Circular Economy: The Next Level of Company Success

André Podleisek · Katharina Luban ·
Roman Hänggi
Editors

Circular Economy: The Next Level of Company Success

A Comprehensive Study on How to Create Value and Improve Operational Excellence Through Circularity

Springer

Editors
André Podleisek
Wettswil, Switzerland

Katharina Luban
Sevelen, Switzerland

Roman Hänggi
Brülisau, Switzerland

ISBN 978-3-031-66405-2 ISBN 978-3-031-66406-9 (eBook)
https://doi.org/10.1007/978-3-031-66406-9

This Springer imprint is published by the registered company Springer Nature Switzerland AG
The registered company address is: Gewerbestrasse 11, 6330 Cham, Switzerland

Preface

Circular Economy (CE) is gaining significant momentum in businesses. Initially, this was motivated by environmental concerns, particularly those related to CO_2 emissions. Recently, there has been a noticeable shift in customer opinions towards environmentally responsible products. This will be a major differentiation factor for companies, further strengthened by evolving regulatory frameworks and requirements. In addition, the scope of CE initiatives has expanded to also include further effects such as resource scarcity and supply chain instability. Furthermore, it has been observed that CE initiatives can result in cost savings for companies and have great potential for innovation.

Contrasting sharply with the prevailing Linear Economy (LE) take–make–dispose approach, CE is shifting the aim towards the use and repeated use of resources in closed loops. This paradigm shift is prompting a fundamental rethinking of our approach to material usage.

We ask ourselves: What must companies do to align with this new way of thinking? Conversations within our group of colleagues and with students revealed that the topic CE needs a deep understanding of potential concepts, as well as a nuanced exploration of feasibility and implementation approaches.

Over the course of two years and more, 20 students and research assistants participated in our research at different stages. The dialogue with our fellow professors and experts at the Eastern Switzerland University of Applied Sciences (OST) further enriched our understanding, bringing diverse perspectives to the challenges of the different approaches.

On top of a comprehensive review of the literature, we collected insights from more than 500 companies on their view on CE and their current implementation status. Additional in-depth case studies provided detailed information on companies and their obstacles to implement CE specifically. We summarise our learning in the Rapperswil Circular Economy Model to guide companies on the path to circularity.

Our research clearly showed that integration of CE is a must for every company. Naturally, not all concepts are similarly relevant and the depth of implementation will vary greatly depending on the nature of the business and its value chain.

Our book not only intends for practice but also adds to the academic discussion of CE. It also helps to establish a solid foundation in our different lectures, where CE has become an integral part.

This book would not have been possible without the work and commitment of many people. We thank the students in the 2021 master class who started the journey of circular economy research in practice with us. They are listed alphabetically: Giuliano Bernard, Morena DiSilvestre, Samuel Folini, Patrick Looser, Sylvan Mächler, Fabio Marti, Odin Peter, Mladen Petraskovic, Mauro Ramirez, Mathilde Schäfer, Philipp Thomann and Raphael Welter. Not all of the above students decided to participate as authors in this book, but their first insights were valuable, and their investigations set the pace for this book.

What is science without colleagues who challenge ideas and provide support where necessary? Therefore, we want to thank our colleagues who were always eager to explore innovative concepts and who supported the students in their first research on CE, namely Prof. Dr. Daniel Keller, Prof. Dr. Daniel Politze, Prof. Dr. Felix Nyffenegger and Prof. Dr. Rainer Bunge.

The second year of master students, who engaged in the case studies, walked the extra mile to become the first authors in the chapters of Part II of this book. Their names can be found in the book, but we especially want to thank them for their commitment and enthusiasm: Nico Bleisch, Denis Hammer, Yannik Keller, Raphael Meyer, Roger Rinderer, Raffael Römer and Reto Weber.

We also express our gratitude to the dedicated team of scientific assistants of the Institute of Product Development and Design (IPEK) at OST who played a pivotal role in the realisation of this book. Their unwavering commitment to excellence, meticulous organisation of inputs, growing expertise with LaTeX and thorough proofreading significantly contributed to the quality and coherence of the final manuscript. A special thanks goes to the research assistant Nicolas Hofer, whose tireless efforts in gathering and organising data during his Master's Thesis, as well as his attention to detail, ensured the accuracy of the scientific content. His commitment to this book and to the topic of CE is truly exceptional.

Two more contributors spontaneously agreed to add their expertise to make our discoveries more useful in practice: Prof. Daniel Schwendemann from OST and Prof. Dr. Ramesh Subramoniam from the University of Texas at Dallas (UTD), who undoubtedly enhanced the overall depth and credibility of this work.

In addition to the professional support received, we are grateful to our friends and family for their patience and support during the writing and editing process. Their understanding and encouragement has played an invaluable role in the realisation of this book.

Rapperswil, Switzerland
December 2024

André Podleisek
Katharina Luban
Roman Hänggi

About This Book

Without any doubt, the current prevailing Linear Economy principle of take-make-dispose is inherently unsustainable: Raw materials are extracted, transformed into products, and discarded as waste after use. The resulting environmental and economic challenges brought together both practitioners and academics to develop more sustainable methods for production and consumption. Especially the concept of Circular Economy has received a lot of attention. But despite many technical solutions, a key question for practical implementation has remained unanswered until now: How can enterprises integrate Circular Economy into their business strategy to leverage success?

This book examines the topic of Circular Economy in an entrepreneurial context. It describes ten options for companies to retain value and identifies drivers and enablers for implementing Circular Economy in businesses.

A survey among more than 500 industry experts revealed the extent to which retention options were implemented in practice. For the first time, relevant drivers and enablers for each retention option were identified. On top of this, case studies of successful circular market offerings shed light on each retention option in detail.

The knowledge gained from research and the further shared experiences of experts that contributed to this book result in a 6-step management model. This model provides guidance on how to implement Circular Economy in organisations as an element of operational excellence and as a source of product, service, and business innovation. Now, every enterprise can start the journey!

As the survey and the cases show: Circular Economy is a promising way to the next level of company success. The book at hand will help organisations to develop their own individual approach that matches their business model. In short, you will find:

- First comprehensive analysis of 10 retention options, their drivers, and enablers.
- Survey of more than 500 companies and in-depth case studies for each retention option.
- Practical research-based implementation model for companies.

Contents

About the Editors

Prof. André Podleisek is a specialist in Sustainability and Quality in Industry. His enthusiasm for environmental protection and Circular Economy began when he studied 20 years ago. Since then, he has been employed by global industrial companies as a Sustainability Officer, where he has changed their way of doing business to incorporate Circular Economy. Recently, he joined the university to pass on his practical sustainability knowledge to the next generation of industrial engineers.

Prof. Dr. Katharina Luban is a distinguished expert in the field of Supply Chain Management. She dedicates her professional endeavours to pioneering modern, technology-driven solutions that enhance logistics operations, procurement activities and product cost. A focal point of her research, teaching and consulting efforts lies in embracing Circular Economy principles, which play a pivotal role in fortifying supply chains for long-term resilience in the ever-evolving landscape of the future.

Prof. Dr. Roman Hänggi is an expert in Production Management, having accrued 25 years of experience in the industry at all levels of hierarchy. His passion for the ongoing development of lean in the digital age continues to motivate him today. He assisted numerous industrial companies in their journey to become operational champions through the combination of lean and digitalisation. His expertise has enabled him to incorporate Circular Economy into the operational excellence path in both practice and research.

Acronyms

AHP	Analytic Hierarchy Process
AI	Artificial Intelligence
ANSI	American National Standards Institute
AR	Augmented Reality
B2B	Business-to-Business
B2C	Business-to-Consumer
BIM	Building Information Modelling
BPA	Bisphenol-A
BPF	Business Performance Factors
C2C	Consumer-to-Consumer
CCU	Carbon Capture Utilisation
CE	Circular Economy
CEBM	Circular Economy Business Models
CHF	Swiss Francs
CLSC	Closed-Loop Supply Chains
cobot	Collaborative Robots
CSR	Corporate Social Responsibility
EEE	Electrical and Electronic Equipment
EIP	Environmental Impact Points
EOL	End-of-Life
EPA	Environmental Protection Agency
EPR	Extended Producer Responsibility
ERP	Enterprise Resource Planning
ESG	Environment, Social and Governance
EU	European Union
EUR	Euro
GBP	Great Britain Pound
GDP	Gross Domestic Product
GWP	Global Warming Potential
I4.0	Industry 4.0
IAM	Independent Aftermarket

IoMT	Internet of Manufacturing Things
IoT	Internet of Things
IPEK	Institute of Product Development, Industrial Engineering and Product Lifecycle Management
IWK	Institute for Materials Science and Plastics Processing
KPI	Key Performance Indicators
LCA	Life Cycle Assessment
LE	Linear Economy
LED	Light-Emitting Diode
LIB	Lithium-Ion Battery
LO	Lista Office
MCP	Manufacturers of Convenience Products
MDP	Manufacturers of Durable Products
MNP	Manufacturers of Non-durable Products
MPC	Manufacturers of Parts and Components
NFC	Near-Field Communication
OE	Original Equipment
OE-CE	Operational Excellence Circularity
OEM	Original Equipment Manufacturer
OES	Original Equipment Service
OST	Eastern Switzerland University of Applied Sciences
PA	Impact Assessment
PaaS	Product as a Service
PAH	Polycyclic Aromatic Hydrocarbons
PBT	Polybutylene Terephthalate
PE	Polyethylene
PET	Polyethylene Terephthalate
PLA	Polylactic acid
PLC	Product Lifecycle
PLM	Product Lifecycle Management
PM-CE	Product/Service/Business Model Innovation Circularity
POM	Poly-Oxy-Methylene
PP	Polypropylene
PRV	Pressure Regulating Valve
PSS	Product Service System
PVC	Polyvinylchloride
R	Retention Option
R0	Refuse
R1	Reduce
R2	Reuse
R3	Repair
R4	Refurbish
R5	Remanufacture
R6	Repurpose
R7	Recycle

R8	Recover
R9	Remine
RA	Reclaimed Asphalt
RAP	Recycled Asphalt Pavements
RFID	Radio Frequency Identification
RIC	Remanufacturing Industries Council
RIME	Repair and Installation of Machinery and Equipment
RL	Reverse Logistics
ROI	Return on Investment
RQ	Research Question
rTPU	Recycled Thermoplastic polyurethane
RW	Retail and Wholesale
SBB	Swiss Federal Railways
SDG	Sustainable Development Goal
SME	Small and Medium-sized Enterprises
SPLM	Smart Product Lifecycle Management
SVC	Service Industry
TA	Time Assessment
TCO	Total Cost of Ownership
TPE	Thermoplastic elastomer
TRA	Theory of Reasoned Action
UNEP	United Nations Environmental Programme
USD	US Dollar
UTD	University of Texas at Dallas
VR	Virtual Reality
WEEE	Waste from Electrical and Electronic Equipment
WERZ	Institut für Wissen, Energie und Rohstoffe Zug
WTB	Wind Turbine Blade
WWF	World Wide Fund for Nature

List of Figures

List of Tables

Chapter 1
Introduction to This Book

Katharina Luban, Nicolas Hofer, and Roman Hänggi

Abstract Without any doubt, the current prevailing Linear Economy principle of take-make-dispose is inherently unsustainable due to its adverse effects on the environment and the depletion of natural resources. In this context, the concept of Circular Economy has gained increasing attention in recent years as a way to address the resulting problems. In this introduction, we explore our perspective on why Circular Economy has evolved into a strategic focal point for corporate management. We introduce the methodologies and approaches that were applied throughout the book and how we derived practical recommendations for companies on their journey towards a Circular Economy.

1.1 Circular Economy as a Guiding Theme of This Book

The current prevailing Linear Economy (LE) operates according to the principle of take-make-dispose: Raw materials are extracted, transformed into products and discarded as waste after use [1, 2]. The result is an increasing waste stream of non-degradable plastic and electrical and electronic equipment waste (e-waste) [3]. On the consumption side we saw an increase of more than 65% in the global amount of domestic material consumption between 2000 and 2019, reaching 95.1 bill. tonnes in 2019 – approximately 12.3 tonnes per person in one year [4]! The global population

K. Luban (✉)
Sevelen, Switzerland
e-mail: katharina.luban@ost.ch

N. Hofer
Zürich, Switzerland

R. Hänggi
Brülisau, Switzerland

A. Podleisek et al. (eds.), *Circular Economy: The Next Level of Company Success*,
https://doi.org/10.1007/978-3-031-66406-9_1

is estimated to reach 9.7 bill. people by 2050 [5] and if consumption remained constant, this population would require two planets worth of resources to maintain living standards [6].

This is a dreadful scenario. The growing demand for resources, the increase in landfill, and waste incineration contribute to 'sharp rises in greenhouse gas emissions, acidification, aquatic eco-toxicity and emissions of smog-forming substances' [3, p. 11]. In addition to this, the extraction and processing of resources also has significant environmental impacts, including air and water pollution and habitat destruction [7–9]. This will have severe effects on ecosystems, which are essential for human health and well-being, but have limited resilience to human impacts [7].

Consequently, it is increasingly discussed and recognised that a finite planet cannot sustain unconstrained economic growth based on resource-intensive activities [1, 7] and the consumption of natural resources has become a major global concern. We see an increasing shift of consumption from non-renewable materials to renewable materials. This shift is supported by innovations that enable the use of renewable materials. In addition, driven by environmental and economic challenges, there has been a growing interest in exploring alternative methods of organising production and consumption. In this context, the concept of Circular Economy (CE) has received increasing attention from both academics and practitioners over the past decade [10–12].

CE advocates for the use and repeated use of resources in closed loops rather than following a linear approach of extraction, use, and disposal. It envisions an industrial system with the ultimate goal of achieving 'the holistic closure of the entire "take, make, waste" cycle' [2, p. 20] reflecting the efficiency of natural ecosystems where each output serves as input for another process [13]. As a result, CE emerges as a promising response to environmental and resource-related challenges, gaining considerable traction in industry, policy making, and academia. Its objective is to improve economic performance and growth without depleting resources beyond the Earth's capacity [7, 14–16].

In addition to this general social challenge, there are many reasons why companies should consider CE as a strategic topic. In this chapter, we discuss some of these considerations. Then we introduce the methodologies and approaches that were applied throughout the book and how we derived practical recommendations for companies on their journey to CE.

This book examines the topic of CE in an entrepreneurial context. It will help organisations develop their own individual approach that matches their business model. The main takeaways are a comprehensive analysis of 10 retention options to close the loop and their specific drivers and enablers. A survey of more than 500 companies sheds light on the current state of industry, and in-depth case studies provide examples and practical insight for each retention option. The result is a practical research-based implementation model for companies.

1.2 Circular Economy as a Strategic Topic in Corporate Management

Due to the exhaustion of finite resources and the growing problem of environmental degradation, companies need to complement their strategies. This chapter will dive into the concept of CE, which has garnered substantial attention as a transformative approach and should not be ignored by company leaders.

Resource Scarcity and Cost Pressures

Fluctuating raw material prices, scarcity and uncertain supply security threaten 'the long-term competitiveness of companies, regions and countries' [17, p. 6]. The COVID-19 crisis and recent conflicts unexpectedly disrupted global supply chains and showed how material shortages can result in production delays, increased costs, and loss of competitiveness [18]. Dependencies such as Europe's high dependence on metal imports (between 75% and 100%) place industrial companies in a very vulnerable position [19]. The strategic and also geopolitical importance of natural resources is today a critical element on the political agenda of all countries. Decisions about investments, political economic programmes, and even military conflicts are influenced by the scarcity and cost pressures that arise from the uncertainty surrounding critical natural resources.

Therefore, it is not surprising that operational programmes, product changes, and even new business models that incorporate CE have gained increasing attention in recent years due to their potential to improve resource efficiency and reduce waste [20]. Adopting the CE principles enables companies to decouple economic growth from resource consumption [2]. It advocates for a shift toward a regenerative model, focussing on the reuse, repair, and recycling of resources to reduce dependence on raw materials. Ultimately, this strategic transition empowers organisations to mitigate the risks of resource scarcity and minimise the costs associated with the purchase of raw materials [21].

Innovation and Differentiation

CE thinking advocates a departure from the traditional linear 'take-make-dispose' approach, emphasising innovation, creativity, and collaboration through circularity. Reimagining product design, business models, and value chains in alignment with circular principles opens opportunities for companies to pioneer new and inventive solutions. By embracing the CE, companies position themselves to stand out in the market, attract environmentally conscious customers, and cultivate a culture of innovation, thus fostering long-term growth and profitability.

These positive results have been supported by several studies, which indicate that companies experience various economic advantages when transitioning to CE. Benefits include a better reputation, increased customer satisfaction, and increased revenues [22–24]. Furthermore, companies benefit from innovations in production and infrastructure, along with new opportunities to earn [12, 25].

Environmental Stewardship and Sustainability

Amid growing apprehensions regarding climate change and environmental decline, companies face increasing pressure from customers, employees, communities, and investors to demonstrate greater accountability. Embracing the principles of CE serves as a testament to the commitment of a company to environmental stewardship and sustainable practices. Through the reduction of waste, the reduction of greenhouse gas emissions and the prioritisation of product designs that emphasise longevity and recyclability, businesses have the opportunity to establish themselves as sustainability pioneers.

This will bestow a competitive edge in comparison to competitors and will foster value creation in a broader sense: Two studies showed that the implementation of CE can generate social value through the creation of jobs and the support of local economies [14, 26]. Additionally, sustainable practices can improve customer loyalty, attract and retain top talent, and improve investor confidence. Involving stakeholders in CE initiatives nurtures trust, cultivates shared value, and establishes lasting relationships, all of which are essential for organisational success in the interconnected landscape of today.

Regulatory Compliance and Risk Mitigation

Governments and regulatory bodies around the world are progressively implementing more stringent environmental regulations and policies. Laws related to Extended Producer Responsibility (EPR) require that producers assume responsibility for the entire lifecycle of their products, including collection, recycling, and proper disposal. These regulations either incentivise or require companies to design products with recyclability and reusability in mind. For example, numerous countries and regions have enforced or proposed bans or restrictions on single-use plastics, such as plastic bags, straws, and packaging. In addition to these, there are numerous other initiatives and regulations that businesses must comply with.

Ultimately, more stringent waste management and recycling regulations will drive the adoption of circular practices, as this shift allows businesses to expand their risk management capabilities. Through the proactive integration of CE principles into their strategies, organisations can proactively stay ahead of regulatory requirements, diminish compliance risks, and mitigate the potential consequences of penalties and reputational damage linked to non-compliance.

Integration of Circular Economy in Strategic Considerations

Although there is widespread recognition of climate change as a problem [16], and despite the evident potential discussed above, along with the support of political decision makers and other national government strategies [27], the acceptance of CE principles by companies and organisations remains relatively modest [27–32]. This observation holds true, even within the realm of industrial companies [33].

The shift from a linear to a circular economy requires a comprehensive long-term approach [34], involving fundamental changes in the existing economic structure and a societal shift in attitudes toward consumption [35]. There is already an observable change in the behaviour and attitudes of consumers and stakeholders, along with the enforcement of more stringent governmental policies and regulations [34, 36]. Now, the industry must also confront and adapt to this transformation.

Companies that choose to participate in this paradigm shift must adjust their business practices accordingly. Embracing Circular Economy is one or maybe the only way to adapt to this change, involving the integration of CE concepts into strategic management decisions. Various authors have proposed ideas on how to do this [37, 38]. In this book, we will elucidate and analyse various options, emphasise the correlation between suggested activities and their outcomes, and offer guidance to company leaders navigating the transformative journey toward a CE.

1.3 Goal and Methodology of This Book

Discussions with academics and professionals, including C-level managers of industrial companies, reveal that the strategies to implement CE vary significantly depending on the specific business domains of the companies. However, there is reason to assume that there are commonalities in the approaches across various sectors. This book is a result of proving this hypothesis, denominating the approaches, and then identifying success factors that are mainly relevant to a certain approach and others that show more relevance to other approaches (see Fig. 1.1).

To answer the question 'Which success factors enable which approach to circular economy?', a procedure was followed that is reflected in the different parts of this book. All steps together result in the Rapperswil Circular Economy Model. The different steps are visually illustrated in Fig. 1.2 and will be elaborated more closely in the remaining part of this chapter.

The activities that resulted in publishing this book started with exploratory and immersive research to get acquainted with the topic. In 2021, 12 master students of the MSc in Business Engineering course at the Eastern Switzerland University of Applied Sciences (OST) analysed more than 30 implemented CE cases during the course of a seminar dedicated to learning about research methods.

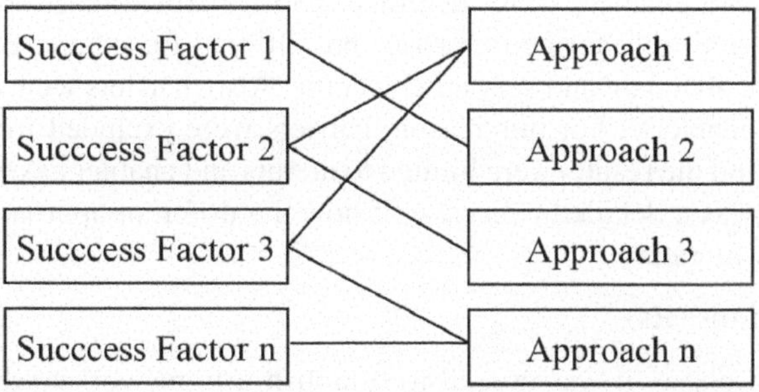

Fig. 1.1 Main question of this book: Which success factors enable which approach to circular economy?

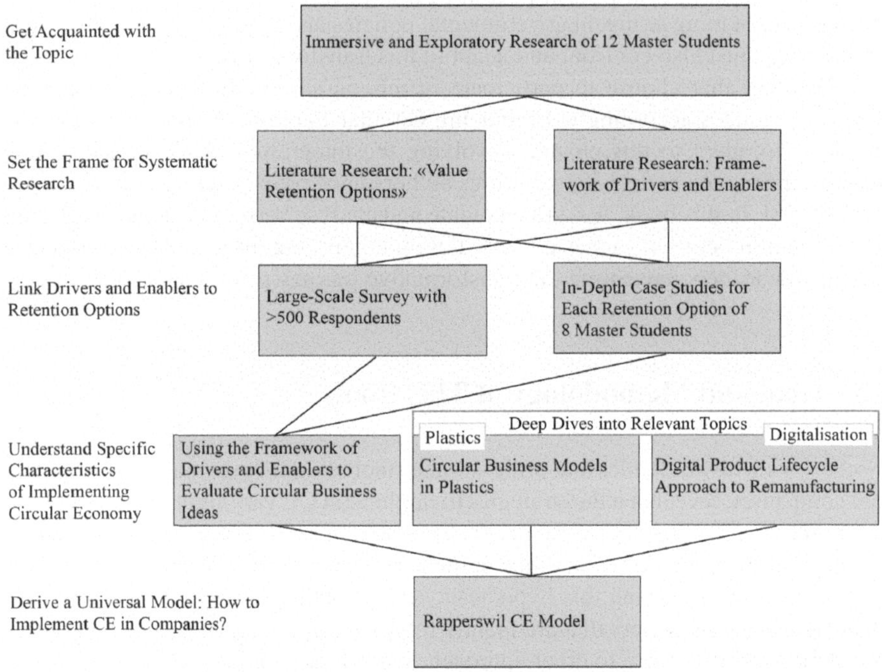

Fig. 1.2 Methodology of this book

Each student identified and analysed two to three industry cases in the field of CE. The task was to work out relevant success factors, namely drivers, enablers, and barriers for each case. Drivers are factors that motivate or inspire the adoption of CE, while enablers help facilitate the achievement of the project. Barriers are factors that prevent or make it difficult to achieve the goals; they have to be overcome in order to succeed. During the seminar, relevant information was obtained by consulting primary and secondary sources. Primary sources were collected during interviews with representatives of some of the companies.

The result of exploring over 30 realised practical cases was astonishing: Many similar drivers and enablers were identified, and it seemed that some drivers and enablers were applicable for certain cases but differed for others. Barriers, on the other hand, were abundant and very case-specific. Also, barriers were often regarded as 'absence of enablers'. For this reason, barriers were excluded from subsequent considerations and the results were limited to drivers and enablers. Another outcome of the seminar was a detailed overview of how the different approaches to CE are classified in the literature.

Retention Options (Rs)

After the initial phase, it was crucial to establish a framework to discuss the different approaches to the implementation of CE in a structured way. Guided by the idea of 'Value Retention', the R-framework of Reike et al. [39] was chosen. Reike

Table 1.1 Ten options to CE that are relevant to industrial companies

Retention options	
Refuse	Remanufacture
Reduce	Repurpose (Rethink)
Reuse (Resell)	Recycle
Repair	Recover (energy or biocycle)
Refurbish	Remine

et al. describe 10 practical Retention Options (Rs)[1] to retain value in products or components by either slowing, narrowing, or closing the loops (see Table 1.1). These 10 options were also the basis of the United Nations Environmental Programme (UNEP) Circularity Programme [41] and describe value retention as practical strategies to reduce the use of virgin materials and avoid waste. Products, components, and materials are kept in a useful state for as long as possible, and they are fed back into the value creation at some stage of product realisation. Due to its practical and comprehensive structure, the R-framework of Reike et al. is very suitable and relevant for industrial enterprises.

Theoretical Framework for Drivers and Enablers of Circular Economy

To establish a comprehensive framework for general drivers and enablers when implementing CE, an extensive and structured review of the literature was carried out as shown in Fig. 1.3. Within the review process, a total of 209 potentially interesting papers were identified by screening two selected databases for peer-reviewed articles according to previously defined search strings. By excluding duplicates and qualitatively assessing titles and abstracts, this number was condensed to 94 results. Finally, a full-text analysis led to a remaining number of 32 highly relevant peer-reviewed papers.

The papers revealed drivers and enablers that were subsequently coded, synonyms were categorised, and categories were clustered. The resulting framework consists of 16 drivers in 5 clusters and 12 enablers in 5 clusters. This resulting framework of drivers and enablers was used in further research to identify those drivers and enablers that play the most significant role in the adoption of CE in the industry.

Survey: Key Drivers and Enablers to Implement Circular Economy in Industrial Companies

A broad industry survey was conducted using a standardised online questionnaire to obtain a good understanding of the current state of implementation of CE in the industry and to verify the general framework of drivers and enablers. The results are based on responses from 512 entities from 32 European countries. As shown in

[1] Throughout this edited volume the terms retention option and retention strategy are used interchangeably, especially when talking about their implementation or adoption. Both variants exist in the literature, e.g., [39, 40] writing about the same aspects.

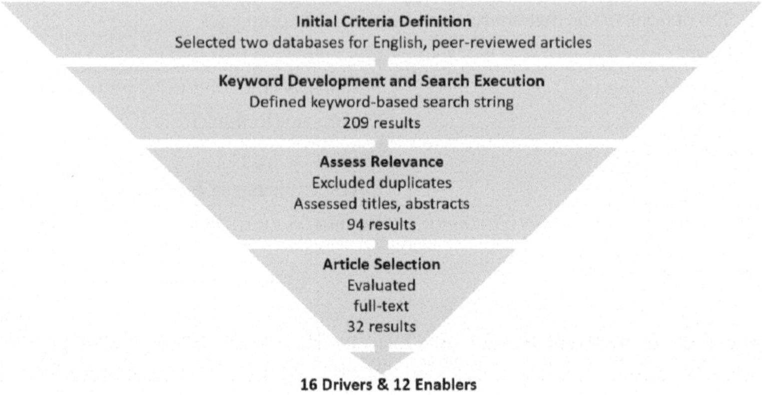

Fig. 1.3 Procedure for identifying the drivers and enablers of Circular Economy

Table 1.2 Participants of the industry survey

		n	Percentage (%)
Position*	CEO/Business owner	223	43.6
	Business Development Manager	73	14.3
	Environmental/Sustainability/Governance Manager	67	13.1
	Other managerial positions	50	9.8
	Other C-Suite Officers	39	7.6
	Technical staff	29	5.7
	Other	27	5.3
	Administration and Finance	3	0.6
Main area of expertise*	Manufacturer of durable products	214	41.8
	Manufacturer of non-durable products	80	15.6
	Service	80	15.6
	Manufacturer of parts or components	49	9.6
	Other	42	8.2
	Retail/Wholesale	32	6.3
	Repair and installation of machinery and equipment	10	2.0
	Manufacturer of convenience products	5	1.0

*For each of these questions, one participant chose not to answer

Table 1.2, most of the participants work in the manufacturing industry, 223 of them holding the position of Chief Executive Officer.

The survey revealed the driving forces behind the adoption of CE by European industrial companies and confirmed the general drivers and enablers that foster the adoption and implementation of CE. At the same time, it was possible to determine differences in the relevance of drivers and enablers according to the approach chosen for implementing CE. This insight leads to the development of a heat map, that is, a table in which the relevance of drivers for certain Rs is visualised by a colour palette from red (not relevant) to dark green (highly relevant). This research is particularly valuable for practitioners seeking guidance in choosing a promising approach to CE and understanding the necessary prerequisites for successful CE implementation.

The findings of the study also strongly indicate that the incorporation of CE into business models can positively affect factors of business performance, such as profit, company image, and manufacturing costs. Retaining value undoubtedly has an economic implication. Many of the responding companies have successfully implemented strategies from the 10 R Framework and have seen positive impacts on their business.

The survey also clearly showed that current CE initiatives are driven by the desire to save money and improve resource consumption. At the moment, only few companies use CE as a strategic approach to develop new products and services. However, once operational processes have benefited from circular approaches, the strategic consideration of CE as a competitive factor will become decisive for the future success of all companies.

Analysis of Successful Industry Use Cases

While the survey gave an indication as to what drivers and enablers are relevant for which Retention Option (R), the commitment and bright minds of a new class of students were engaged to find use cases that illustrate the Rs: In the fall semester of 2022 another class of prospective masters in Business Engineering set out to identify and analyse specific use cases of implemented CE that had not yet been described in the literature. Each student was assigned the task of finding a suitable case for a specific R. The students used sources on the Internet and conducted several interviews with company representatives to gather relevant information in order to develop a deep understanding of specific cases. To do so in a standardised way, they all applied the before-mentioned framework of drivers and enablers to assess the use cases. A special way of quality assurance was implemented when six students from the first year volunteered to support and challenge the resulting use case descriptions. They became second and third authors of the respective articles in this book.

The use case analyses served to illustrate examples of successful implementation of each of the 10 Rs, which will help deepen the knowledge and understanding of possible approaches to CE and their most prevalent drivers and enablers. The case study methodology also revealed insights into the complex phenomenon of 'success' in the context of implementing CE in companies. In the end, the resulting case studies not only highlight the driving forces of success and describe enablers that supported the implementation of the cases, but also make statements about the perceived impact and outcome of the implemented cases.

Using the Framework of Drivers and Enablers of Circular Economy to Evaluate Circular Business Ideas

For the acceptance of CE in companies, the measurable success is crucial. This insight was shown by the survey and the case studies. However, the strategic importance of

CE in the development of new products, services and business models should not be neglected, although it might be the a later step.

One of our earlier findings was that saving costs is not the most significant driver when developing new circular products or business models, but it is when using CE approaches to improve operations. Rather, the involvement of customers will ensure the successful implementation of new CE-based products, services, or business models. When talking about new circular products and services, the drivers and enablers of CE can balance resulting costs, as well as resulting inconveniences from a customer's point of view.

To compare different circular ideas, a systematic, holistic, and unbiased assessment of drivers, enablers, and costs is essential. First, an analysis based on the general drivers of CE sharpens the understanding of the business case and supports the decision about the most appropriate R with respect to this strategic dimension. Then, we propose to perform a utility analysis based on the general framework of drivers and enablers. We show how to do this and how this can be used to differentiate between several circular ideas by evaluating their concepts and costs.

We used a circular business idea to illustrate our method. In a first step, the drivers were analysed and led to the conclusion that the high-value retention option *Reuse* would be a good strategy to pursue. The example of reusable food packaging served as a demonstration of how to bring transparency to the assessment process of circular ideas.

When using a similar procedure, businesses will gain a deeper understanding of how to best leverage the drivers and enablers of their ideas, and they can also use the procedure to compare any CE idea or business model with the solutions of competitors.

Rethinking and Positioning Plastics for Circular Economy

Despite legislative efforts, a significant proportion of plastic materials continue to contribute to the accumulation of waste. This not only poses environmental challenges, but also, on the positive side, represents untapped economic opportunities. The intricacies of plastic types and the corresponding recycling methods are multifaceted. Gaining a better understanding of these elements will provide invaluable insights into the use, reuse, and recycling of plastics. Professor Daniel Schwendemann, a leading figure in the domain of plastics processing, provides insight into key considerations and challenges inherent to plastics. He suggests moving from a general rejection of plastics to fostering a culture of responsible plastic use and effective *Recycling*.

Digital Transformation of Remanufacturing

For many years, *Remanufacturing* was largely an afterthought for manufacturers, who often overlooked the opportunity associated with reclaiming their own products. This began to change during the 1990s and into the early 2000s, when data visibility increased across organisations, fostering transparency and efficiency in return operations. The push for data visibility led companies to harness the data recovered from the returned products. Professor Dr. Ramesh Subramoniam has been working in the field of supply chain, manufacturing, and data since many years and has used his experience to develop a framework for a digitised Product Lifecycle (PLC) integrated with business phases including remanufacturing stages. He points out that future success in *Remanufacturing* will hinge on companies adopting a mindset open to change, backed by a robust technology infrastructure to facilitate the transition.

The Rapperswil Circular Economy Model

All findings and learning from our research are summarised in the final chapter. For companies that want to include CE in their strategy and seek guidance on how to address the implementation of circularity, we propose the Rapperswil Circular Economy Model. The basis is a six-step management model that starts with understanding the current situation, then setting vision and goals, setting ambition levels, defining possible actions, including prioritisation, developing a roadmap, and finally measuring results and adjusting as needed (see Fig. 1.4).

To understand the current state of the company with respect to CE, relevant drivers and retention strategies must be identified and aligned, which can be done using a structured list. Then, the actual level of circularity must be objectively rated according to the two dimensions, 'operational excellence' and 'product/service/business model'. These two dimensions of circularity were derived from our research because retention strategies work differently for internal activities and for outside or market activities. For the rating, we propose a quantitative approach.

Once the current level of circularity has been identified, the level of ambition must be set. To do so, many considerations come into play that are individual to each company. Therefore, the book provides many examples of where ambition can go, so the reader receives valuable input to think about.

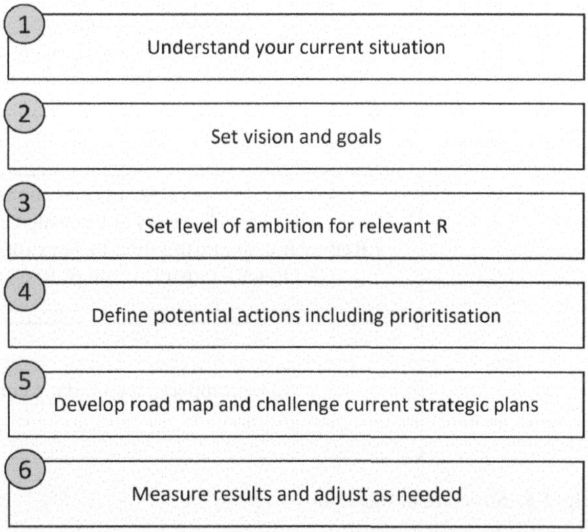

Fig. 1.4 6-step management model for the implementation of the Rapperswil CE Model

Ambition levels are noted on the same chart as the current level of circularity. When comparing the current level with the level of ambition, a gap will be identified that needs to be closed by defining concrete actions. All circularity actions must become an integral part of the company's transformation plan. To round out the model, a last step defines the continuous review of the current state, the ambitions, and the actions that will lead companies on the path of circularity.

1.4 Structure of This Book

In order to present the contents of this book in a balanced and targeted way, three parts were created (see Fig. 1.5).

Part I starts with an introduction to the topic of Circular Economy and the development of two basic frameworks (Chap. 3). The first framework introduces possible

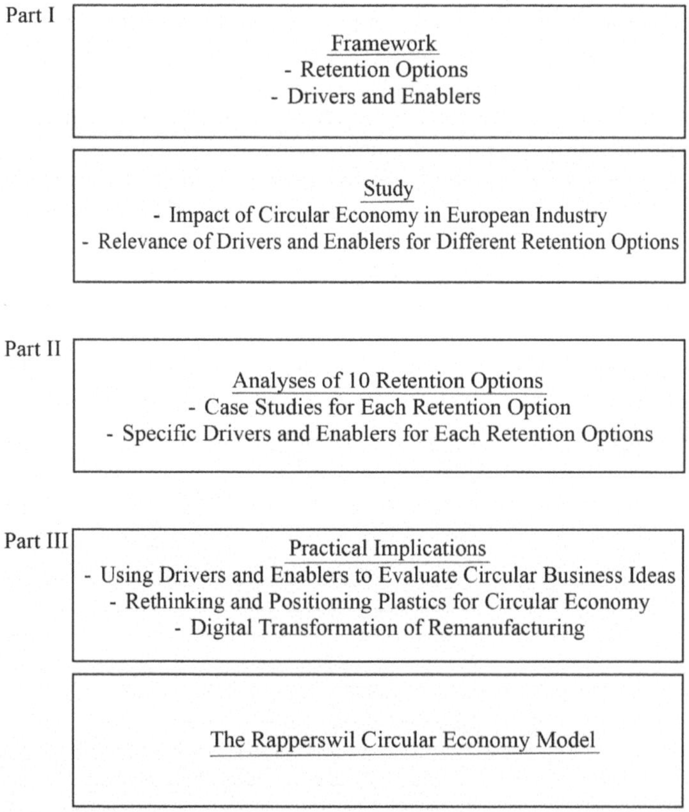

Fig. 1.5 Structure of this book

approaches to CE. Based on comprehensive literature analyses and using the concept of a Value Hill, 10 Rs were identified that are the most applicable to industrial companies either involved or about to engage in CE (see Sect. 3.1). Further review of literature resulted in a framework for Drivers and Enablers of CE, consisting of 16 drivers in 5 clusters and 12 enablers in 5 clusters (see Sect. 3.2). The two frameworks were used in a quantitative survey (Chap. 4) with 512 respondents from European industrial companies to highlight the forces driving certain Rs, and to show enablers that were most relevant to the respondents. The findings of the study also strongly indicate that the incorporation of CE into business models can positively affect factors of business performance, such as profit, company image and manufacturing costs.

Part II of the book contains a review of use cases for each of the R (Chaps. 6–14). Master students in Business Engineering identified use cases, tested the framework of drivers and enablers, and derived insights about the outcome and impact that the implementation of R had on the respective case companies. All cases were presented in the same structure, resulting in a unified representation of all the cases under consideration. An exception are the chapters on *Repurpose* (Chap. 11) and *Remine* (Chap. 14). These follow a slightly different approach due to limitations in available use cases and limited access to interview partners. Instead, drivers and enablers were evaluated using information obtained from the literature.

Part III is dedicated to the practical implications of implementing CE in companies. A *Reuse* case serves as an example on how to evaluate ideas and business models with the help of the driver-enabler-framework (Chap. 16). Plastics is one of the most relevant industries for *Recycling* and therefore we added an expert view on the use, reuse and final recycling of plastics in Chap. 17. Today, companies make data-driven decisions and data is an important enabler of CE. *Remanufacturing* can be a way to monetise the data collected in all stages of the PLC. Our author validates this approach of a digital PLC with case studies based on his previous experience working for major companies in various industries (Chap. 18).

All findings are concluded in Chap. 19. This final chapter summarises the link between retention options, drivers, and enablers. Then it uses these findings and the practical implications to derive the Rapperswil Circular Economy Model. This model proposes a concrete six-step implementation path for a specific company, shares examples on how to evaluate the current status of circularity, set ambition levels, and derive actions to start the transformation.

References

1. Ellen MacArthur Foundation. (2013). *Towards the circular economy: Economic and business rationale for an accelerated transition*.
2. Lacy, P., Long, J. & Spindler, W. (2020). *The circular economy handbook: Realizing the circular advantage*. London, UK: Palgrave Macmillan. ISBN: 978-1-349-95967-9.
3. Bringezu, S. et al. (2017). *Assessing global resource use: A systems approach to resource efficiency and pollution reduction (A Report of the International Resource Panel)*. ISBN: 978-92-807-3677-9.

4. United Nations. (2022). *The Sustainable Development Goals Report, 2022.*
5. United Nations. (2019). *How certain are the United Nations global population projections?* Issue: 2019/6.
6. De Morais, L., Pinto, D. & Cruz-Jesus, F. (2021). Circular economy engagement: Altruism, status, and cultural orientation as drivers for sustainable consumption. *Sustainable Production and Consumption, 27*, 523–533. ISBN: 2352-5509.
7. Desing, H. et al. (2020). A circular economy within the planetary boundaries: Towards a resource-based, systemic approach. *Resources, Conservation and Recycling, 155*, 104673. ISBN: 0921-3449.
8. Islam, A. et al. (2020). Advances in sustainable approaches to recover metals from E-waste - A review. *Journal of Cleaner Production,* **244**, 118815. ISBN: 0959-6526.
9. Niinimäki, K. et al. (2020). The environmental price of fast fashion. *Nature Reviews Earth & Environment, 1*, 189–200. ISBN: 2662-138X.
10. Geissdoerfer, M., Savaget, P., Bocken, N. M. & Hultink, E. J. (2017). The Circular Economy – A new sustainability paradigm? *Journal of Cleaner Production, 143*, 757–768. ISBN: 0959-6526.
11. Ghisellini, P., Cialani, C. & Ulgiati, S. (2016). A review on circular economy: The expected transition to a balanced interplay of environmental and economic systems. *Journal of Cleaner Production, 114*, 11–32. ISBN: 0959-6526.
12. Murray, A., Skene, K. & Haynes, K. (2017). The circular economy: An interdisciplinary exploration of the concept and application in a global context. *Journal of Business Ethics, 14*, 369–380. ISBN: 0167-4544, 1573-0697.
13. Del Río, P., Kiefer, C. P., Carrillo-Hermosilla, J. & Könnölä, T. (2021). *The circular economy: economic, managerial and policy implications*. Cham: Springer International Publishing. ISBN: 978-3-030-74791-6.
14. European Commission & Directorate-General for Environment. (2020). *Circular Economy Action Plan – For a cleaner and more competitive Europe*. Publications Office of the European Union. https://data.europa.eu/doi/10.2779/05068.
15. European Commission. (2020). *A new Circular Economy Action Plan For a cleaner and more competitive Europe* Issue: COM/2020/98.
16. Langen, S. et al. (2021). Promoting circular economy transition: A study about perceptions and awareness by different stakeholders groups. *Journal of Cleaner Production, 316*. https://doi.org/10.1016/j.jclepro.2021.128166.
17. Mont, O., Plepys, A., Whalen, K. & Nußholz, J. L. K. (2017). *Business model innovation for a Circular Economy: Drivers and barriers for the Swedish industry - the voice of REES companies*. Mistra REES.
18. O'Connor, M., Zinnerman, J., Anastas, P. & Plata, D. (2016). A Strategy for material supply chain sustainability: Enabling a circular economy in the electronics industry through green engineering. *ACS Sustainable Chemistry & Engineering, 4*, 5879–5888. ISBN: 2168-0485.
19. European Commission & Industry Directorate-General for Internal Market, Entrepreneurship and SMEs. (2018). *Raw materials scoreboard 2018 – European innovation partnership on raw materials*. Publications Office. https://data.europa.eu/doi/10.2873/08258.
20. Lüdeke-Freund, F., Gold, S. & Bocken, N. M. P. (2019). A review and typology of circular economy business model patterns. *Journal of Industrial Ecology, 23*, 36–61. ISBN: 1088-1980, 1530-9290.
21. Agyemang, M. et al. (2019). Drivers and barriers to circular economy implementation: An explorative study in Pakistan's automobile industry. *Management Decision, 57,* 971–994. ISBN: 0025-1747.
22. Gusmerotti, N. M., Testa, F., Corsini, F., Pretner, G. & Iraldo, F. (2019). Drivers and approaches to the circular economy in manufacturing firms. *Journal of Cleaner Production, 230*, 314–327. ISBN: 0959-6526.
23. Matsumoto, M., Yang, S., Martinsen, K. & Kainuma, Y. (2016). Trends and research challenges in remanufacturing. *International Journal of Precision Engineering and Manufacturing-Green Technology, 3*, 129–142. ISBN: 2288-6206, 2198-0810.

24. Milios, L. & Matsumoto, M. (2019). Consumer perception of remanufactured automotive parts and policy implications for transitioning to a circular economy in Sweden. *Sustainability, 11*. https://doi.org/10.3390/su11226264.
25. Rizos, V. et al. (2016). Implementation of circular economy business models by small and medium-sized enterprises (SMEs): Barriers and enablers. *Sustainability, 8,* 1212. ISBN: 2071-1050.
26. Rizos, V. et al. (2019). Identifying the impact of the circular economy on the fast-moving consumer goods industry: Opportunities and challenges for business, workers and consumers: Mobile phones as an example. ISBN: 978-92-830-4572-4.
27. Rizos, V. & Bryhn, J. (2022). Implementation of circular economy approaches in the electrical and electronic equipment (EEE) sector: Barriers, enablers and policy insights. *Journal of Cleaner Production, 338,* 130617. ISBN: 0959-6526.
28. Circle Economy. (2020). *The circularity gap report 2020*. The Netherlands: Amsterdam. https://www.circularity-gap.world/2020.
29. Fonseca, L., Domingues, J., Pereira, M., Martins, F. & Zimon, D. (2018). Assessment of circular economy within Portuguese organizations. *Sustainability, 10*, 2521. ISBN: 2071-1050.
30. Letchumanan, L. T. et al. (2022). Analyzing the factors enabling green lean six sigma implementation in the industry 4.0 era. *Sustainability, 14*, 3450. https://doi.org/10.3390/su14063450.
31. Tura, N. et al. (2019). Unlocking circular business: A framework of barriers and drivers. *Journal of Cleaner Production, 212*, 90–98. ISBN: 0959-6526.
32. Kirchherr, J. et al. (2018). Barriers to the circular economy: Evidence from the European Union (EU). *Ecological Economics, 150*, 264–272. ISBN: 0921-8009.
33. Kristoffersen, E., Blomsma, F., Mikalef, P. & Li, J. (2020). The smart circular economy: A digital-enabled circular strategies framework for manufacturing companies. *Journal of Business Research, 120*, 241–261. ISBN: 0148-2963.
34. Van Buren, N., Demmers, M., van der Heijden, R. & Witlox, F. (2016). Towards a circular economy: The role of Dutch logistics industries and governments. *Sustainability, 8*, 647. ISBN: 2071-1050.
35. Gillabel, J., Manshoven, S. & Grossi, F. (2021). *Business Models in a Circular Economy* Eionet Report ETC/WMGE 2021/2 (European Topic Centre on Waste and Materials in a Green Economy).
36. Hart, J., Adams, K., Giesekam, J., Tingley, D. D. & Pomponi, F. (2019). Barriers and drivers in a circular economy: The case of the built environment. *Procedia CIRP, 80*, 619–624. ISBN: 2212-8271.
37. Blomsma, F. et al. (2019). Developing a circular strategies framework for manufacturing companies to support circular economy-oriented innovation. *Journal of Cleaner Production, 241*, 118271. ISBN: 0959-6526.
38. Potting, J., Hekkert, M., Worrell, E., & Hanemaaijer, A. (2017). *Circular Economy: Measuring innovation in the product chain PBL publication number: 2544*. The Hague: PBL Netherlands Environmental Assessment Agency.
39. Reike, D., Vermeulen, W. J. & Witjes, S. (2018). The circular economy: New or Refurbished as CE 3.0? – Exploring controversies in the conceptualization of the circular economy through a focus on history and resource value retention options. *Resources, Conservation and Recycling, 135*, 246–264. ISBN: 0921-3449.
40. Diaz, A., Reyes, T. & Baumgartner, R. J. (2022). Implementing circular economy strategies during product development. *Resources, Conservation and Recycling, 184*, 106344. ISBN: 0921-3449.
41. UNEP. (2019). *Circularity* UNEP - United Nations Environment Programme. Retrieved September 21, 2022 from http://www.unep.org/circularity.

Part I
The Retention of Value in a Circular Economy

Chapter 2
Introduction to Part I

Katharina Luban

Abstract This chapter is a brief introduction to Part I of this book. It describes the methodology that was applied and the results that can be expected in this part of the book.

Theoretical Basis of this Book

After an initial exploratory research phase, the real work of this book began with a systematic analysis of the literature. On the one hand, it was necessary to categorise the approaches that companies can choose when implementing Circular Economy (CE). The second goal was to get a clear understanding of the drivers that drive companies to think about and implement CE in the first place and to identify the enablers that support implementation.

Value Retention Options

Companies can engage in CE using different approaches. First of all, these approaches need to be classified. Forming categories for different approaches is a fundamental cognitive and organisational strategy that helps individuals and groups make sense of complex information, facilitate communication, and promote effective decision-making.

Guided by the idea of 'Value Retention', a framework for different approaches was chosen that describes value retention in products or components by either slowing, narrowing, or closing the loops. Products, components, and materials are kept in a useful state for as long as possible, and they are fed back into the value creation at some stage of product realisation. 10 options are presented that describe value retention as practical strategies to embrace CE.

Theoretical Framework for Drivers and Enablers of Circular Economy

Despite many discussions about the benefits of implementing CE, the factors that drive companies to do so have not been systematically analysed. Neither were the factors that enable the implementation and make the cases successful.

K. Luban (✉)
Sevelen, Switzerland
e-mail: katharina.luban@ost.ch

A. Podleisek et al. (eds.), *Circular Economy: The Next Level of Company Success*,
https://doi.org/10.1007/978-3-031-66406-9_2

Therefore, in this chapter, a comprehensive framework of drivers and enablers is developed by means of an extensive and structured review of the literature. During the review process, a criterion-based selection and, finally, a full-text analysis produced a list of drivers and enablers that were coded, categorised, and clustered. The resulting framework consists of 16 drivers in 5 clusters and 12 enablers in 5 clusters.

Survey: Key Drivers and Enablers to Implement the Circular Economy in Industrial Companies

The resulting retention options, on the one hand, and the framework of drivers and enablers, on the other, were used in further research. The aim was to identify those drivers and enablers that play the most significant role in the adoption of the 10 different retention options in the industry.

To obtain this insight, a quantitative survey of the actual state of implementation of CE was carried out among European industrial companies. 512 respondents from 32 European countries responded to a standardised online questionnaire.

The survey confirmed the driving forces behind the adoption of CE and it was possible to determine differences in the relevance of drivers and enablers depending on the approach chosen for implementing CE. This research is particularly valuable for practitioners seeking guidance in choosing a promising approach to CE and understanding the necessary prerequisites for a successful implementation.

The findings of the study also strongly indicate that the incorporation of CE into business models can positively affect factors of business performance, such as profit, company image, and manufacturing costs. Retaining value undoubtedly has an economic implication. Many of the responding companies have successfully implemented strategies from the 10R framework and saw positive impacts on their business.

Chapter 3
Approaches, Drivers and Enablers of Circular Economy – A Theoretical View

Nicolas Hofer, André Podleisek, Roman Hänggi, and Katharina Luban

Abstract Circular Economy aims to decouple economic growth from resource consumption, offering businesses new ways to thrive in a sustainable way. This is achieved through resource efficiency and value-retaining practices. To accelerate the adoption of these practices, this chapter provides an overview of value retention options relevant to industrial companies. A detailed literature review was carried out to determine the key drivers and enablers of these value retention options, with particular emphasis on the European industry. This resulted in the identification of 16 driving and 12 enabling factors grouped into a framework, accompanied by an overview of the challenges companies face when transitioning to circular practises.

3.1 The Circular Approach to Value Retention

This chapter will provide an overview of the different Retention Options (Rs) that companies can adopt to implement Circular Economy (CE).

We focus on those value retention options that are relevant to manufacturing companies. First, the general functional aspect of the Rs, reducing the use of virgin materials and reducing waste in production, will be explained using the Value Hill model. Next, an overview of various Rs and R Frameworks will be provided, leading to the adoption of a specific R Framework in this study. The hierarchy of Rs will also be discussed, along with the identification of three distinct value propositions. The

N. Hofer (✉)
Zürich, Switzerland
e-mail: nicolas.hofer@ost.ch

A. Podleisek
Wettswil am Albis, Switzerland

R. Hänggi
Brülisau, Switzerland

K. Luban
Sevelen, Switzerland

A. Podleisek et al. (eds.), *Circular Economy: The Next Level of Company Success*,
https://doi.org/10.1007/978-3-031-66406-9_3

chapter will conclude by linking the Rs to different types of waste, using a practical example to illustrate the connections.

3.1.1 Methodology and Literature on the Implementation of Circular Economy

The literature on CE with respect to manufacturing is presented chronologically. Starting in 2015, Tukker [1] reviewed Product Service Systems (PSSs) to show their implications from a product and sustainability perspective. PSSs are business models in which companies maintain ownership of their products by renting, sharing, or leasing them out. This gives the company greater control over the lifecycle of the product and facilitates the application of CE practices.

In 2016, Lewandowski [2] reviewed existing Circular Economy Business Models (CEBM). CEBMs are specific types of business models that aim to close the loop on resource use by reusing, refurbishing, and recycling materials and products to create value by using resources in multiple cycles. The review aimed to identify gaps in the literature and propose a new framework for a better implementation of CE. Implementation strategies were also reviewed by Lieder and Rashid [3], but with a broader angle and paying particular attention to the manufacturing industry. They found that the existing literature rarely touched on the implementation level. Matsumoto et al. reviewed trends and challenges in research and development with a particular focus on remanufacturing [4]. They pointed out that most of the value in End-of-Life (EOL) products can be exploited by remanufacturing them, resulting in potentially high economic margins.

Swain [5] reviewed research and development on technologies to recover and recycle lithium from Lithium-Ion Batteries (LIBs) used in electric vehicles. The driving factor for this study was the growing demand for this metal and the low recovery and recycling rate. Swain concluded that technology must be further developed to make recovery and recycling more environmentally friendly and more economically attractive.

At the end of a product's life, Reverse Logistics (RL) and Closed-Loop Supply Chains (CLSC) play an important role in waste management. RL involves collecting and transporting used products and materials from their final destination back to the manufacturer or other designated point in the supply chain. CLSC extends this process by reintroducing these materials into production, creating a circular economy where waste is minimised and resources are conserved. In 2018, Islam and Huda [6] reviewed and summarised the relevant literature on RL and CLSC to provide an overview of the state of the art in these areas. The review examined the challenges and opportunities related to RL and CLSC, as well as strategies to optimise these processes. The authors identified logistical, economic, and environmental challenges and emphasised effective planning and coordination as key factors for successful implementation. Overall, the review provides insights into how companies can navigate the complexities of RL and CLSC and can serve as a helpful resource when looking to optimise these processes as part of the R-adoption efforts.

Due to increasing interest in CE, in 2019, several reviews on the role of digital technologies in supporting CE were published. Ávila-Gutiérrez et al. [7] conducted a review of the literature on the use of digital technologies in CE, focusing on the various types of technology that can be used to support CE initiatives and the various challenges and opportunities associated with their adoption. Kerin and Pham [8] focused on the various ways in which these technologies can be used to facilitate the transition to CE and the various challenges and opportunities associated with their adoption. Manavalan and Jayakrishna [9] reviewed the use of digital technologies in the fashion industry. Finally, Ren et al. [10] reviewed the literature on the use of digital technologies in the recycling industry.

Similarly, several studies have examined different aspects of CEBMs. Pieroni et al. [11] focused on the interfaces and application areas of CEBMs and how they can be integrated into different sectors and industries, while Lüdeke-Freund et al. [12] examined how CEBMs can reduce the extraction of raw materials and the generation of waste. To assess the effectiveness of CEBMs in achieving waste reduction and resource efficiency, circular performance assessment methods are necessary. Sassanelli et al. [13] conducted a review of these methods, examining the various approaches that have been proposed, as well as their strengths and limitations. Additionally, Saidani et al. [14] reviewed indicators for CE, which can be used by practitioners to set appropriate targets and track their progress in implementing CEBMs and transitioning to CE.

A similar review was conducted in 2020 by Kristensen and Mosgaard [15]. They focused specifically on the CE indicators for companies and products, identifying 30 indicators to measure progress. In the same year, Acerbi and Taisch [16] reviewed the literature on the adoption of CE practices in the manufacturing sector and identified current and possible future research directions in the literature.

This was continued in 2021 by Bjornbet et al. [17], who reviewed only case studies to measure progress toward CE of manufacturing companies. The list of relevant studies is growing year by year. The last publication to be mentioned here was conducted by Rahman et al. [18], who reviewed how 10 different actions can be applied to reduce the waste stream of EOL LED lamps.

While the aim of achieving waste reduction and improving efficiency is shared among the aforementioned reviews, other studies concentrate on identifying the factors that contribute to a successful transition to CE. These reviews are then systematised.

3.1.2 *The Concept of Value Retention*

As outlined in the Introduction to this Book, there are limits to the resources available and the emission-absorbing capacity of our planet. Therefore, human activities must be within biophysical boundaries to ensure the viability of future generations [19].

In a resource-based context, CE aims to extract as much value as possible from the resources used and keep the materials in the loop as long as possible.

In this sense, Rs are placed at the heart of CE because they are regarded as the 'how-to' in minimising the inputs of the resources used and the waste produced [20]. Rs are the practical approach to achieving circularity that can be applied to the 'concept and design' phase as well as the 'production and use' phase of a Product Lifecycle (PLC) [21]. Their general objectives are to design out waste, reduce the need for virgin materials and retain the intrinsic value of resources by slowing, narrowing, or closing the loops, i.e., keeping products, components, and materials in a useful state for as long as possible.

However, the deployment of Rs also makes sense from a business perspective. Manufacturing companies typically strive to increase the value of their products while simultaneously decreasing the cost of production. Since reducing waste throughout production is one way to reduce production costs, the deployment of Rs can have a significant impact on achieving these objectives. Therefore, it can free up capital that can be allocated to more valuable aspects of the product, such as improving the quality or features of the product, ultimately leading to an increase in the perceived value of the products for customers. The functionality of Rs can be demonstrated by examining the PLC, which is analysed in both a linear and a circular context using the Value Hill model. This model is structured around three PLC stages, namely *pre-use*, *use*, and *post-use*.

The Linear Value Hill

The Value Hill in Fig. 3.1 shows the linear PLC. In this model, the focus is generally on maximising short-term profit, and little attention is paid to the long-term sustainability of the product or the resources used to make it.

In the *pre-use* phase, virgin material extraction takes place first. Then it is refined for manufacture and transformed into parts via a mould or tool. Then the parts are assembled into products and are ready for retail. Desing et al. [19] pointed out that every conversion process to convert raw material into a finished product requires different types of energy and resources (e.g., electrical, thermal, labour, or transportation). At the same time, different forms of waste are produced (e.g., scrap and various emissions). On the one hand, this creates costs. On the other hand, value is added at every step to meet customer requirements. The product reaches its maximum value when it is ready for sale and as such is at the top of the Value Hill.

During the *use* phase, the product value starts to decrease due to wear and tear or other factors. Once the product is no longer useful, it is discarded as waste and sent to a landfill or incinerated. This marks the end of the product's lifecycle and represents the loss of its value as it has descended to the bottom in the *post-use* phase.

This also applies to the parts, components, and materials of which the product is made, meaning that all their value is also lost. As a result, all products produced in the Linear Economy (LE) must go through all stages of the *pre-use* phase, which requires constant input of virgin materials and energy and leads to all forms of waste in the *pre-use* and *post-use* phases. This is a waste of resources and reduces economic and environmental efficiency.

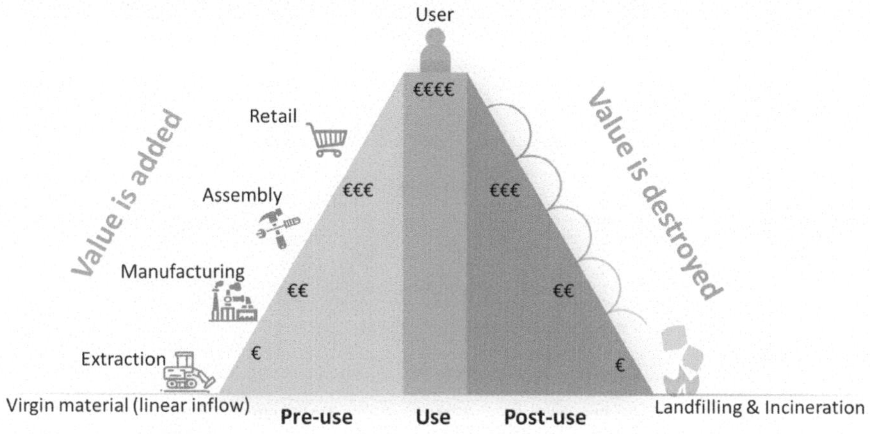

Fig. 3.1 The Linear Value Hill (own representation adapted from [22])

In contrast, using Rs aims to keep resources in use as long as possible and to recover and regenerate products and materials at the end of their useful life.

The Circular Value Hill

In the Circular Value Hill (Fig. 3.2), value is also added as the product moves uphill in the *pre-use* phase. Likewise, the maximum value is reached at the top of the hill in the *use* phase. However, when the product cascades down in the *post-use* phase, the Rs aim to slow or close this cycle and attempt to capture the added value of the product beyond its use phase by conserving all or parts of resources already used [22]. Therefore, in a CE, the lifecycle of a product does not end and instead is constantly extended or renewed.

Figure 3.2 illustrates this concept through three arrows, two *slowing the loop* and the last *closing the loop*. The upper arrow represents the *looping of products*, the middle arrow represents the *looping of components*, and the lower arrow represents the *looping of materials*. These loops enable the retention of value, as opposed to value being lost, by redirecting products, components, and materials back to the *pre-use* phase. The closer the loop is to the top of the hill, the more value can be retained, as it bypasses more stages of the pre-use phase, which helps prevent waste and emissions generated in earlier stages. This can lead to cost savings in production from a holistic value chain view.

Rs that enable the upper two loops *slow the flow* of resources by extending or intensifying the use period of products and components [23]. Rs that enable the looping of materials (bottom loop) build the bridge between post-use and manufacturing. By returning the material and allowing it to be used for new applications, they *close the loop* of resources [23].

Since a product's lifecycle is usually understood as the period from production to disposal, use cycle extending and use cycle renewing can be used to describe the process more accurately. This conceptual change highlights the key differences between

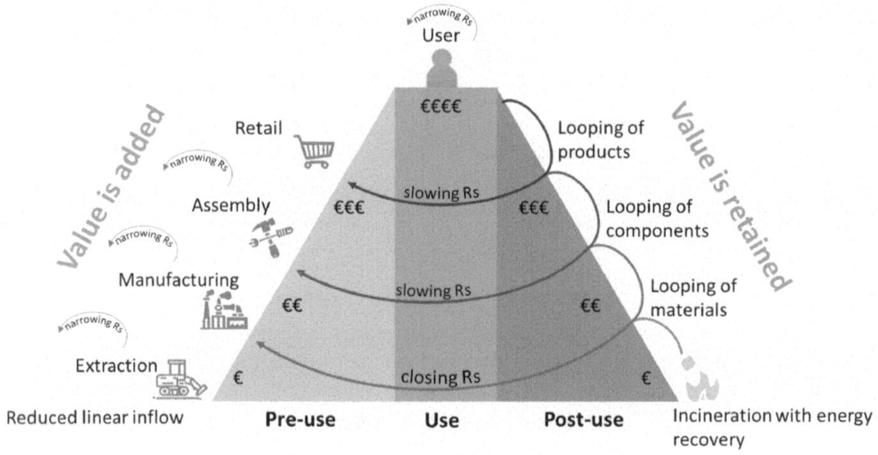

Fig. 3.2 The Circular Value Hill (own representation adapted from [22])

linear economy and circular economy, with the latter prioritising the continuation of resource use rather than disposal.

In theory, Rs would allow an indefinite looping of materials. However, complete closure of cycles is hardly possible due to different mechanisms such as dispersion, dilution, contamination, process losses, and due to the mechanical properties of the materials [19].

However, these Rs reduce the amount of virgin material inflow and the production of waste in the *pre-use* phase while minimising the waste that reaches the bottom of the Value Hill in the *post-use* phase. In that case, controlled and regulated waste incineration to recover energy can avoid the pollution otherwise caused by uncontrolled waste burning and landfilling.

In addition to what has been said, one more category of Rs has yet to be addressed, those that *narrow the flow*. Narrowing Rs aim to reduce resource use, improve the efficiency of production processes, and ultimately decouple value creation from virgin material consumption [23, 24]. In this regard, they are to be applied at any stage of the *pre-use* phase and help to eliminate waste before it is even created. Due to the fact that they address the *initial use* of resources, in contrast to *looping* them, they do not primarily retain value. Instead, they can be considered value-adding options.

While waste management holds a central position in CE, it is not a novel idea. These narrowing Rs can include strategies such as Lean Management, increased material and energy efficiency, and reuse of materials or waste heat in the process. Lean Management has been focused on waste reduction since its inception. This understanding was not motivated by circularity, but rather by cost reduction and operational excellence [25]. Nevertheless, the attitude is the same.

Although resource efficiency enhancement options are not exclusive to the circular Value Hill, their circular potential unfolds if they reduce the inflow of non-renewable virgin materials by utilising looped materials instead or substituting them with environmentally friendly ones altogether.

Value Cycle in Product Lifecycle Management

While the Value Hill is considering the lifecycle of a product, PLC management looks at the creation and management of products from the product's market lifecycle, as described by Nyffenegger et al. [26], see Fig. 3.3. This engineering-driven approach proposes a data concept throughout the lifecycle process of a product: Using Product Lifecycle Management (PLM) to combine the horizontal information perspective of the life of a discrete product with the vertical perspective, that is, the lifecycle of the product offering as such, for cross-functional improvement [27]. This combination helps to identify where to address which CE concept in a PLM. We will address this in more detail in Sect. 3.1.5.

3.1.3 Hierarchy of 10 Retention Options

With the functional aspects discussed, the focus is shifted to the specific Rs. In the literature, various authors have proposed many different Rs. Reike et al. identified 38

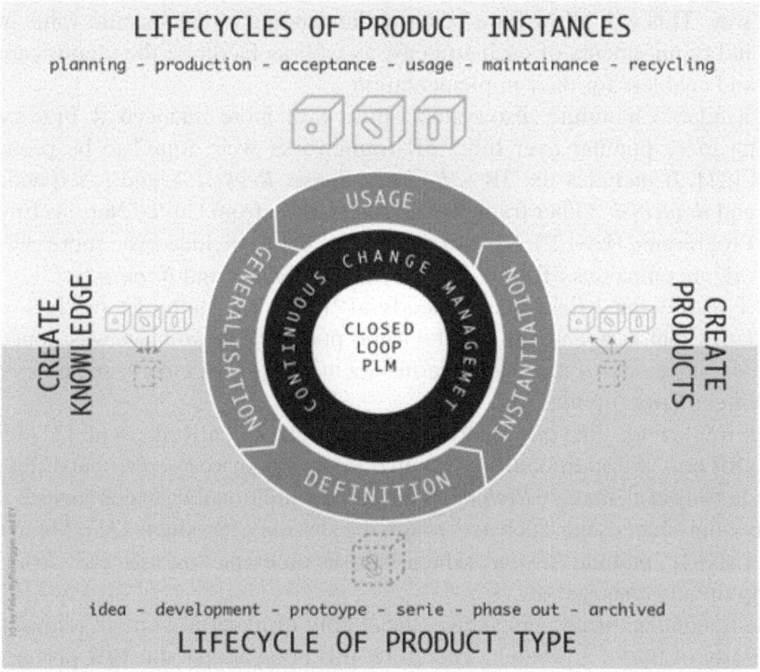

Fig. 3.3 Product Lifecycle Management [26]

different strategies in their review of the literature. They are: 're-assembly, re-capture, reconditioning, recollect, recover, recreate, rectify, recycle, redesign, redistribute, reduce, re-envision, refit, refurbish, refuse, remarket, remanufacture, renovate, repair, replacement, reprocess, reproduce, repurpose, resale, resell, re-service, restoration, resynthesize, rethink, retrieve, retrofit, retrograde, return, reuse, reutilise, revenue, reverse and revitalize' [21, p. 253]. They also exemplified that Rs are listed in various combinations as R Frameworks.

An example of an R Framework is the **3R**, which consists of *Reduce*, *Reuse*, and *Recycle*. This framework is often used as a foundation for waste management and sustainability efforts and was found to be the most prominent in 2016 [28]. The implication of this framework is that the Rs are defined in a broad sense. *Reuse*, for example, as used by Bakker et al., is defined as 'using a product or component again for the same purpose for which it was conceived' [29, p. 11]. The absence of a clear distinction between *direct reuse*, i.e., without the need to restore product functionality, and *indirect reuse*, with the need to restore product functionality first, leaves room for multiple retention options to be included under the term *Reuse*. Similarly, the definition also does not clarify the degree of restoration needed for *Reuse*.

Given this implication, Bakker et al. has listed product life-extending options, specifically *Repair*, *Refurbishment*, and *Remanufacturing*, as options for *Reuse* while not considering *direct reuse* at all. To eliminate uncertainty, it is important to clearly distinguish between different circular retention options and define them in a more precise way. This will allow for a better understanding of the specific value propositions and requirements of each strategy, as well as facilitate the identification of drivers and enablers for their implementation.

The available literature also reflects this, with more nuanced R Frameworks becoming more popular over time. **5R**-frameworks were found to be prominent in 2018 [21]. It includes the 3R – *Reduce*, *Reuse*, *Recycle* – and is expanded by *Repair* and *Refurbish*. Other frameworks, like the one from United Nations Environmental Programme (UNEP), have expanded the list to include even more. UNEP's framework encompasses **5R** plus *Refuse*, *Remanufacture* and *Repurpose*.

UNEP also defined *Reuse* more clearly as 'the using again of a product, object or substance that is not waste, for the same purpose for which it was conceived, without the necessity of repair or refurbishment' [30]. This clearly separates direct from indirect *Reuse* options.

Other frameworks like the one from Potting et al. [31] or Reike et al. [21] include UNEP's **8R** plus two additional ones. Both frameworks share *Recover* but differ in the second. Potting et al. include *Rethink* as the second additional R, which focuses on the intensity of product usage, such as through the sharing of products [31]. Meanwhile, Reike et al. [21] include *Remine*, which focuses on extracting materials from EOL products already disposed of.

From a manufacturing perspective, the supply of material is more relevant than the intensity of use of a product. Therefore, this book adopts the **10R** proposed by Reike et al. [21]. A descriptive overview of the Rs and commonly used synonyms is provided in Table 3.1. It is important to remember that different R Frameworks may

Table 3.1 Overview of the retention options and their description

R# option	Description
R0 Refuse	Rejecting virgin or hazardous materials in product design [21] or eliminating physical products by offering the same function or combined functions. Typically enabled by radically different products, technologies, or both [31]. Synonyms: reinvent, prevent
R1 Reduce	Eliminating waste in production by using less material per unit produced, as well as reducing the input of natural resources and energy in general [21, 31]
R2 Reuse	Extending the existing use cycle of products or parts by using them again for the same purpose for which it was conceived without reconditioning [21]. It can involve a minimum amount of condition monitoring (cleaning, repackaging), but involves no disassembly. Synonyms: redistribution, resell
R3 Repair	Extending the existing use cycle of products or parts by maintenance or replacing broken parts to retain or restore the original functionality [21]. Synonyms: Corrective, condition-based, predictive, prescriptive maintenance
R4 Refurbish	Extending to new use cycles of products or parts by repairing or replacing defective components of a multi-component product to restore functionality that may be inferior to the original specification [32]. It can also refer to upgrading the product with newer/advanced components [21] and may involve partial disassembly. Synonyms: recondition, retrofit, refresh, remodel
R5 Remanufacture	Extending to new use cycles of products or parts and using them for the original function with at least the same Original Equipment Manufacturer (OEM) performance specification and quality [32]. It is a standardised industrial process that involves complete disassembly, check, cleaning, and reassembly. Synonyms: rebuild, overhaul, remake
R6 Repurpose	Extending to new use cycles of products and parts by using them for other purposes than their original function [21, 31]. Synonyms: alternate use
R7 Recycle	Extending the material lifespan by processing mixed streams of products to obtain the same high-grade or lower-grade quality [32]. It involves technological equipment to shred, melt, and capture materials, and the structure of the product is not maintained [21]. Recycled material is often referred to as secondary material that can be reintroduced into production for the same initial purpose or alternative ones
R8 Recover	Extracting elements or materials from end-of-life compounds that are not recycled, and recovery of energy through incineration [21]. Synonym: waste-to-energy
R9 Remine	Extracting valuable materials (e.g., gold, copper) stored in landfills for revalorisation purposes [21]. Synonyms: landfill mining, urban mining, ocean cleanup

prioritise certain Rs over others and can include other retention options beyond the ones listed. The choice of which R Framework to use depends on the specific goals and needs of a company or organisation.

The Rs in Table 3.1 were ordered from the most desirable/preferred (R0) to the least (R9) from a CE perspective, based on their environmental effects and the inherent value they retain. In addition, the number can be seen as a representation of the amount of effort put into it, or the 'length' of its respective circular process; hence, we talk about short, medium and long loop Rs, as shown in Fig. 3.4.

Short Loop Rs (R0 and R1)

Short loop Rs narrow the flow of resources. They are in the pre-use phase of the Value Hill and preferred, because they aim to mitigate the need for the extraction of virgin material and waste from the outset. This is achieved by completely eliminating the use of certain or all materials as Refuse (R0) or by using fewer materials through efficiency and innovation as Reduce (R1). Therefore, they impact product design and production processes, making them effective in reducing the dependence on virgin materials, increasing operational efficiencies, and reducing the environmental impact of products. This is also considered the value proposition they offer to manufacturing companies [33]. They are placed next to *production* in Fig. 3.4.

Medium Loop Rs (R2 to R6)

Medium loop Rs slow down the flow of resources. Reuse (R2) and Repair (R3) extend the existing use cycle of products or parts while Refurbish (R4), Remanufacture (R5), and Repurpose (R6) extend it to a new use cycle. This is also illustrated in Fig. 3.4 where R2 and R3 have their starting and ending point at *use*, while R4 to R6 have their starting point at *end of use* and lead back to *production*.

R2 does not require additional resources to extend the use cycle. R3, R4 and R5, on the other hand, require more work and resources incrementally and are therefore placed further down with higher numbers. R6 is a particular case because by repurposing a product, the intended product's function is lost and is generally replaced by a new product. For example, by repurposing used truck tyres as fenders for impact protection, the flow of resources to manufacture tyres does not decrease. If they were remanufactured instead, the required resources are reduced. Thus, R6 is the least circular option of this group, as the manufacturing of a new product requires new resources and produces more waste in the pre-use phase.

In contrast to short loop Rs, medium loop Rs allow manufacturers to make use of residual value of products or parts by renewing their use cycle [33]. These strategies are placed in the middle of the hierarchy because they slow the flow of resources and do not fully reduce the dependency on virgin materials.

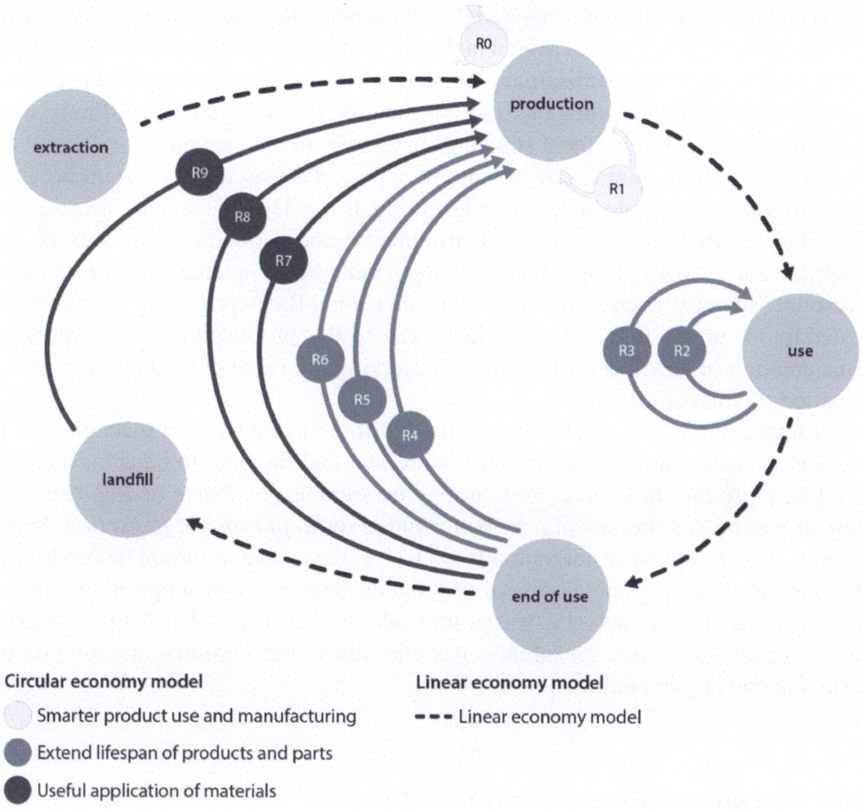

Fig. 3.4 Retention strategies in the circular economy model from the perspective of manufacturing companies (own representation based on [21])

Long Loop Rs (R7 to R9)

Long loop Rs 'denote traditional waste management activities' [21, p. 256] and are responsible for closing the cycle of materials. They enable the flow of materials from *end of use* directly back to the *production* stage (Fig. 3.4). The exception is the bottom option Remine (R9), which, according to its definition, is the extraction of materials already disposed of in landfills. Because Recycle (R7) and Recover (R8) are strategies that prevent waste from ending up in landfills, they are ranked above Remine – while Remine addresses waste that has already been created rather than preventing waste from being generated in the first place.

Compared to short and medium loop Rs, waste management activities require that the products be shredded, melted, or other processes are used to break them down into materials or particles. Because the structure of the product is destroyed, only the value of the material remains. Furthermore, these strategies are usually highly energy-consuming processes that often require expensive technological equipment, which can exceed the material value obtained [21]. This being the case, some consider them only marginally better in environmental and economic efficiency than landfilling EOL products [34]. Due to this, many scholars agree that these are the least preferred Rs and should be applied when there are no other alternatives [35]. Accordingly, they represent the three bottom-most strategies in the hierarchy presented.

Despite this, long loop Rs are an important part of a closed-loop system because they help prevent valuable materials from getting lost in landfills and use them again. This can contribute to the overall environmental and economic efficiency of the system. It is also worth noting that long loop Rs can be complementary (or, in cases of resource scarcity, even a prerequisite) to short loop Rs by providing the materials needed for the production of new products. Their value proposition can be expressed as the creation of new forms of value by preserving the value of the materials of the discarded products [33].

It is important to note that the suitability of different strategies for managing EOL products depends on the condition of the product and the goal to be achieved. For example, a product in good condition may be suitable for *Reuse* or *Refurbish*. In contrast, a damaged product may be better suited for *Repurpose* or *Recycling*. In the long run, the strategies are not mutually exclusive; they can and should be combined. All strategies have the common goal of reducing resource consumption and transforming discarded products, i.e. waste, into valuable resources. This helps conserve natural resources, reduce greenhouse gas emissions, and minimise the amount of waste that ends up in landfills.

3.1.4 Strategic Options Driven by Rs

Material Reduction Strategies

The Rs included in this group aim to reduce the use of virgin materials and the generation of waste generated during production. As such, they affect product design and production processes. Companies that are successful in these strategies can decrease their demand for virgin materials, increase operational efficiency, and reduce their environmental impact [33].

- *R0 Refuse* – refusing the use of materials by dematerialising/eliminating physical products by providing the function through a radically different offer, e.g. music from physical CD to electronic mp3.
- *R1 Reduce* – decreasing virgin material usage in product design by using renewable or recycled materials instead.

Product Use Cycle Extending Strategies

These strategies slow the flow of resources in the economy and allow companies to extract value from products by extending their use cycle [33]. This means that they aim to extend the period during which a product performs its original function.

- *R2 Reuse* – reusing, redistributing, or reselling a product that is still in good condition (minimum amount of condition monitoring, such as cleaning and repackaging, can apply).
- *R3 Repair* – corrective or condition-based maintenance (replacing defective parts) to extend a product's useful life.

Product or Part Use Cycle Renewing Strategies

These strategies also slow the flow of resources and allow companies to extract value from products or parts thereof [33] that have reached the end of their lifecycle. When these strategies are implemented, the lifecycle is renewed to a new use cycle.

- *R4 Refurbish* – upgrading a product with newer or advanced components or restoring the functionality of a multi-component product by replacing defective parts.
- *R5 Remanufacture* – rebuilding a product or its parts. It is an industrial process that involves the total disassembly of products, followed by checking, cleaning, replacing, or reworking the parts where necessary.
- *R6 Repurpose* – 0using discarded products or their parts for a purpose other than the original.

Material Processing Strategies

These Rs are waste management techniques to close the material cycle. This involves the processing of discarded products or parts into materials that can be used and reused in the production of the same or other products to create new forms of value [33].

- *R7 Recycle* – involves processing materials to obtain the same material with equal or lower quality.
- *R8 Recover* – recover energy from processing materials through incineration (waste-to-energy), or with bio-chemical, chemical or kinetic operations.
- *R9 Remine* – extraction of valuable materials (e.g., gold, copper, plastic) already disposed of in landfills, waterways or nature.

3.1.5 *Retention Option Throughout the Product Lifecycle – A Process View*

Adding to the conceptual outline in Sect. 3.1.2 of incorporating circular economy thinking into the product development process or PLM, we dive into the framework developed by Nyffenegger et al. (see Fig. 3.3). This framework distinguishes between the innovation process as a *lifecycle of product definitions* and the product's lifespan as a *lifecycle of product instances*. It focusses on a process view that is crucial for addressing all strategic options relevant to CE. Reike et al. suggest linking circular retention strategies R0 to R9 with the product definition lifecycle [21]. The elements of corporate sustainability strategy, sustainability design guidelines & principles, and sustainability design implementation must be integrated into the product definition lifecycle through the respective retention options R0 to R9. This approach ensures that the needs of a CE are addressed in every phase of product definition.

Furthermore, we expand this perspective by including the production process (product creation) and the lifecycle of the product instances. Here, the retention strategies can also significantly enhance circular economy practices. Our research indicates that the relevant drivers and enabling factors (see Sect. 4.2) play a crucial role in defining and continuously improving the PLC process.

The PLM framework of Nyffenegger et al. proposes creating a digital twin of each unique product instance. This greatly facilitates learning throughout the lifecycle of any product instance. This closed-loop thinking, as emphasised by Nyffenegger et al., is essential for learning and continuous improvement, which are vital for circular economy practices. The knowledge generated serves as a foundation for further advancements in sustainability with each new product generation. This learning will be incorporated into the corporate sustainable strategy. Figure 3.5 summarises key inputs for circular economy in the PLC process and proposes a circular economy extended PLC model.

3.1.6 *The Value of Waste in a Circular Economy*

What is the monetary aspect of waste, assuming that waste is valuable? Lacy et al. [36] assign it to be worth 4.5 trill. USD globally, which is contained in four categories of waste: wasted resources, capacity, lifecycles, and value. A definition for each type of waste connected with the value propositions presented previously is shown together with the applicable R in Table 3.2. It also shows that, with the **10R**-framework, each type of waste is addressed. In this sense, companies that implement multiple Rs have greater potential to minimise their own waste and capitalise on the waste of others.

The fashion industry provides a good example to demonstrate the four types of waste. Fast fashion relies on customers' recurring consumption and impulse purchases [37]. This pushed mass production with large quantities and low prices and has led customers to buy more, wear each item less, and dispose of them after little

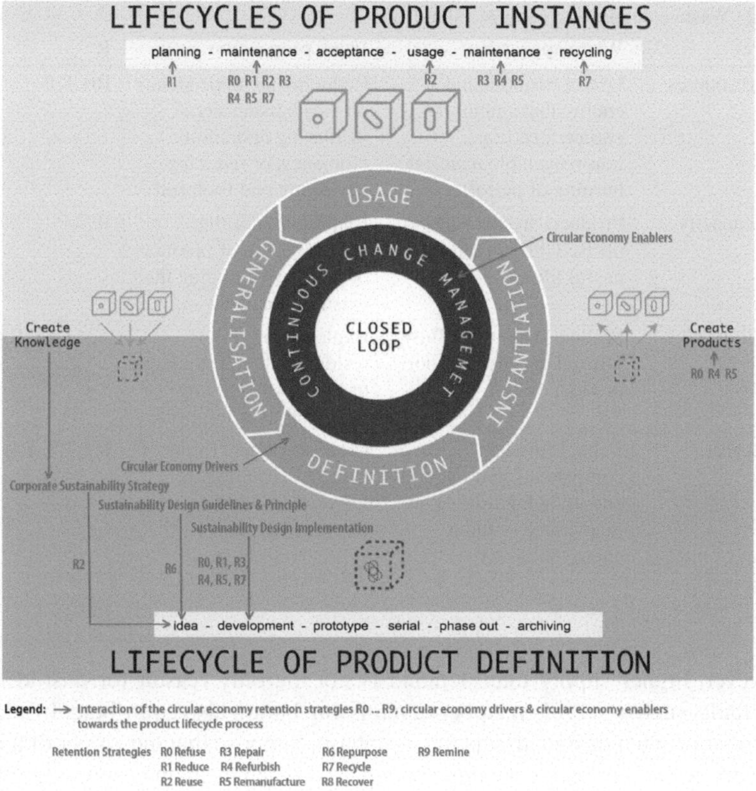

Fig. 3.5 CE integrated into product lifecycle process (CE extended PLM) based on [21, 26]

use [38]. In this scenario, the disposal of flawless clothing represents *wasted capacity*. Europe's second-hand apparel market capitalises on that waste with the *Reuse* strategy. It is estimated that the market value was nearly 17 bill. EUR in 2022 and is expected to increase to 38 bill. EUR by 2032 [39].

Some items, in addition, are never even sold in the first place and are incinerated or thrown away [37]. If one would imagine that the entire supply chain would be owned by the same company, the production of these unused items only generated costs without returning any profit. This represents three types of waste. First, it is *wasted capacity* because the items have not been worn. Second, it is *wasted resources* due to overproduction, and third, it is *wasted value* because both the functional and the material value are lost by landfilling it.

The New York Times [40] reported that the Swedish fashion giant H&M has a deadstock (new unsold items designated as waste) worth 4.3 bill. USD, and The Times [41] reported that the British luxury brand Burberry incinerated 28 mill. GBP worth of deadstock in 2018. In addition, textiles waste disposal comes with additional costs. For the British economy, it is estimated to be 82 mill. GBP a year [42].

Table 3.2 Waste categories to capture value from

Category	Waste appearance	Value proposition	R
Wasted resources	Use of material and energy that cannot be regenerated (e.g., non-recyclable material, burning of petrol)	Reducing the dependency on virgin materials, increasing operation efficiency, or reducing environmental footprint	R0, R1
Wasted capacity	Products are not fully utilised throughout their useful life	Exploitation of the residual value of products or parts by extending their existing use cycle	R2
Wasted lifecycles	Products cannot be fully utilised because of poor product design	Exploitation of the residual value of products or parts by extending* or renewing their use cycle	R3*, R4, R5, R6
Wasted value	Value lost when components and materials end up in landfills or are incinerated without energy recovery	Creation of new forms of value by retaining the material value of discarded products	R7, R8, R9

However, higher supply than demand is not the only reason for unsold items. Design faults such as a broken zipper or a missing button also contribute. In addition to that, poor product design affects the use phase. Suppose that the zipper of a jacket breaks and, due to the design, it cannot be repaired. Then the item becomes worthless, even if the rest of it is still in good condition. As such, this represents the category of *wasted lifecycles*.

There is a high potential for Rs throughout the value chain, which is not exclusively the case for fashion products. For example, the disposal of products with 'unused functional potential' [43, p. 6] is a widespread phenomenon in the Electrical and Electronic Equipment (EEE) sector, one of the fastest growing today [44]. This is mainly driven by the short PLCs, the constant modernisation of electronic devices, and a society that demands the latest technologies [45]. A particular example is the almost annual release of a new iPhone [46], accompanied by customers queuing up the day before the sale [47], ready to dispose of the 'old' but still functioning one as e-waste. Of the 12 mill. tonnes of electronic waste produced in Europe annually, the material value alone is estimated at 12.8 bill. EUR [48]. However, most materials are not recycled or recovered [49].

Considering that the European industry could save 630 bill. EUR annually through the better use of resources [50], the concept of circulation should resonate with an industry driven by the idea of achieving economic value. The potential benefits of implementing Rs are numerous, and their inclusion will be necessary for manufacturers to compete on a global level in the long term [51].

Many companies have already implemented Rs, such as DESSO and Hewlett–Packard, and have benefited from waste reduction, technological developments, and cost savings. DESSO, which was acquired by Tarkett in 2015, for example, makes use of several Rs. The Dutch carpet manufacturer has been in business for nearly a century and has successfully reduced its demand for virgin material to make yarn. In 2008 the company started to take back any used carpets and invested in developing carpet recycling technology. The challenge was to recycle yarn without degrading it. Since the company succeeded, recycled yarn has become a new raw material in their carpets. By 2011, 60% of the company's carpets were manufactured using only recycled material [52]. The company also saves resources by using modular carpet designs to enable easier carpet repair. They only have to swap broken parts instead of the entire carpet flooring. This saves time, reduces resource use, and creates less waste.

Hewlett–Packard is also active on various fronts. For example, they aim to be more efficient and sustainable in their packaging design. By redesigning the packaging, Hewlett–Packard reduced the amount of packaging material used per product and also made it reusable. Furthermore, they added secondary materials and replaced hard-to-recycle materials with recyclable substitutes. Hewlett–Packard claims that this saved them 12 mill. USD in 2014 [53].

There are many other examples of companies that have successfully implemented strategies from the 10R framework and seen positive impacts on their business (e.g., [54–56]. Despite these successes, not all companies opt to embrace these strategies. This observation leads to an important inquiry: What motivates certain companies to implement Rs, while others choose not to? The factors contributing to this decision will be explored in the next chapter.

3.2 Developing a Framework of Drivers and Enablers

The research is based on a comprehensive review of the abundant literature on circular economy, aiming to identify and analyse key drivers and enablers. The findings of the review will be used to construct a theoretical framework specifically for drivers and enablers that influence the adoption of Rs by manufacturing companies. The chapter starts with the description of the research methodology and then presents and discusses the driving factors of CE. This is followed by the identification of enabling factors, which are discussed alongside barriers for a clearer understanding of the results.

For a summary of these findings and a discussion on the research limitations, please refer to Sect. 3.3.

3.2.1 Methodology and Literature Review on Drivers, Enablers, and Barriers

In recent years, there has been a significant increase in the number of scientific articles published on CE. A search for review articles with the keyword 'circular economy' using the Web of Science database returned over 800 articles published in 2021, compared to just 59 in 2017 and 5 in 2014. The sample of review articles presented here is divided into two groups: those related to CE and the manufacturing industry and those focused on the drivers, enablers, or barriers to adopting CE practices.

A keyword search was conducted using two multi-publisher databases (Web of Science and ProQuest) to identify relevant literature. Using two databases is considered appropriate to provide a comprehensive understanding of a topic [57] and allows for the repeatability of the search process. The literature identification process was divided into two steps: the initial keyword search and the subsequent screening of the resulting articles. This led to the inclusion of the sample in the review. The complete process can be seen in Fig. 3.6.

Literature Identification

In the literature identification process, a string of keywords was applied using Boolean logic to both databases and searched for in the titles, abstracts, and keywords of the papers. Multiple search strings were tested and discussed with three professors for applicability. The final search string is presented in Table 3.3. It includes the terms 'Circular Economy' and 'Closed Loop', the latter of which is commonly cited as a synonym of CE. These terms were separated by the *OR* function to return

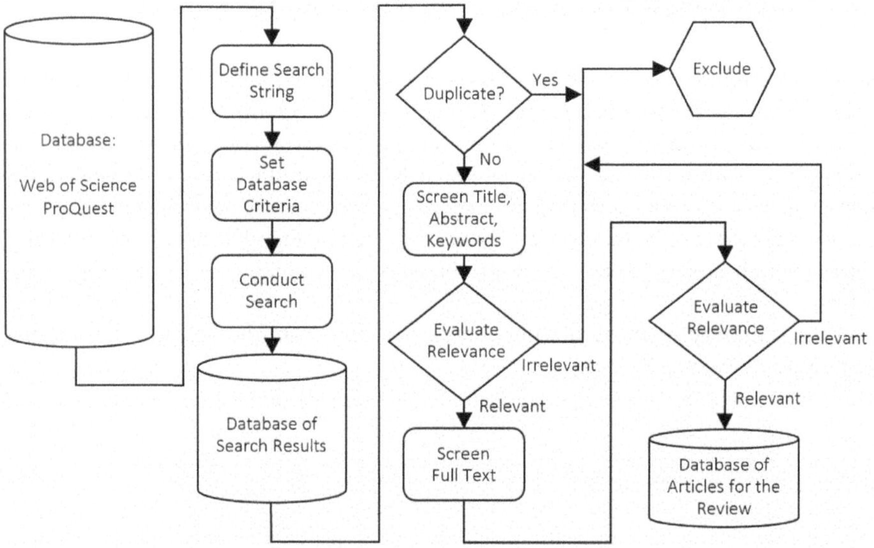

Fig. 3.6 Process for identifying relevant literature

Table 3.3 Boolean literature search string and results

Search string	Database	# Results
(((Circular Economy) OR (Closed Loop)) AND (European Union OR EU OR Europe*) AND (Survey OR Case OR Report) AND (Retention OR Refuse OR Reduce OR Reuse OR Repair OR Refurbish OR Remanufacture OR Repurpose OR Recycle OR Recover OR Remine) AND (Driver* OR Enabler* OR Opportunity* OR Accelerator* OR Constraint* OR Practice* OR Success Factor* OR Obstacle* OR Challenge* OR Inhibitor*))	Web of Science	157
	ProQuest	52
Total number of papers found		209
Duplicates excluded		26
Rejected cause of title, keywords, and abstract		89
Rejected after scanning of text		62
Total number of papers included in the review		32

papers that included either of them. The search string was then combined with the strategy names of the 10Rs using the *AND* operator to further narrow the search to papers related to the R Framework. The strategy names were separated by the *OR* operator. The string was further refined by including the terms 'European Union,' 'EU,' and 'Europe*' to focus on the geographical location of interest. The asterisk was used as a wildcard symbol to match words starting with the same letters but with different endings. The same applies to the included terms 'Driver*,' 'Enabler*,' and 'Barrier*' and their various other denotations, as found in previous literature reviews (see Sect. 3.2.1). Because the absence of an enabler is often perceived as a barrier, 'Barrier' was included in the search. To minimise the findings based on theoretical data, the string also included the terms 'Survey', 'Report', and 'Case' (as in case study) to find papers based on empirical data. In addition to that, four criteria for the database search were set:

1. Only articles (books, white papers, etc. were excluded)
2. Only peer-reviewed
3. Only English language
4. Published in scholarly journals

The final search was carried out on May 31, 2022, and a total of 209 publications were found. After removing duplicates, 183 unique articles remained.

Screening

The remaining papers were then evaluated in two steps. In the first step, the titles, keywords, and abstracts of the papers were read, and exclusion criteria were applied. The criteria were the following: the paper does not relate to CE; the paper focuses only on the mining, bio-, agri- or food industry; the paper is specific to a region or country not in Europe; the paper focuses on the impact of CE on the social dimension;

the paper deals with performance measures; the paper is not in English. As a result, a total of 89 articles were rejected from further evaluation.

In the second step, the text of the remaining papers was screened for relevance to the topic. A set of inclusion criteria were applied to decide if the article would be included in the final sample. The inclusion criteria were the following: literature discusses drivers, enablers, and/or barriers; literature is based on empirical data. Any paper that did not deal with the topic of interest was excluded from the sample.

This resulted in a total of 32 peer-reviewed papers, which are included in the review. Of these, 25 papers are relevant for the drivers, and 20 are relevant for the enablers, with overlaps.

Review

To understand why some manufacturing companies have implemented Rs while others have not, it is necessary to explore the factors that drive companies to adopt Rs, the enabling factors that facilitate their practice, and the barriers that may hinder their adoption. This provides an understanding of the difficulties and possibilities that companies face when trying to put Rs into action and can help to recognise approaches that can facilitate the successful implementation of these techniques. The following literature review was conducted to identify the drivers, enablers, and barriers of R adoption by industrial companies, with a specific focus on the European industry.

The published reviews on drivers, enablers, and barriers are shown in Table 3.4. The overview includes the title, author, year, and applied perspective. The named studies reviewed drivers and barriers to transitioning to a CE in different industry sectors, including the agri-food supply chain, the fashion industry and eco-innovation. They also covered various perspectives such as government, organisation, supplier, and consumer. For example, Aloini [61] and van Eijk [65] targeted political decision-makers, while Tan et al. [58] used the perspective of business leaders. Others, such as Govindan and Hasanagic [63] focused on finding factors that influence the implementation of CE in a supply chain context using a multi-perspective approach. These studies have identified a range of drivers and barriers to transitioning to a CE, including economic, technological, regulatory, cultural, and behavioural factors.

Many authors use multiple categories to classify these factors, helping to identify areas where progress must be made [66]. For example, de Jesus and Mendonça [64] used four categories for drivers (technological; social and culture; institutional and regulatory; economic, market, and financial), and Aloini et al. [61] used seven categories (technological; supply chain; social; organisational; institutional, environmental; economic).

Our research revealed that most reviews address only drivers and barriers and focus less on enablers. This is explained by the fact that the authors use 'driver' and 'barrier' as metonyms. As such, the term *driver* is often used as a substitute for enablers, opportunities, or accelerators, while the term *barrier* represents constraints, obstacles, challenges, or inhibitors (e.g., [58–60, 62, 64]). It is also worth noting that barriers and enablers can be perceived as mirror images in some cases,

Table 3.4 Overview of previous literature reviews on drivers, enablers, or barriers

Title	Author	Year	Applied perspective
Transitioning to a Circular Economy: A Systematic Review of Its Drivers and Barriers	Tan, J., Tan, F. Ramakrishna, S. [58]	2022	Reviewed existing literature for drivers and barriers to CE from a business leader's perspective
Drivers and Barriers Towards Circular Economy in Agrifood Supply Chain: A review	Mehmood, A., Ahmed, S., Viza, E., Bogush, A., Ayyub, R. [59]	2021	Reviewed existing literature for drivers and barriers to CE from an agri-food supply chain perspective
Can Fashion Be Circular? A Literature Review on Circular Economy Barriers, Drivers, and Practices in the Fashion Industry's Productive Chain	de Aguiar Hugo, A., de Nadae, J., Lima, R. [60]	2021	Reviewed existing literature to find and categorise the drivers and barriers to CE from the perspective of the fashion industry
Driving the Transition to a Circular Economic Model: A Systematic Review on Drivers and Critical Success Factors in Circular Economy	Aloini, D., Dulmin, R., Mininno, V., Stefanini, A., Zerbino, P. [61]	2020	Reviewed existing literature to classify drivers and critical success factors (enablers) to support practical implementations and to drive policymakers in their CE agenda
Unlocking Circular Business: A Framework of Barriers and Drivers	Tura, N., Hanski, J., Ahola, T., Ståhle, M., Piiparinen, S., Valkokari, P. [62]	2019	Reviewed previous literature on CE and combined it with results from an empirical case study to develop an integrative framework for drivers and barriers from a CE business perspective
A Systematic Review on Drivers, Barriers, and Practices towards Circular Economy: A Supply Chain Perspective	Govindan, K., Hasanagic, M. [63]	2018	Reviewed existing literature for drivers, enablers, and barriers to CE from different perspectives (governmental, organisational, supplier, and consumer) to construct a conceptual framework
Lost in Transition? Drivers and Barriers in the Eco-Innovation Road to the Circular Economy	de Jesus, A., Mendonça, S. [64]	2018	Reviewed academic and non-academic literature on drivers and barriers to CE from an eco-innovation perspective in which the European Union (EU) has been highlighted
Barriers & Drivers towards a Circular Economy	van Eijk, F. [65]	2015	Reviewed literature on barriers and drivers to CE through multiple lenses, including geographic and value chain approaches to policy-orientated drivers

where the absence of an enabler is seen as a barrier. As a result, enablers are often listed as drivers when present or as barriers when absent. Although the terms are closely related, there is a subtle difference. When looking at the terminology of the three terms, *drivers* can be described as factors that motivate or inspire progress, development, and innovation, while *enablers* help to facilitate the achievement of an undertaking. *Barriers* are factors that prevent or make it difficult to achieve a goal and must be overcome to succeed. In light of this, only two reviews address enablers separately from drivers and barriers, once termed critical success factors [61] and once key practices [63].

To have a complete understanding of all the factors that contribute to the success of companies transitioning to CE, it is necessary to analyse the drivers, enablers, and barriers that impact the process. As a result, two Research Questions (RQs) were formulated to guide this review in relation to the specific Rs:

- RQ1: What drives European industrial companies to implement Rs?
- RQ2: Which factors enable European industrial companies to practise Rs successfully?

3.2.2 Drivers to Implement Retention Strategies in Manufacturing Companies

Progress, development, and innovation do not take place on their own [67]. A force behind it pushes for these advances, namely the driver. Drivers can be created intentionally or originate from circumstances. Hence, this chapter will answer the first guiding RQ by investigating the drivers that force or encourage European manufacturing companies to adopt R practices.

The drivers were first categorised based on their similarity and meaning and then inductively grouped into five clusters. A code was also assigned to each driver to facilitate reference in later chapters. A total of 16 drivers were identified and are shown in Table 3.5 with their corresponding sources. The clusters are as follows:

- *Environmental* – drivers in this cluster are due to the need to reduce environmental pollution or the increased demand for resources.
- *Institutional* – drivers in this cluster are intentionally established by institutions such as laws and regulations.
- *Stakeholder* – in this cluster, the driving reasons are due to stakeholders that require the company to respond.
- *Financial* – this cluster includes drivers that increase or secure a company's economic wealth and competitiveness.
- *Responsibility* – drivers in this cluster are socially motivated.

Table 3.5 Factors that drive European manufacturing companies to adopt Rs

Cluster	Code	Driver	Context	Sources
Environmental	D.E1	Resources	Decrease the resources used in production	[62, 68–76]
	D.E2	Footprint	Decrease product carbon footprint	[62, 69, 71, 77, 78]
	D.E3	Risk	Resilience against resource-related risks (scarcity, supply challenges, etc.)	[69, 71, 75–77, 79]
Institutional	D.I1	Legal	Comply with legal requirements and norms	[62, 68, 69, 71, 72, 74, 75, 80–83]
	D.I2	Punitive	Avoid punitive measures, fines	[72, 76, 83–85]
	D.I3	Juridification	Anticipate or meet future legal obligations	[68, 69, 82, 86]
	D.I4	Tax	Taxation benefits, access to supportive funds and subsidies	[72, 76, 85]
Stakeholder	D.S1	Image	Improve corporate image, legitimacy, or credibility	[62, 68, 69, 71, 74, 79, 80, 83, 87, 88]
	D.S2	Customer	Adapt to changing consumer behaviour, meet customer expectations, increase customer satisfaction	[68, 69, 71, 74–78, 80, 82, 86, 87]
	D.S3	Practices	Meet the standards and follow the best practices of competitors	[68, 74, 76, 83, 89]
Financial	D.F1	Volatility	Price volatility of raw material	[51, 71]
	D.F2	Savings	Cost saving opportunities	[62, 68, 69, 74, 87, 89]
	D.F3	Competitiveness	Competitiveness through innovation and business development	[62, 68–70, 74, 77, 78, 80, 83, 87, 89, 90]
	D.F4	Value	Create value from waste and production side streams	[62, 72, 73, 77, 79, 82]
Responsibility	D.R1	CSR	Social and ethical responsibility	[68, 72, 74, 76, 79]
	D.R2	Promoter	Become a promoter of systemic change	[69]

Environmental Drivers

It is argued that short-term profits mainly dominate the LE [91]. The growing population and a middle class with increasing purchasing power [92] are good market prerequisites for manufacturing companies. After all, while the population doubled between 1970 and 2020 [93], economic growth quadrupled [94]. The disproportionate economic growth rate can be attributed to mass production at low costs, the inundation of modern society with non-essentials [95], and consumer behaviour that tends to replace products at an accelerated rate [43]. The ever-increasing demand for natural resources needed to sustain production and meet consumer demand is a problem in itself. Another issue is the negative impact of current production processes and waste landfilling on the environment [37].

An estimated 70% of toxic and hazardous chemicals in the environment stem from e-waste [96], and 20% of water pollution is attributed to the release of dangerous dying and treatment chemicals from the textile industry [42]. Moreover, textile manufacturing uses annually 98 mill. tonnes of non-renewable resources (e.g., oil to produce synthetic fibres) and 93 bill. m^3 of water [42] while releasing half a mill. tonnes of microplastic into oceans [97]. This makes the industry responsible for some of the most severe environmental impacts [37, 60].

The effect of the linear economy and the need to reduce waste and pollution while being more responsible for resource consumption is a driving reason for many companies to practice Rs. Three investigated companies by Manninen et al. [78] have decreased the environmental impact of their products by making use of short loop R, medium loop R, and long loop R. By doing so, they saved on virgin raw materials, reduced energy and water consumption in production, and increased the process utilisation rate. From that, the first driver is derived. Companies start practising Rs because they want to decrease the resources used in production. Gusmerotti et al. [71] also report that companies want to decrease their products' carbon footprint to become more sustainable and therefore start to implement Rs.

Other reasons companies have started practising Rs have emerged due to globalisation. Europe's dependency on imports of various resources has made the economy susceptible to export restriction measures by the supplier countries [49]. China, for example, provides 98% of the rare earth elements used in the European Union (EU). Such dependencies can severely affect various industries, especially Europe's EEE sector, which relies on critical and rare earth metals for further development.

Furthermore, due to the COVID-19 pandemic, supply chains have been disrupted and manufacturers had to rethink and reinvent business practices regardless of size or industry sector to ensure survival [88]. As a result of the pandemic, concerns are growing 'about the security of outsourcing supply and firm profitability' [97, p. 872]. Supply disruptions are also due to increasing resource scarcity for certain materials. That additionally drives up market prices. In fact, prices have already doubled since the beginning of this century [91]. Tura et al. [62] argue that circular practices can contribute to becoming less dependent on materials and mitigate vulnerability to fluctuations in the prices of virgin materials. A survey of 821 participating companies established that resource-related risks and economic advantages are the main drivers for companies to adopt Rs [71]. Stewart and Niero [75] came to the same conclusion by examining 46 sustainability reports from companies in the fast-moving consumer goods industry. In addition, their research shows that short and long loop Rs are the most practised ones reported. Furthermore, concerns about resource scarcity and supply drive interest in technical solutions on how to recover them [89]. The driver, in this case, is resilience against resource-related risks.

Institutional Drivers

In response to environmental degradation, European governments are enforcing different laws and policies that aim to reduce the environmental impact, extend product lifetime, and reduce waste [65, 98]. For example, the Waste from Electrical and Electronic Equipment (WEEE) Directive (2012/19/EU) sets collection, recycling, and recovery targets for electrical goods [99], or the Directive 2008/98/EC on waste, which sets measures to protect the environment and human health by preventing or reducing waste generation [100]. Such directives aim to involve all economic stakeholders in an efficient operating system to recover the raw materials contained in EOL products. Therefore, government initiatives have become a driving force for companies to implement Rs, especially when they are mandatory.

Companies start to practice circular strategies, knowingly or unknowingly, to avoid sanctions and penalties [63, 68, 69], like reducing their packaging waste or stopping to use of certain hazardous materials. However, it may only motivate companies to make the minimum effort [79] to keep their operating licences and maintain their legitimacy before the law, customers, and competitors. As an interviewee of Bansal and Roth stated: 'We will do what we need to do legally' [68, p. 725]. Therefore, legal measures such as laws and policies force companies to engage with Rs and are essential in developing CE in the industry.

Amendments in the directives, such as increasing collection, recycling, and reuse targets, further push companies' involvement. For example, by 2018, the collection rate for e-waste needed to be 45% of the EEE on the market. From 2019, it increased to 65% [84]. Companies anticipating future legal obligations are more likely to increase their involvement [69] to minimise possible adverse consequences and costly capital refits that could affect business operations [68, 86]. Thus, they are driven to improve resource efficiency and find new ways to reduce waste. Juridification has triggered multiple companies to start with Rs. For example, Giama and Papadopoulos [70] report on a French manufacturing company that started to reduce waste by using recycled glass in its products, although it is not yet mandatory. It resulted in savings of virgin material and a 20% reduction in energy consumption.

Another government strategy is to create favourable conditions for companies to implement and practice Rs. Taxation benefits, supportive funds, and subsidies [63] have been cited as driving forces in multiple papers. Dytianquin et al. [85] report that one Dutch construction company investigated, practices *Refuse* because circular projects are being subsidised.

Punitive measures for 'linear-relate' activities are also being set in place. For example, there is a landfill tax for residual waste in Serbia. At the same time, the disposal of recyclables remains free of charge [72]. Such measures can influence companies' decisions when considering which materials to use in product design. The increasing cost of waste disposal and regulations on what is legally allowed to be landfilled can also push companies to make efforts to find ways to reduce and/or reuse their production waste.

Avoiding punitive measures, such as sanctions, penalties, and increased disposal costs, is another driver that emerged in the literature due to legal requirements. This can motivate companies to adopt strategies from the 10R framework in order to avoid these negative consequences.

Stakeholder Drivers

The term 'stakeholder' is used here to refer to society, customers, and competitors. Since the negative environmental impacts of industries are becoming more transparent, consumer awareness is increasing. This pressures companies to change and could lead to 'reputational risk' [42, p. 22], which can impact profits if they do not respond. The pressure is especially on manufacturers that have already attracted blame 'for the environmental impact of products caused by the frequency of sales-driven replacement cycles' [79, p. 424]. Engaging in more visible activities to improve corporate environmental reputation is a common strategy to avoid bad publicity. It motivates some companies to start or increase their involvement with Rs [62, 88]. Not least, having a green image has become a business strategy and, thereby, a driver for many companies [80]. An example of this is presented by Ekokem, a waste management company. The company increased its recycling and reuse rate to increase its public image because it is normatively valued through certifications [83]. Another driver closely related to this is derived from Gusmoretti et al. [71]. They report that company managers have taken the initiative to offer products with lower environmental footprints to target environmentally sensitive customers. To do that, the companies either reduce the resources used or make use of other Rs. Also, Calzori et al. and Rizos et al. mention that companies are motivated to adopt Rs due to the preference of customers for green products [69, 87]. In return, these companies often benefit from additional competitive advantages. Therefore, adapting to changing consumer behaviour to meet customer expectations and increase customer satisfaction has become a driving reason for companies to adopt R practices.

As more companies implement Rs, new standards are set that can put pressure on other companies, particularly those with similar customer groups or operating in close proximity, to follow suit [89]. These companies are often subject to the same social norms and regulations and need to adopt and follow the best practices of their competitors to remain competitive, even if it involves additional costs [83]. For example, fashion brands such as Filson, REI, and Eileen Fisher promote sustainable consumption by offering durable clothing, encouraging consumers to invest in high-quality items that can be repaired rather than thrown away [101]. If customers start to demand such offerings, other companies have to meet these standards and follow their competitors' best practices. Hence, it can be a driving factor that pushes companies to implement Rs.

Responsibility Drivers

Some organisations adopt Rs 'despite the perceived additional costs involved' [79, p. 423]. They either practice Rs within their organisation or at least support other companies to practice them by giving access to their waste. They do this for ethical reasons, such as reducing the negative impacts of pollution on public health and animals [72]. These motivations, which can be referred to as Corporate Social Responsibility (CSR), can include the desire to avoid wasting valuable resources and to promote systemic change [69]. In an interview, a representative of a socially driven company stated: 'it's about being a good environmental citizen, about being responsible. There's nothing wrong with doing good' [68, p. 725].

Financial Drivers

'Jealousy, competitiveness, call it what you like. That is what drives the organisation. It is greed and competition' [68, p. 725]. Economic success is a salient driver for companies to implement Rs. According to Cole et al. [79], many manufacturers have recognised that their EOL products contain a value that can be retained using Rs. Therefore, more manufacturers are taking the initiative to develop recovery strategies to have customers return the EOL products so that materials and parts can be *Reused*, *Repaired*, *Refurbished*, etc. According to Milios and Matsumoto [102], *Remanufacturing* large products is 40–60% lower in cost than manufacturing new ones. Moreover, it effectively decreases the products' environmental footprint.

The employment of Rs offers several benefits for companies beyond the possibility of reducing material costs [87]. It also creates new income and business opportunities [28, 103]. For example, the implementation of landfill bans has increased the demand for waste utilisation solutions [62], providing new business opportunities for companies.

The desire to identify and take advantage of other market opportunities can motivate companies to experiment with and develop new business models centred on Rs [69]. This can involve expanding the company's services, such as offering *Repair* services, which can provide a more stable revenue stream than just selling products [4]. Other companies, on the other hand, focus on using materials as efficiently as possible [62], which drives the development of technology for Rs. Those who succeed in these endeavours can gain a competitive advantage. A survey of 99 Portuguese companies conducted by Fonseca et al. [77] found that the main motivators for companies to engage with CE include having a sustainable business model, accessing new markets and consumers, improving environmental performance, and creating environmental value.

As interest in recycled materials and various forms of production waste grows, the potential for companies to create value from them as a secondary source of income increases. A recent survey by Ratner et al. [73] found that more than 20% of the European companies participating in the study are already valorising their waste by selling it to other companies. For example, demand for Polyethylene Terephthalate (PET) waste, often used to replace cotton in the fibre industry, is rising and driving its value on international markets [82]. Additionally, Gusmerotti et al. [71] found that circular companies, which engage in R practices, tend to perform better in terms of turnover, employees, and clients compared to companies that operate linearly. It is not surprising that companies that practice Rs are considered the main beneficiaries of the CE [20, 104].

3.2.3 Enablers and Barriers to Implement Retention Strategies in Manufacturing Companies

In this review, enablers are factors that facilitate the successful implementation and practice of Rs. They may not directly cause success, but they contribute to creating better conditions for it to occur. On the other hand, barriers are factors that hinder or obstruct the implementation and practise of Rs. It is important to note that enablers can also be challenged by their own barriers and their absence is often perceived as a barrier in itself [66]. Therefore, barriers are discussed analogously to enablers to provide more context to answer the second RQ.

A total of 12 enablers are highlighted and summarised in Table 3.6. Similar to the drivers they were first categorised according to their similarity and meaning and then inductively appointed to five clusters:

- *Firm* – refers to the company's internal factors that will ease the transition, like available capital and committed company culture.
- *Value chain* – refers to enablers that enhance collaboration, which is necessary for the quality of materials, products, and parts.
- *Technological* – refers to the technology required to practise Rs.
- *Customers* – refers to enablers that are associated with customer demand for products produced with Rs.
- *Material supply* – includes enablers that address the supply of materials, parts, or products.

Enablers and Barriers in the Firm

There are several internal barriers that can hinder a company's transition to circular practice, including hierarchical structures and silos between departments [111, 112], risk-averse management [58], weak leadership [111], lack of management support [62], and lack of commitment from top management [74]. Risk-averse management has been identified as particularly problematic, as it can lead to a reluctance to participate in circular initiatives, a focus on short-term profits, and a lack of willingness to share knowledge and technology [58]. Kirchherr et al. [113] also identified the hesitant company culture as one of the main barriers to the implementation of Rs and related it to circular projects that do not produce immediate financial returns. In addition, measuring the long-term benefits of circular activities is said to be challenging [80]. As a result, managers must adopt a long-term perspective in order to recognise the benefits of circular practices.

Table 3.6 Enabling factors for European manufacturing companies to practice Rs

Cluster	Code	Enabler	Context	Sources
Firm	E.F1	Culture	Committed company environment and culture	[74, 87]
	E.F2	Capital	Available capital or financial support to transition	[74, 105]
Value chain	E.V1	Partnership	Build closer relationships/ invest in collaboration and partnerships	[33, 69, 75, 78, 87, 90, 105, 106]
	E.V2	Transparency	Enhance information sharing and transparency, make use of digital technologies and standardise information	[69, 106, 107]
Technological	E.T1	Research	Experiment and research	[33, 106, 108]
	E.T2	Mutuality	Joint decision-making (customers and suppliers)	[69, 106]
	E.T3	R-Technology	Innovate and apply technology that supports circular strategies	[62, 74, 82, 106]
	E.T4	Outsourcing	Possibility of outsourcing the activity	[33]
Customers	E.C1	Engagement	Increase customer engagement and awareness, provide incentives	[33, 68, 75, 79, 87, 88, 105, 106, 109]
Material supply	E.M1	Logistics	Available reverse logistic network	[79]
	E.M2	Collection	Create simple collection and take-back systems	[72, 79, 81, 110]
	E.M3	Symbiosis	Engage in industrial symbiosis	[69, 90, 106]

The commitment of managers and employees is essential for the successful implementation of circular practices. This is supported by research conducted by Rizos et al., who found in their study of 30 European Small and Medium-sized Enterprises (SMEs) that 'the most frequently mentioned enabler is the company culture of the staff and manager' [87, p. 11]. In addition, companies that are committed to sustainability often integrate it into their mission and vision, which can help facilitate the implementation of circular practices [74]. Startups have an advantage in this regard, as they have more flexibility in shaping their organisational structure, strategy, and culture.

The transition from a linear to a circular production system can be challenging due to the high initial investment, time, and labour required for activities such as redesigning the shop floor, managing reverse logistics, and improving distribution planning [87, 105]. Financial constraints and the high upfront costs associated with these efforts can be significant barriers for companies [87, 105, 113].

Companies surveyed by Rizos et al. [87] refer to special funds available to aid the implementation of circular solutions. These funds include grants from EU or other government or public support programmes. This can help companies overcome the financial barriers associated with the implementation of Rs [105, 114]. Unfortunately, small companies often face barriers when it comes to obtaining grants and funding due to the administrative burden and the time needed to carefully assess these opportunities [105]. In addition, these funding opportunities tend to be primarily geared toward medium and large enterprises, which can hinder companies with limited employee and management capacity. The survey conducted by Rizos et al. found that 20% of the respondents perceive the added administrative burden of practising circular strategies as a barrier [87].

Other researchers have also noted that it can be difficult for companies to allocate resources to the development of new technologies or business models that support the Rs [33, 105]. This difficulty is often increased by the lack of availability of in-house know-how, which is crucial for understanding and implementing R-Technologies (technology that supports circular strategies) effectively. The relatively new nature of many R-supporting solutions also means that there is a lack of successful case studies [62] and best-practice examples [105]. There is also a lack of demonstrations of new technologies or processes to improve resource efficiency. This uncertainty can make investors and companies hesitant to invest due to the associated economic risks and the perceived gap in internal expertise required to navigate these novel solutions.

Enablers and Barriers in the Value Chain

In addition to financial and 'perceived' technological barriers, the lack of support from supply and demand networks (value chain) can also pose a challenge to the implementation of circular practises [87]. As Rizos et al. found, this is particularly true for SMEs, which often have low bargaining power in their supply chains.

Resistance is also experienced due to other stakeholders. Driven by the threat of possible cannibalisation [33], competing companies discourage promoting circular products to protect their sales. These manufacturers are concerned that environmental requirements will reduce their profit margins and that strong reuse markets impact their sales of new items [79]. Others who embrace CE and have found innovative technical solutions, such as the Spanish company Mahonitas, which creates recyclable shoes, have protected their knowledge through patenting [88].

Approximately one third of SMEs that have participated in the survey of Rizos et al. view networking as a key enabler for Rs. This can involve collaboration with like-minded companies through supply chain partnerships or networking groups [87].

The complexity of modern product design and material mixtures, as well as the lack of standardisation and rapid technological advancement, can also make it difficult for products to be *repaired*, *refurbished*, *remanufactured* or *recycled* [66, 115].

This challenge is exacerbated if information about the product and its materials is not readily available. As such, the use of digital technologies to facilitate efficient information sharing and specifically to create more transparency can help overcome this barrier [87].

Many authors have also noted that companies face challenges or do not implement Rs because their suppliers or customers do have a missing or incomplete understanding of the benefits of circulating materials [87, 105]. Therefore, they are reluctant to exchange knowledge and information about their products and the materials contained [33]. The lack of cooperation and the dependency on the supply chain can be a severe challenge for companies that want to practise Rs. To efficiently circulate materials, parts, and products, it is necessary to engage with suppliers and customers [73]. Companies must develop close relationships with other supply chain actors and communicate their needs and reasons. Transparency is considered the key [62].

Enablers and Barriers Around Technology

Vermunt et al. report that knowledge and technology are often lacking, particularly for short loop circular practises [33]. However, Sá et al. found that the lack of technological solutions is not always the main barrier in implementing circular practices, as companies can sometimes find alternative solutions through collaboration or by adapting existing technologies [74]. They also note that the lack of solutions is considered significantly less of a barrier when comparing technicians' perceptions to those of executives and managers. They concluded that hierarchical superiors might not know about, value, or consider solutions, arguably for financial reasons [74]. In any case, the availability of technology and technical know-how is an important enabling factor for implementing circular practises.

Technical barriers are considered minor by Kirchherr et al. [113] based on the results of 208 survey participants and 47 expert interviews. Instead, when specialised technology for practising an R is accessible (R-Technology), the lack of technical know-how has been reported to hinder companies from changing their operations. However, only 10% of the sample in the survey by Rizos et al. mentioned the know-how of individuals within the company as an enabling factor [87]. Companies that overcame such difficulties claimed that they developed technological know-how by experimenting or outsourcing technical activities altogether [33].

Salvador et al. [106] suggest that innovative solutions are found in close collaboration with suppliers and customers. Other authors also proposed the idea that joint decision-making encourages innovation and often leads to sustainable solutions [69]. Another enabler mentioned is collaboration with research institutes [78]. Keeping up to date with published research can also help. For example, Quinones et al. [108] showed how waste could be reduced in the design phase by using a Building Information Modeling (BIM) tool that proposes alternative building elements.

Enablers and Barriers Concerning Customers

Companies that strive to practise Rs often face challenges due to inadequate market signals. For example, the low price of virgin materials [113] decreases the demand for recycled secondary materials, i.e. looped materials, and makes it difficult to establish a market for them. Improving the efficiency of *recycling* processes through technology is necessary to improve the cost-effectiveness of the process [33]. Until these technological advances are made, the higher price of secondary materials also leads to higher product prices, making products less attractive to price-sensitive customers. In addition, there is often a general misconception that products made from waste or with extended use cycles are of poorer quality, which can further discourage demand for these products. To address these challenges, strategies such as introducing durability criteria or extending warranty times can be implemented to increase cultural acceptance of recycled products and make it easier for consumers to understand the benefits of these products [116].

Enablers and Barriers to Material Supply

Multiple authors have identified barriers in laws and policies that can hinder the implementation of R practises in firms. Van Buren et al. [91] criticised that some legal frameworks are based on linear economic principles and do not support CE concepts. Andersen [84] and Vermunt et al. [33] also point out that certain environmental laws prohibit the reuse of certain waste materials as resources, such as recycled plastics in medical equipment [82]. These restrictions can make it difficult for companies to practice R0, especially since the number of suppliers of circular materials is still limited [33]. Furthermore, the Waste from Electrical and Electronic Equipment (WEEE) Directive (2012/19/EU), which mainly promotes R7 [84], has been criticised for misaligned targets that hinder the adoption of medium loop Rs.

Companies interviewed by Cole et al. [79] also mentioned that the existing infrastructure for waste management, the lack of RL, and the higher cost of handling products for medium loop R contribute to a preference for recycling, which is seen as a simpler and cheaper option.

The lack of a 'concrete, coherent and strict legislative framework' [87, p. 4] has also been identified as a challenge for companies. For example, EU waste legislation prohibits the transportation of certain waste materials across borders, and there is no consensus on what these materials are [91]. This can make it difficult for companies to practise Rs and import waste and also limits the availability of circular materials, parts, or products [33, 73, 82].

Concerns about the quality and quantity of circular materials are also common among companies practising Rs [82]. Collaboration and partnerships with waste management and reprocessing firms, as well as industrial symbiosis, can help address these concerns. Specifically, industrial symbiosis, as mentioned by Calzolari et al. [69], is a way to ensure the quality and supply of materials. Connecting different industries and enabling them to share resources such as waste, water, and energy, can create more efficient and sustainable systems by reducing waste and conserving resources. It can also help companies access new markets and create new business opportunities. It is often facilitated by intermediaries, such as government agencies or non-profit organisations, bringing companies together and facilitating resource exchanges. While industrial symbiosis can help companies improve their resource efficiency and reduce their environmental impact, it also creates interdependencies that can affect the risk profile of participating companies. Therefore, it is important for companies to carefully consider the potential risks and benefits of engaging in industrial symbiosis before entering into such arrangements.

Consumers also play a role in the supply of materials, which can be challenging. An interviewee by Cole et al. stated: 'if it fits in a bin, it will go in a bin' [79, p. 421], and thus the product is not returned. Consumers must be more involved and informed about the goals of a company to participate actively. However, their participation is often lacking [75]. Hewlett-Packard, for example, has a hardware remanufacturing and refurbishment programme called HP Renew. They solved this problem using strategies such as trade-in, cash-return, and leasing to encourage their customers to return products [53]. The key is to make it as easy as possible for customers to return products.

Therefore, incentivising product returns and providing easy return logistics can help encourage customers to participate in circular practices. However, these efforts require good support and infrastructure for RL [117], which is often lacking. Another enabling factor in this regard, which allows more control over quality and supply, is to use the reverse logistic system of retailers [79]. Overall, logistics is a key enabler for turning waste flows into resources, and, unsurprisingly, partnering with logistics providers was mentioned to facilitate the process [66, 117].

3.3 Summary and Research Limitations

3.3.1 Summary

The circular approach to value retention is focused on minimising the use of virgin materials and reducing waste in production, to extract as much value as possible from resources, and to keep products and materials in use for as long as possible. This approach has both environmental and economic benefits, as it reduces the costs associated with waste and the constant input of new materials.

To successfully adopt a circular approach, it is essential to identify the value proposition of a product or service, which encompasses both the intrinsic value of the product and the additional value that comes with the implementation of circular strategies. By understanding the value proposition, businesses can better design and market their products for circularity.

Using the Value Hill allows companies to visualise their position in the pre-use phase, identify the value contained in their products, and potential retention options to benefit from in CE. The ten proposed Rs are ordered according to their impact on the environment and their level of retention of value. Those Rs on top of the hill are preferred due to their greater potential to reduce the use of virgin materials and prevent waste. At the same time, they aim to maintain the highest level of value (complete products) compared to those Rs at the bottom of the hill, which focus only on materials or energy. In summary, short loop Rs, referring to *Refuse* and *Reduce*, aim to eliminate the need to extract virgin material and reduce waste from the beginning of PLC. Medium loop Rs, such as *Reuse* and *Repair*, extend the existing use cycle of products or parts. Long loop Rs, like *Recycling*, *Recovery*, and *Remine*, involve traditional waste management activities that break down products into materials, particles, or energy that can be returned to the production stage.

Companies in Europe are motivated to implement Rs for a variety of reasons. Sixteen driving factors (referred to as Drivers) have been identified in the literature review and have been arranged into five clusters:

- *Environmental drivers* include the desire to decrease resource consumption and carbon footprint, as well as the negative environmental impact of current production processes.
- *Stakeholder drivers* include the pressure of consumers and competitors to adopt CE practices and improve the corporate environmental reputation, as well as the desire to meet customer expectations and follow best practices.
- *Social drivers* include ethical motivations such as corporate social responsibility and the desire to promote systemic change, not least to minimise negative impacts on public health and animals.
- *Financial drivers* are related to the desire for economic success through cost savings, competitive advantages, and new markets, as well as recognition of value in EOL products.
- *Institutional drivers* include the need to comply with legal requirements and avoid punitive measures, as well as the desire to benefit from government incentives.

Of the multiple internal and external factors that can enable or hinder a company's transition to circular practises, a total of 12 principal Enablers have been grouped into the five categories:

- *Firm enablers* show the internal aptitude of the company.
- *Value Chain enablers* highlight cooperation and information exchange.
- *Technological enablers* address realisation from a technology and capability perspective.
- *Customer enablers* are linked to customer loyalty and engagement.
- *Material Supply enablers* point to structure and roles in the supply chain.

The involvement of the value chain is particularly important for the successful implementation of Rs. For companies to reuse their products, it is necessary to increase customer engagement and have a functioning RL to return EOL products. Financial barriers can be overcome through financial support or alternative financing options, and technological barriers can be addressed through research, experimentation, and collaboration with suppliers and customers. However, regulatory barriers, such as the lack of legal requirements and successful business cases, reflected in the absence of the respective driver, can make risk-averse managers reluctant to participate in circular initiatives. A committed company culture, openness to industrial symbiosis, and the aforementioned partnerships and transparency help facilitate the transition to a circular future.

3.3.2 Research Limitations

The limitation of this study includes the fact that only one reviewer analysed the papers. Furthermore, the search string was very specific and other relevant literature could use different keywords, which was not identified because of that fact. In addition, the literature identification process of the study was limited, as it did not look for terminology that overlaps with the concept of circular economy, such as 'industrial ecology', 'industrial symbiosis', or 'cradle-to-cradle'. Future research should consider adding these terms to their database search.

References

1. Tukker, A. (2015). Product services for a resource-efficient and circular economy – a review. *Journal of Cleaner Production, 97,* 76–91. ISSN: 09596526.
2. Lewandowski, M. (2016). Designing the Business Models for Circular Economy – Towards the Conceptual Framework. *Sustainability, 8,* 43. ISSN: 2071-1050.
3. Lieder, M. & Rashid, A. (2016). Towards circular economy implementation: A comprehensive review in context of manufacturing industry. *Journal of Cleaner Production, 115,* 36–51. ISSN: 09596526.
4. Matsumoto, M., Yang, S., Martinsen, K. & Kainuma, Y. (2016). Trends and research challenges in remanufacturing. *International Journal of Precision Engineering and Manufacturing-Green Technology, 3,* 129–142. ISSN: 2288-6206, 2198-0810.
5. Swain, B. (2017). Recovery and recycling of lithium: A review. *Separation and Purification Technology, 172*, 388–403. https://doi.org/10.1016/j.seppur.2016.08.031
6. Islam, M., & Huda, N. (2018). Reverse logistics and closed-loop supply chain of Waste Electrical and Electronic Equipment (WEEE)/E-waste: A comprehensive literature review. *Resources, Conservation and Recycling, 137*, 48–75. https://doi.org/10.1016/j.resconrec.2018.05.026
7. Ávila-Gutiérrez, M. J., Martín-Gómez, A., Aguayo-González, F. & Córdoba-Roldán, A. (2019). Standardization framework for sustainability from circular economy 4.0. *Sustainability, 11,* 6490. ISSN: 2071-1050.
8. Kerin, M. & Pham, D. (2019). A review of emerging industry 4.0 technologies in remanufacturing. *Journal of Cleaner Production, 237.* https://doi.org/10.1016/j.jclepro.2019.117805
9. Manavalan, E., & Jayakrishna, K. (2019). A review of Internet of Things (IoT) embedded sustainable supply chain for industry 4.0 requirements. *Computers & Industrial Engineering, 127*, 925–953. https://doi.org/10.1016/j.cie.2018.11.030
10. Ren, S., Zhang, Y., Liu, Y., Sakao, T., Huisingh, D., & Almeida, C. M. V. B. (2019). A comprehensive review of big data analytics throughout product lifecycle to support sustainable smart manufacturing: A framework, challenges and future research directions. *Journal of Cleaner Production, 210*, 1343–1365. https://doi.org/10.1016/j.jclepro.2018.11.025
11. Pieroni, M., McAloone, T., & Pigosso, D. (2019). Business model innovation for circular economy and sustainability: A review of approaches. *Journal of Cleaner Production, 215,* 198–216. https://doi.org/10.1016/j.resconrec.2019.104426
12. Lüdeke-Freund, F., Gold, S. & Bocken, N. M. P. (2019). A review and typology of circular economy business model patterns. *Journal of Industrial Ecology, 23,* 36–61. ISSN: 1088-1980, 1530-9290.
13. Sassanelli, C., Rosa, P., Rocca, R., & Terzi, S. (2019). Circular economy performance assessment methods: A systematic literature review. *Journal of Cleaner Production, 229*, 440–453. https://doi.org/10.1016/j.jclepro.2019.05.019

14. Saidani, M., Yannou, B., Leroy, Y., Cluzel, F. & Kendall, A. (2019). A taxonomy of circular economy indicators. *Journal of Cleaner Production, 207,* 542–559. ISSN: 09596526.
15. Kristensen, H. & Mosgaard, M. (2020). A review of micro level indicators for a circular economy - moving away from the three dimensions of sustainability? *Journal of Cleaner Production, 243*. https://doi.org/10.1016/j.jclepro.2019.118531
16. Acerbi, F. & Taisch, M. (2020). A literature review on circular economy adoption in the manufacturing sector. *Journal of Cleaner Production, 273,* 123086. ISSN: 0959-6526.
17. Bjornbet, M., Skaar, C., Fet, A. & Schulte, K. (2021). Circular economy in manufacturing companies: A review of case study literature. *Journal of Cleaner Production, 294*. https://doi.org/10.1016/j.jclepro.2021.126268
18. Rahman, S., Pompidou, S., Alix, T., & Laratte, B. (2021). A review of LED lamp recycling process from the 10 R strategy perspective. *Sustainable Production and Consumption, 28,* 1178–1191. https://doi.org/10.1016/j.spc.2021.07.025
19. Desing, H., et al. (2020). A circular economy within the planetary boundaries: Towards a resource-based, systemic approach. *Resources, Conservation and Recycling, 155,* 104673. ISSN: 0921-3449.
20. Kirchherr, J., Reike, D. & Hekkert, M. (2017). Conceptualizing the circular economy: An analysis of 114 definitions. *Resources, Conservation and Recycling, 127,* 221–232. ISSN: 0921-3449.
21. Reike, D., Vermeulen, W. J. & Witjes, S. (2018). The circular economy: New or refurbished as CE 3.0? – Exploring controversies in the conceptualization of the circular economy through a focus on history and resource value retention options. *Resources, Conservation and Recycling, 135,* 246–264. ISSN: 0921-3449.
22. Achterberg, E., Hinfelaar, J. & Bocken, N. (2016). *Master circular business with the value hill.*
23. Bocken, N. M. P., de Pauw, I., Bakker, C. & van der Grinten, B. (2016). Product design and business model strategies for a circular economy. *Journal of Industrial and Production Engineering, 33,* 308–320. ISSN: 2168-1015, 2168-1023.
24. Kristoffersen, E., Blomsma, F., Mikalef, P. & Li, J. (2020). The smart circular economy: A digital-enabled circular strategies framework for manufacturing companies. *Journal of Business Research, 120,* 241–261. ISSN: 0148-2963.
25. Hänggi, R., Fimpel A., & Siegenthaler, R. (2021). LEAN Production – einfach und umfassend: Ein praxisorientierter Leitfaden zu schlanken Prozessen mit Bildern erklärt. Berlin, Heidelberg: Springer Berlin Heidelberg. https://doi.org/10.1007/978-3-662-62702-0
26. Nyffenegger, F., Hänggi, R. & Reisch, A. (2018). In P. Chiabert, A. Bouras, F. Noël & J. Ríos (Eds.) *Product lifecycle management to support industry 4.0*(pp. 358–366). Cham: Springer International Publishing. ISBN: 978-3-030-01614-2.
27. Diaz, A., Reyes, T. & Baumgartner, R. J. (2022). Implementing circular economy strategies during product development. *Resources, Conservation and Recycling, 184,* 106344. ISSN: 0921-3449.
28. Ghisellini, P., Cialani, C. & Ulgiati, S. (2016). A review on circular economy: The expected transition to a balanced interplay of environmental and economic systems. *Journal of Cleaner Production, 114,* 11–32. ISSN: 0959-6526.
29. Bakker, C., Wang, F., Huisman, J. & den Hollander, M. (2014). Products that go round: Exploring product life extension through design. *Journal of Cleaner Production, 69,* 10–16. ISSN: 095-96526.
30. UNEP. *UNEP Circularity Platform. Reuse*. Retrieved December 17, 2023 from https://buildingcircularity.org/reuse/
31. Potting, J., Hekkert, M., Worrell, E., & Hanemaaijer, A. (2017). Circular economy: Measuring innovation in the product chain. *PBL publication number: 2544*. The Hague: PBL Netherlands Environmental Assessment Agency.
32. Blomsma, F., et al. (2019). Developing a circular strategies framework for manufacturing companies to support circular economy-oriented innovation. *Journal of Cleaner Production, 241,* 118271. ISSN: 0959-6526.

33. Vermunt, D., Negro, S., Verweij, P., Kuppens, D. & Hekkert, M. (2019). Exploring barriers to implementing different circular business models. *Journal of Cleaner Production, 222,* 891–902. ISSN: 0959-6526.
34. Kerr, W. & Ryan, C. (2001). Eco-efficiency gains from remanufacturing. *Journal of Cleaner Production, 9,* 75–81. ISSN: 09596526.
35. Rashid, A., Asif, F. M., Krajnik, P. & Nicolescu, C. M. (2013). Resource conservative manufacturing: An essential change in business and technology paradigm for sustainable manufacturing. *Journal of Cleaner Production, 57,* 166–177. ISSN: 09596526.
36. Lacy, P., Long, J. & Spindler, W. (2020). *The circular economy handbook: Realizing the circular advantage*. London: Palgrave Macmillan UK. ISBN: 978-1-349-95967-9.
37. Niinimäki, K., et al. (2020). The environmental price of fast fashion. *Nature Reviews Earth & Environment, 1,* 189–200. ISSN: 2662-138X.
38. Morgan, L. R. & Birtwistle, G. (2009). An investigation of young fashion consumers' disposal habits. *International Journal of Consumer Studies, 33,* 190–198. ISSN: 14706423, 14706431.
39. Future Market Insights. (2022). *Europe Second-hand Apparel Market*. Retrieved December 15, 2022 from https://www.futuremarketinsights.com/reports/europe-second-hand-apparel-market.
40. Paton, E. (2018). H&M, a fashion giant, has a problem: $4.3 billion in unsold clothes. *The New York Times*. ISSN: 0362-4331. Retrieved June 11, 2022 https://www.nytimes.com/2018/03/27/business/hm-clothes-stock-sales.html.
41. Ellson, A. (2018). Luxury brands including Burberry burn stock worth millions. *The Sunday Times*. ISSN: 0140-0460. Retrieved April 11, 2022 from https://www.thetimes.co.uk/article/luxury-brands-burning-stock-worth-millions-zxxscjcmj.
42. Morlet, A., et al. (2017). A new textiles economy: Redesigning fashion's future. *Ellen MacArthur Foundation* .
43. Mont, O., Plepys, A., Whalen, K. & Nußholz, J. L. K. (2017). *Business model innovation for a Circular Economy: Drivers and barriers for the Swedish industry - the voice of REES companies*. Mistra REES.
44. O'Connor, M., Zinnerman, J., Anastas, P. & Plata, D. (2016). A strategy for material supply chain sustainability: Enabling a circular economy in the electronics industry through Green engineering. *ACS Sustainable Chemistry & Engineering, 4,* 5879–5888. ISSN: 2168-0485.
45. Abalansa, S., Badr El Mahrad, Icely, J. & Newton, A. (2021). Electronic waste, an environmental problem exported to developing countries: The GOOD, the BAD and the UGLY. *Sustainability, 13.* Place: Basel Publisher: MDPI AG, 5302.
46. Hasnain, Q. (2021). *Every iPhone release in chronological order*. Retrieved June 5, 2022 from https://screenrant.com/iphone-apple-release-chronological-order-smartphone/.
47. Cakebread, C. (2017). We got in line at 4 AM to wait for an iPhone X - Here's why some people love 'the wait'. *Business Insider.* Retrieved June 5, 2022 from https://www.businessinsider.com/the-iphone-x-launched-today-and-i-waited-7-hours-to-buy-it-2017-11.
48. Forti, V., Balde, C. P., Kuehr, R., & Bel, G. (2020). *The Global E-waste Monitor 2020: Quantities, flows and the circular economy potential (United Nations University (UNU)/United Nations Institute for Training and Research (UNITAR)- co-hosted SCYCLE Programme*. International Telecommunication Union (ITU) & International Solid Waste Association. Bonn/Geneva/Rotterdam: ISWA. ISBN: 978-92-808-9115-7.
49. European Commission & Industry Directorate-General for Internal Market, Entrepreneurship and SMEs. (2018). *Raw materials scoreboard 2018 – European innovation partnership on raw materials*. https://data.europa.eu/doi/10.2873/08258 (Publications Office).
50. Greenovate! Europe. (2012). *Guide to resource efficiency in manufacturing: Experiences from improving resource efficiency in manufacturing companies*. Europe INNOVA.
51. Kamble, S. S., Belhadi, A., Gunasekaran, A., Ganapathy, L. & Verma, S. (2021). A large multi-group decision-making technique for prioritizing the big data-driven circular economy practices in the automobile component manufacturing industry. *Technological Forecasting and Social Change, 165.* Place: New York Publisher: Elsevier Science Ltd., 1. ISSN: 00401625.

52. Desso. (2015). *As a Cradle to Cradle® company, we go beyond what most people understand by sustainability.* Retrieved April 14, 2022 from https://vloeren.projecten.tarkett.nl/nl_NL/node/onze-duurzaamheidsstrategie-omgezet-in-acties-1831.
53. Hewlett Packard Co. (2014). *HP 2014 Living Progress Report.*
54. Bundgaard, A. M. & Huulgaard, R. D. (2019). Luxury products for the circular economy? A case study of Bang & Olufsen. *Business Strategy and the Environment, 28,* 699–709. ISSN: 0964-4733, 1099-0836.
55. Mishra, J. L., Chiwenga, K. D. & Ali, K. (2021). Collaboration as an enabler for circular economy: A case study of a developing country. *Management Decision, 59,* 1784–1800. ISSN: 0025-1747.
56. Ranta, V., Keranen, J. & Aarikka-Stenroos, L. (2020). How B2B suppliers articulate customer value propositions in the circular economy: Four innovation-driven value creation logics. *Industrial Marketing Management, 87,* 291–305. ISSN: 0019-8501.
57. Green, B. N., Johnson, C. D. & Adams, A. (2006). Writing narrative literature reviews for peer-reviewed journals: Secrets of the trade. *Journal of Chiropractic Medicine, 5,* 101–117. ISSN: 15563707.
58. Tan, J., Tan, F. J., & Ramakrishna, S. (2022). Transitioning to a circular economy: A systematic review of its drivers and barriers. *Sustainability, 14,* 1757. https://doi.org/10.3390/su14031757
59. Mehmood, A., Ahmed, S., Viza, E., Bogush, A. & Ayyub, R. (2021). Drivers and barriers towards circular economy in agri-food supply chain: A review. *Business Strategy and Development, 4,* 465–481. ISSN: 2572-3170.
60. De Aguiar Hugo, A., de Nadae, J. & da Silva Lima, R. (2021). Can fashion be circular? A literature review on circular economy barriers, drivers, and practices in the fashion industry's productive chain. *Sustainability.* https://doi.org/10.3390/su132112246
61. Aloini, D., Dulmin, R., Mininno, V., Stefanini, A. & Zerbino, P. (2020). Driving the transition to a circular economic model: A systematic review on drivers and critical success factors in circular economy. *Sustainability, 12.* Place: Basel Publisher: MDPI AG, 10672.
62. Tura, N., et al. (2019). Unlocking circular business: A framework of barriers and drivers. *Journal of Cleaner Production, 212,* 90–98. ISSN: 0959-6526.
63. Govindan, K. & Hasanagic, M. (2018). A systematic review on drivers, barriers, and practices towards circular economy: A supply chain perspective. *International Journal of Production Research, 56,* 278–311. ISSN: 0020-7543, 1366-588X.
64. De Jesus, A. & Mendonça, S. (2018). Lost in transition? Drivers and barriers in the Eco-innovation road to the circular economy. *Ecological Economics, 145,* 75–89. ISSN: 09218009.
65. Van Eijk, F. (2015). *Barriers & Drivers towards a Circular Economy - Literature Review A-140315-R-Final.* Acceleratio.
66. Hart, J., Adams, K., Giesekam, J., Tingley, D. D. & Pomponi, F. (2019). Barriers and drivers in a circular economy: The case of the built environment. *Procedia CIRP, 80,* 619–624. ISSN: 2212-8271.
67. OECD. (2009). In *Working out change* (pp. 115–145). Paris: OECD. ISBN: 978-92-64-06715-8.
68. Bansal, P. & Roth, K. (2000). Why companies go green: A model of ecological responsiveness. *Academy of Management Journal, 43,* 717–736. ISSN: 0001-4273, 1948-0989.
69. Calzolari, T., Genovese, A. & Brint, A. (2021). The adoption of circular economy practices in supply chains-An assessment of European Multi-National Enterprises. *Journal of Cleaner Production, 312,* 127616. ISSN: 0959-6526.
70. Giama, E. & Papadopoulos, A. (2020). Benchmarking carbon footprint and circularity in production processes: The case of stonewool and extruded polysterene. *Journal of Cleaner Production, 257.* https://doi.org/10.1016/j.jclepro.2020.120559
71. Gusmerotti, N. M., Testa, F., Corsini, F., Pretner, G. & Iraldo, F. (2019). Drivers and approaches to the circular economy in manufacturing firms. *Journal of Cleaner Production, 230,* 314–327. ISSN: 0959-6526.
72. Ilić, M. & Nikolić, M. (2016). Drivers for development of circular economy – A case study of Serbia. *Habitat International, 56,* 191–200. ISSN: 01973975.

73. Ratner, S., Gomonov, K., Lazanyuk, I. & Revinova, S. (2021). Barriers and drivers for circular Economy 2.0 on the firm level: Russian case. *Sustainability, 13,* 11080 https://doi.org/10.3390/su131911080.
74. Sá, M. M., et al. (2022). Integration of the circular economy paradigm in companies from the Northwest of the Iberian Peninsula. *Sustainability, 14*, 7940. https://doi.org/10.3390/su14137940
75. Stewart, R. & Niero, M. (2018). Circular economy in corporate sustainability strategies: A review of corporate sustainability reports in the fast-moving consumer goods sector. *Business Strategy and the Environment, 27,* 1005–1022. ISSN: 09644733.
76. Kumar, V., Sezersan, I., Garza-Reyes, J. A., Gonzalez, E. D. & Moh'd Anwer AL-Shboul. (2019). Circular economy in the manufacturing sector: Benefits, opportunities and barriers. *Management Decision, 57,* 1067–1086. ISSN: 00251747.
77. Fonseca, L., Domingues, J., Pereira, M., Martins, F. & Zimon, D. (2018). Assessment of circular economy within Portuguese organizations. *Sustainability, 10,* 2521. ISSN: 2071-1050.
78. Manninen, K., et al. (2018). Do circular economy business models capture intended environmental value propositions? *Journal of Cleaner Production, 171,* 413–422. ISSN: 09596526.
79. Cole, C., Gnanapragasam, A., Cooper, T., & Singh, J. (2019). An assessment of achievements of the WEEE Directive in promoting movement up the waste hierarchy: Experiences in the UK. *Waste Management, 87*, 417–427. https://doi.org/10.1016/j.wasman.2019.01.046
80. Kapetanopoulou, P. & Tagaras, G. (2011). Drivers and obstacles of product recovery activities in the Greek industry. *International Journal of Operations & Production Management, 31,* 148–166. ISSN: 0144-3577.
81. Nowakowski, P., Krol, A., & Mrowczynska, B. (2017). Supporting mobile WEEE collection on demand: A method for multi-criteria vehicle routing, loading and cost optimisation. *Waste Management, 69*, 377–392. https://doi.org/10.1016/j.wasman.2017.07.045
82. Paletta, A., Leal, W., Balogun, A., Foschi, E. & Bonoli, A. (2019). Barriers and challenges to plastics valorisation in the context of a circular economy: Case studies from Italy. *Journal of Cleaner Production, 241.* ISSN: 0959-6526.
83. Ranta, V., Aarikka-Stenroos, L., Ritala, P. & Mäkinen, S. J. (2018). Exploring institutional drivers and barriers of the circular economy: A cross-regional comparison of China, the US, and Europe. *Resources, Conservation and Recycling, 135,* 70–82. ISSN: 0921-3449.
84. Andersen, T. (2022). A comparative study of national variations of the European WEEE directive: Manufacturer's view. *Environmental Science and Pollution Research, 29*, 19920–19939. https://doi.org/10.1007/s11356-021-13206-z
85. Dytianquin, N., Gregersen-Hermans, J., Kalogeras, N., J van Oorschot & Ritzen, M. (2021). Circularity in selected EU Countries: The case of construction and demolition industry. *IOP Conference Series: Earth and Environmental Science, 855.* ISSN: 1755-1307.
86. Jora, O., Patruti, A., Iacob, M. & Sancariuc, D. (2020). "Squaring the Circle"-The disregarded institutional theory and the distorted practice of packaging waste recycling in Romania. *Sustainability, 12.* https://doi.org/10.3390/su12229440
87. Rizos, V., et al. (2016). Implementation of circular economy business models by small and Medium-Sized Enterprises (SMEs): Barriers and enablers. *Sustainability, 8,* 1212. ISSN: 2071-1050.
88. Vatamanescu, E., Dabija, D., Gazzola, P., Cegarro-Navarro, J. & Buzzi, T. (2021). Before and after the outbreak of Covid-19: Linking fashion companies' corporate social responsibility approach to consumers' demand for sustainable products. *Journal of Cleaner Production, 321.* https://doi.org/10.1016/j.jclepro.2021.128945
89. Deutz, P., Baxter, H., Gibbs, D., Mayes, W., & Gomes, H. (2017). Resource recovery and remediation of highly alkaline residues: A political industrial ecology approach to building a circular economy. *Geoforum; Journal of Physical, Human, and Regional Geosciences, 85*, 336–344. https://doi.org/10.1016/j.geoforum.2017.03.021
90. Stumpf, L., Schöggl, J. P. & Baumgartner, R. J. (2021). Climbing up the circularity ladder? - A mixed-methods analysis of circular economy in business practice. *Journal of Cleaner Production, 316*. https://doi.org/10.1016/j.jclepro.2021.128158

91. Van Buren, N., Demmers, M., van der Heijden, R. & Witlox, F. (2016). Towards a circular economy: The role of Dutch logistics industries and governments. *Sustainability, 8,* 647. ISSN: 2071-1050.
92. Mason, A., Lee, R. & members of the NTA Network. (2022). Six ways population change will affect the global economy. *Population and Development Review, 48,* 51–73. ISSN: 0098-7921, 1728-4457.
93. Roser, M., Ritchie, H. & Ortiz-Ospina, E. (2013). World Population Growth. *Our World in Data.*
94. Gill, K. & Verma, I. (2021). Circular economy: A review of global practices and initiatives with special reference to India. *FOCUS : Journal of International Business, 8,* 187–205. ISSN: 2347-4459, 2395-258X.
95. Liu, L. & Ramakrishna, S. (2021). *An introduction to circular economy*. Singapore: Springer Singapore. ISBN: 9789811585098 (2022).
96. Islam, A., et al. (2020). Advances in sustainable approaches to recover metals from E-waste - A review. *Journal of Cleaner Production, 244,* 118815. ISSN: 0959-6526.
97. Huynh, P. H. (2022). Enabling circular business models in the fashion industry: The role of digital innovation. *International Journal of Productivity and Performance Management, 71,* 870–895. Place: Bradford Publisher: Emerald Group Publishing Limited. ISSN: 17410401.
98. Peiro, L. T., Polverini, D., Ardente, F. & Mathieux, F. (2020). Advances towards circular economy policies in the EU: The new Ecodesign regulation of enterprise servers. *Resources, Conservation and Recycling, 154.* https://doi.org/10.1016/j.resconrec.2019.104426
99. *Directive 2012/19/EU of the European Parliament and of the Council of 4 July 2012 on waste electrical and electronic equipment (WEEE) (recast).* https://eur-lex.europa.eu/eli/dir/2012/19/oj.
100. *Directive 2008/98/EC of the European Parliament and of the Council of 19 November 2008 on waste and repealing certain Directives (Text with EEA relevance).* http://data.europa.eu/eli/dir/2008/98/oj/eng.
101. Diddi, S., & Yan, R.-N. (2019). Consumer perceptions related to clothing repair and community mending events: A circular economy perspective. *Sustainability, 11,* 5306. https://doi.org/10.3390/su11195306
102. Milios, L. & Matsumoto, M. (2019). Consumer perception of remanufactured automotive parts and policy implications for transitioning to a circular economy in Sweden. *Sustainability, 11.* https://doi.org/10.3390/su11226264
103. Su, B., Heshmati, A., Geng, Y. & Yu, X. (2013). A review of the circular economy in China: Moving from rhetoric to implementation. *Journal of Cleaner Production, 42,* 215–227. ISSN: 09596526.
104. Geissdoerfer, M., Savaget, P., Bocken, N. M. & Hultink, E. J. (2017). The circular economy – A new sustainability paradigm? *Journal of Cleaner Production, 143,* 757–768. ISSN: 0959-6526.
105. Uvarova, I., Atstaja, D., & Korpa, V. (2020). Challenges of the introduction of circular business models within rural SMES of EU. *International Journal of Economic Sciences, 9,* 128–149. https://doi.org/10.20472/ES.2020.9.2.008
106. Salvador, R., et al. (2021). Circular economy strategies on business modelling: Identifying the greatest influences. *Journal of Cleaner Production, 299,* 126918. ISSN: 09596526.
107. Ranta, V., Aarikka-Stenroos, L. & Väisänen, J.-M. (2021). Digital technologies catalyzing business model innovation for circular economy - Multiple case study. *Resources, Conservation and Recycling, 164,* 105155. ISSN: 0921-3449.
108. Quinones, R., Llatas, C., Montes, M. & Cortes, I. (2021). A Multiplatform BIM-Integrated Construction Waste Quantification Model during Design Phase. The Case of the Structural System in a Spanish Building. *Recycling, 6.* https://doi.org/10.3390/recycling6030062
109. Okorie, O., Salonitis, K., Charnley, F. & Turner, C. (2018). A systems dynamics enabled real-time efficiency for fuel cell data-driven remanufacturing. *Journal of Manufacturing and Materials Processing, 2.* https://doi.org/10.3390/jmmp2040077

110. Aras, N., Boyaci, T. & Verter, V. (2004). The effect of categorizing returned products in remanufacturing. *IIE Transactions, 36,* 319–331. ISSN: 0740-817X, 1545-8830.
111. Liu, Y. & Bai, Y. (2014). An exploration of firms' awareness and behavior of developing circular economy: An empirical research in China. *Resources, Conservation and Recycling, 87,* 145–152. ISSN: 09213449.
112. Pheifer, A. G. (2017). *Barriers & Enablers to circular business models.* White Paper. Brielle (The Netherlands): ValueC. http://valuec.eu/wp-content/uploads/2017/04/Whitepaper_ValueC_Barriers_Enablers_to_CBM.pdf.
113. Kirchherr, J., et al. (2018). Barriers to the circular economy: Evidence from the European Union (EU). *Ecological Economics, 150,* 264–272. ISSN: 0921-8009.
114. Ali, S. S., et al. (2017). *Understanding the nature and impact of the business angels in funding research and innovation: Final report.* Mannheim: Zentrum für Europäische Wirtschaftsforschung (ZEW). http://hdl.handle.net/10419/181907.
115. Zacho, K. O., Mosgaard, M. & Riisgaard, H. (2018). Capturing uncaptured values - A Danish case study on municipal preparation for reuse and recycling of waste. *Resources, Conservation and Recycling, 136,* 297–305. ISSN: 0921-3449.
116. Morseletto, P. (2020). Targets for a circular economy. *Resources, Conservation and Recycling, 153,* 104553. ISSN: 0921-3449.
117. Galvão, G. D. A., Evans, S., Ferrer, P. S. S. & de Carvalho, M. M. (2022). Circular business model: Breaking down barriers towards sustainable development. *Business Strategy and the Environment, 31,* 1504–1524. ISSN: 0964-4733.

Chapter 4
Implementation of Circular Economy in European Industrial Companies – A Survey

Nicolas Hofer, André Podleisek, Roman Hänggi, and Katharina Luban

Abstract The concept of circular economy promotes the reuse of resources in a closed-loop system rather than the traditional linear model of extraction, use, and disposal. At the core of circular economy are ten different retention options. Their function is to retain resources and prolong the use cycle of products and components. The objective of this study is to investigate the implementation and impact of circular economy within various industries. The study addresses critical questions: What is the extent of adoption of circular economy in industries and how does it influence business performance, including profitability, company image, and manufacturing costs? A key part of the research is to identify the predominant drivers and enablers of circular economy adoption and their relationship to ten individual retention options, as well as the factors motivating their future implementation. The results are based on a large-scale survey conducted with 512 participants. The findings of this study provide data-based insights into current trends, drivers, and enabler associated with the ten retention options.

N. Hofer (✉)
Zürich, Switzerland
e-mail: nicolas.hofer@ost.ch

A. Podleisek
Wettswil am Albis, Switzerland

R. Hänggi
Brülisau, Switzerland

K. Luban
Sevelen, Switzerland

A. Podleisek et al. (eds.), *Circular Economy: The Next Level of Company Success*,
https://doi.org/10.1007/978-3-031-66406-9_4

4.1 Research Design

Research in the field of Circular Economy (CE) has been growing fast, as shown by the number of published papers that nearly doubled each year from 2019 to 2021 [1].

Although the adoption of CE principles by companies has gained increasing attention in recent years, a small percentage of companies have actually adopted them, as reported by the OECD [2]. Additionally, a survey conducted by Masi et al. [3] found that 37% of participants, primarily from the European manufacturing sector, were not aware of the concept of CE. Furthermore, the data presented in the study by Sá et al. [4] suggests that most practitioners do not primarily associate the concept of CE with the Retention Options (Rs). Despite this lack of knowledge or awareness, various company-external drivers, such as institutional or customer pressure, can influence the decision to adopt Rs. This implies that companies might engage in specific R initiatives, just as other companies that incorporate these practices as a conscious effort to support the CE. In response to this hypothesis, our research seeks to explore and contrast the trends in Rs adoption between conscious and unconscious CE practitioners. We want to understand the relevance of this topic in practice.

The survey was carried out throughout December 2022 and distributed to practitioners from a range of sectors. Closed questions were used that aimed to determine which of the 10 Rs proposed by Reike et al. [5] are applied in the industry and the *drivers* and *enablers*, as found in Sect. 3.2 that support their practice. Consequently, the survey looked into these factors separately for each retention option.

The results of this analysis provide valuable insights into the factors that influence the adoption and implementation of CE practices within a company. By understanding these drivers and enablers, organisations can better tailor their strategies and initiatives to effectively implement the principles and practices of CE. This information can also assist policymakers in identifying and addressing any barriers or challenges that may be preventing the broader adoption of CE practices. The analysis shall help to accelerate the transition towards a CE.

Beyond identifying which retention options are commonly used in the industry and the drivers and enablers that support their use, the survey has additional objectives. One of the objectives is to test the assertion that Rs are also incorporated by industry practitioners without recognising themselves as part of CE. Another one is to determine which retention strategies are planned to be implemented in the future and identify the main drivers behind that decision. Additionally, the survey aims to discover the impact of CE practices on various Business Performance Factors (BPF) as perceived by companies that have already incorporated CE into their business model. Furthermore, to enhance the practical relevance of the findings, this study endeavours to evaluate the impact of CE on BPF by conducting an inter-industry comparison, as industries are known to possess distinct characteristics. This can help to provide additional insights to motivate others to adopt CE principles.

In summary, the survey is designed to answer the following research questions:

- What is the extent of adoption of circular economy in businesses?
- Which retention options are commonly implemented, and which are less common?
- How does the implementation of retention options influence business performance?
- How does the implementation of retention options differ between practitioner groups?
- What are the most common drivers for engaging in CE and can they be associated with certain retention options?
- What are the most common enablers for engaging in CE and can they be associated with certain retention options?
- Which motivational factors drive the future implementation or extension of retention options?

4.2 Study Focus and Aspects Surveyed

Govindan and Hasanagic [6] argue that companies are driven primarily by the pursuit of profits before considering environmental impacts. Camilleri [7] also suggests that practitioners are more likely to engage in CE initiatives if they see value in them. This suggests that economic benefits may be one of the most important factors in the adoption of Rs. However, Gusmerotti et al. [8] point out that economic efficiency is strongly related to shareholder expectations and that companies are mainly driven to respond to environmental issues.

The study therefore looks at the impact of CE principles on BPFs. The analysis begins with an overall evaluation of the impact, followed by a comparison of feedback across different industries and levels of CE integration. The objective is to provide a comprehensive understanding of CE's effects and uncover potential trends and insights that can assist organisations when considering CE strategies.

In a second step, the analysis turns its focus on the trend of Rs, examining which ones are currently being practiced by companies and which ones are planned to be implemented in the future. We expand this analysis by highlighting the differences between participants who have actively integrated CE into their business model and those who have not.

Finally, we analyse the drivers and enablers within the larger context of CE. We also compare how these various factors influence specific Rs and identify where they excel. Additionally, we analyse motivational factors similar to the drivers for companies that do not currently practice a specific Rs but intend to adopt them in the future.

4.2.1 Retention Options, Drivers and Enablers

The study encompasses the 10 R included in the framework developed by Reike et al. [5]. The retention options are ranked in order according to their potential to promote circularity in a closed-loop system, with the highest circularity potential in Refuse (R0) and declining circularity until Remine (R9). The Rs are grouped into four categories of strategies: material reduction, product use cycle extending, product and part use cycle renewing, and material processing strategies. The categorisation of the 10 R into these four strategy categories is based on their primary function and value proposition.

Multiple scholars have studied the drivers and enablers behind the adoption and implementation of CE [4, 8–16]. A sample of Italian manufacturing companies was studied by Gusmerotti et al. [8] to determine the most influential factors driving the adoption of the CE paradigm. Rizos et al. [15] analysed 30 case studies of European small and medium enterprises to increase understanding of the barriers and enablers companies encounter when implementing circular business models. Institutional drivers, as well as barriers, were explored by Ranta et al. [9] in a cross-regional comparison of China, the United States, and Europe. Calzolari et al. [11] conducted a study examining 50 Corporate Social Responsibility reports from French and German companies in various industries, including manufacturing, service, energy, and agri-food, in order to understand the extent of circular practices, the level of involvement of supply chain partners, and the drivers behind their adoption. In a recent study, Sá et al. [4] analysed the integration of CE in companies located in the northwest corner of Spain and Portugal. They compared responses from managers, executives, and technicians from the industry and service sectors. Their findings indicated that environmental reasons are the primary motivation for adopting CE practices.

However, according to several authors, there is a need for further empirical studies on the role of drivers and enablers, as much of the current research on these factors is either theoretical or based on a small number of cases [3, 14, 17, 18]. It is also important to note that the listed studies have focused on examining the drivers and enablers for the CE as a whole rather than exploring the factors that promote the adoption of individual Rs. To our knowledge, no prior research has investigated whether there are variances in the drivers and enablers that facilitate the adoption of specific Rs. An approach that examines these differences can provide a better understanding in this regard. Therefore, this study aims to address this by conducting a quantitative online survey focusing on the European industry.

In Sect. 3.2, a framework for drivers and enablers was created by synthesising information from multiple literary sources. It involved identifying the drivers and enablers to adopt CE, categorising them according to their similarity and meaning, and grouping them into several clusters. The drivers were organised into clusters based on their origin, while the enablers were grouped based on the area where they had an impact. The results for the drivers are presented in Table 4.1, and the enablers are listed in Table 4.2.

It should be noted that the study did not consider all the factors: driver D.F4 and enablers E.F1, E.F2, E.T4, and E.C1 were excluded from the study. The explanation for this can be found in the following sections. As a result, the survey includes 15 drivers and 8 enablers.

4.2.2 *Concept of the Drivers in Question*

The mainstream literature identifies a set of benefits of putting Rs into action and the circumstances that have emerged, which act as drivers. This has resulted in faster adoption of CE practices and initiatives than in the past [19].

Environmental drivers emerged due to the effects globalisation, extensive material consumption, and production processes have on the environment [20]. The adoption of Rs is thereby driven by the need to decrease the resourcess used, reduce the environmental footprint and mitigate resource-related risks [21, 22].

Institutional drivers are due to the implementation of laws and policies by European governments to reduce waste and extend the lifetime of products in response to environmental degradation [23, 24]. The European Commission also published action plans to promote the CE [25, 26]. These measures can drive companies to adopt Rs in order to avoid sanctions and penalties or because they anticipate future legal obligations [6, 11, 27]. Governments also implemented favourable conditions through taxation benefits, supportive funds, and subsidies [6] to promote the adoption of Rs.

As consumers become more aware of the harmful environmental effects of industries, some companies feel compelled to improve their environmental image to mitigate reputational risk [28]. Moreover, having a green image has become a business strategy, which drives the adoption of Rs by practitioners to appeal to environmentally conscious customers [11, 15]. In addition, this dynamic has spurred technological advancements in areas where solutions were previously non-existent and established new benchmarks. This pressures other companies to follow best practices to remain viable [29]. These aspects form the *Stakeholder drivers*.

Financial drivers are due to the goal of economic success and the recognition that End-of-Life (EOL) products contain value that can be retained, and cost savings can be achieved when implementing Rs [15]. Moreover, the employment of Rs can create opportunities for new business and income, driven by the demand for waste utilisation solutions created by landfill bans [30]. Companies are also driven to experiment with and develop new business models centred around Rs to identify and take advantage of market opportunities and gain a competitive advantage [11].

The driver D.F4 focuses on generating value from waste and production side streams by selling them to other companies [31]. However, from a company's own perspective, this approach does not encourage the internal adoption of retention options. Instead, it provides financial incentives to exchange waste materials and stimulates a market for these resources, which promotes the implementation of Rs by other organisations. Consequently, while D.F4 is included in the initial framework, it has been excluded from this study in order to concentrate on drivers that more directly influence a company's decision to implement a specific R themselves.

There are companies that adopt circular strategies for ethical reasons, such as to reduce the negative impacts on public health and animal health [32] or because they want to promote systemic change [11]. These company-internal motivational drivers are found in the cluster *Responsibility*.

Table 4.1 Drivers propelling European companies to adopt Rs

Cluster	Code	Driver	Context
Environmental	D.E1	Resources	Decrease the resources used in production
	D.E2	Footprint	Decrease products' carbon footprint
	D.E3	Risk	Resilience against resource-related risks (scarcity, supply challenges, etc.)
Institutional	D.I1	Legal	Comply with legal requirements and norms
	D.I2	Punitive	Avoiding punitive measures
	D.I3	Juridification	Anticipate or meet future legal obligations
	D.I4	Tax	Taxation benefits, access to supportive funds, and subsidies
Stakeholder	D.S1	Image	Improve corporate image, legitimacy, or credibility
	D.S2	Customer	Adapt to changing consumer behaviour to meet customer expectations and increase customer satisfaction
	D.S3	Practices	Meet the standards and follow the best practices of competitors
Financial	D.F1	Volatility	Price volatility of raw material
	D.F2	Savings	Cost saving opportunities
	D.F3	Competitiveness	Competitiveness through innovation and business development
	D.F4*	Value	Create value from waste and production side streams
Responsibility	D.R1	CSR	Social and ethical responsibility
	D.R2	Promoter	Become a promoter of systemic change

*Excluded from the study

4.2.3 Concept of the Enablers in Question

In close relation to the drivers stand the enablers. Enablers are factors that facilitate the implementation and practice of Rs in response to the motivations or incentives provided by the drivers. They can be internal or external factors that also help companies overcome difficulties when adopting Rs. Examples of internal enablers include financial resources and a corporate culture that values sustainability. External enablers include partnerships with other companies or organisations, access to technical expertise, and availability of Reverse Logistics (RL) networks.

While company culture (E.F1) plays an important role as an internal enabler [15] and can greatly influence an organisation's approach to challenges and changes in the business environment, it is not included in this study for several reasons. Company culture is unique to each organisation, making it difficult to establish a generalised framework to assess its impact on the adoption of Rs. Moreover, company culture is often deeply ingrained and dependent on specific departments and their distinct norms and values, which adds complexity to the evaluation process.

Additionally, previous studies have indicated that company culture is the least challenging aspect of implementing Rs within organisations [4, 14, 15]. By focusing on less established enablers, the study can provide more valuable insights and actionable recommendations for organisations seeking to adopt Rs. Consequently, excluding company culture from this study allows a more targeted analysis of the enablers influencing the implementation of Rs.

To implement new practices, organisations must invest in resources such as training, technology, and infrastructure. Similarly, the availability of capital or funding (E.F2) is essential to finance CE initiatives. Previous research has highlighted the significance of financial support through funding in companies' adoption of R practices [4, 13]. The European Investment Bank and other financial institutions have introduced a dedicated line of financing for CE initiatives, aiming to invest at least 10 bill. EUR by 2023 [33]. However, Fonseca et al. [13] found that organisations more frequently finance the adoption of Rs through their own savings and the benefits generated by implementing these practices. Given that access to capital is a crucial prerequisite for any undertaking and multiple studies have already confirmed its importance for CE initiatives, as well as the widespread availability of funding opportunities for CE initiatives in Europe, the enabler E.F2 is excluded from this study.

To effectively practice Rs, it has been stated that it is necessary for companies to establish close relationships with other supply chain actors and increase transparency [15, 30], for example, about the materials used in the design of the product to make *recycling* more efficient. This is supported by using digital technologies and by standardising the information shared.

Companies that have implemented circular practices and faced technological barriers, such as a lack of knowledge or technological solutions, have found success in developing their technological know-how through experimentation or in collaborating with suppliers, customers, and research institutes [4, 34]. To identify innovative solutions, keeping up to date with published research was also cited as the process that has made practising Rs possible.

Outsourcing (E.T4) is an enabling activity for companies that lack the necessary resources or expertise to complete the task themselves. Companies looking to build their own internal capabilities to practice an R would seek other strategies rather than relying on outsourcing. Therefore, it is excluded from the study for similar reasons as D.F4.

Due to the use of quantitative methodology in the survey, the enabler engagement (E.C1) was not included in the research. This is justified by the reason that our quantitative method is not the most effective way to understand which specific factors enable customer engagement as a qualitative approach would. Furthermore, customer-related factors emerged in previous studies as one of the lowest difficulties to overcome in implementing CE strategies [4, 14]. Therefore, it was decided to exclude it and to give space to other factors that are more directly related to the prerequisites necessary to implement Rs.

Enablers to overcome difficulties about the quality and quantity of materials and products needed to practice an R are included in the cluster material supply. Logistics and collection systems, as well as industrial symbiosis and collaboration with waste

Table 4.2 Enablers that make the practices of Rs appealing

Cluster	Code	Enabler	Context
Firm	E.F1*	Culture	Committed company environment and culture
	E.F2*	Capital	Available capital or financial support to transition
Value Chain	E.V1	Partnership	Build closer relationships/invest in collaboration and partnerships
	E.V2	Transparency	Enhance information sharing and transparency, make use of digital technologies and standardise information
Technological	E.T1	Research	Experiment and research
	E.T2	Mutuality	Joint decision-making (customers and suppliers)
	E.T3	R-Technology	Availability of technical and technological know-how in-house
	E.T4*	Outsourcing	Possibility of outsourcing the activity
Customers	E.C1*	Engagement	Increase customer engagement and awareness, provide incentives
Material Supply	E.M1	Logistics	Available reverse logistic network
	E.M2	Collection	Create simple collection/take-back systems
	E.M3	Symbiosis	Engage in industrial symbiosis

*Excluded from the study

management and reprocessing firms, can help address these challenges and enable the successful practice of retention strategies [35, 36].

4.3 Survey Methodology and Sample Description

A web-based research survey was conducted to gather the necessary data for the analysis. Electronic questionnaires are widely used these days and align with internationally accepted scientific standards. They also offer several benefits, such as anonymity for participants, efficiency, and cost-effectiveness. The survey tool provided by Tivian, UNIPARK [37], was utilised for designing the questionnaire and collecting responses. Data analysis was conducted partly with UNIPARKs integrated analysis tool Reporting+ from Questback and partly using Excel.

In the current study, the methodology includes a description of the questionnaire design, followed by the data collection process and ends with presenting the outcome and description of the sample included in further analysis.

4.3.1 Questionnaire Design

The questionnaire was created in English using the information presented in Chap. 3. Due to the objective of the study, the questionnaire must include the 10 retention options (see Table 3.1 in Sect. 3.1.3), each requiring questions about drivers and enablers. To simplify the answering process for the respondents, closed questions were used. They offer a predefined set of options to choose from, making them more efficient and less time consuming compared to open questions [38]. Closed questions are also less prone to bias, as they do not allow respondents to express their own opinions or interpretations [39]. However, this is also considered a drawback, as it is limited in capturing the full range of opinions and providing detailed insights. On the other hand, using open questions would have made the answer process overwhelming and time consuming for the participants.

The survey was designed to avoid bias in the results by not imposing mandatory conditions on participants. This means that participants were allowed to move on to the next question without providing an answer if they chose to do so. This approach ensures that the answers are not influenced by the desire to complete the survey. Instead, responses are more likely to reflect the honest opinions of participants. This is important to ensure the validity and accuracy of the survey results and has been done in other surveys [14]. However, the drawback of this method is that it can lead to a higher rate of missing data if questions are skipped intentionally or accidentally.

To ensure the validity and reliability of the questionnaire, three professors and two practitioners from the manufacturing industry pre-tested it (one from the Netherlands and one from Spain). This led to the identification and correction of minor issues and ambiguities in the questions and helped to ensure that the final version was clear and effective in gathering the intended data.

The questionnaire was structured into three primary sections: demographics, circular economy effects, and circular economy practices.

Demographics

The demographics section aimed to collect information on respondents' positions, company industry, employee count, headquarters location, and primary market of operation. The company industry choices provided were designed to cover a broad range of sectors and to categorise companies according to the nature of their output. To facilitate the decision-making process for participants, examples were provided along with each choice. The options included:

- Manufacturers of Durable Products (MDP) – Companies that produce long-lasting goods, such as machinery, furniture, vehicles, electrical and electronic equipment, appliances, etc.
- Manufacturers of Non-durable Products (MNP) – Companies that manufacture goods with a shorter lifespan, such as chemicals, cleaning products, cosmetics, textiles, apparel, etc.
- Manufacturers of Parts and Components (MPC) – Companies that produce components or parts for other products, such as automotive parts, replacement parts, etc.
- Manufacturers of Convenience Products (MCP) – Companies that produce goods designed for consumption, such as foods, beverages, tobacco, etc.
- Repair and Installation of Machinery and Equipment (RIME) – Companies that specialise in the repair, maintenance, and installation of machinery and equipment.
- Retail and Wholesale (RW) – Companies involved in the sale of goods to consumers or other businesses, including supermarkets, department stores, and distributors.
- Service Industry (SVC) – Companies that primarily offer services to customers or other businesses, including consulting, financial services, public service, IT, etc.
- Other

Circular Economy Effects

The section 'circular economy effects' was designed to explore the integration of CE in a company's business model and its impact on eight BPFs. It also indirectly sought to gauge respondents' familiarity with the CE concept. The response options for the BPFs were intentionally kept broad, including positive impact, no impact, negative impact, and unknown. Using these broad categories, rather than a traditional Linkert scale, allows for a clearer understanding of the general impact direction without delving into subjective nuances of intensity.

Circular Economy Practices

The section 'circular economy practices' focused on the specific Rs. It was further divided into four subsections corresponding to the classification of Rs in Sect. 3.1.4: Material reduction strategies, product use cycle extending strategies, product or part use cycle renewing strategies, and material processing strategies. These subsections aimed to identify which Rs companies were already practising, which ones they intend to adopt in the future, and which ones were not of interest to them. Depending on the selection, follow-up questions were tailored to inquire about the drivers and enablers for each R if the company was already practising it or the primary reason behind their intention to adopt it. Conversely, these questions were bypassed if a specific R was not of interest. To ensure a shared understanding of the different R terminologies, their definitions were included at the beginning of each question related to Rs.

Using closed questions and incorporating conditions that bypass questions based on prior responses, the questionnaire was designed for efficiency and ease of completion. The design aspect also ensured that each survey page remained clear and uncluttered, minimising the chance that respondents missed a question. For further scholarly inquiry or methodological verification, copies of the questionnaire are available from the authors upon formal request.

4.3.2 Data Collection

Responses to the survey were collected during the month of December 2022. To gather as many responses as possible, the survey was distributed through multiple channels, including email, Linkedin, Surveycircle, and personal contacts. In order to identify potential contacts to be reached out to by mail, the study focused on companies in the manufacturing industry located in Europe. LinkedIn SalesNavigator was used to search for companies that met these criteria. The search was narrowed down to focus on employee positions, including CEO, business owner, business development manager, and sustainability manager. These individuals were selected because they were believed to have a good understanding of their company's practices and future direction. The survey was then distributed through email and LinkedIn messages. More than 7,000 individual companies were reached through this method. In addition, reminder emails were sent in the middle of December.

The second approach was to share the survey directly on Linkedin in various groups related to the circular economy. In addition, it was promoted by university professors, fellow students, and personal contacts. These individuals were also asked to forward the invitation to further contacts. Furthermore, the survey was published on the platform Surveycircle. This helped to increase visibility and reach a wider audience.

Due to the distribution method, it is impossible to calculate an accurate response rate, as the total number of individuals who received the survey link is unknown. However, a total of 858 visits to the survey site were recorded. Of these, 548 proceeded to the final page of the questionnaire. This accounts for a completion rate of 64%. Most of the dropouts (20%) were recorded in the first two sections.

The high dropout rate among the survey participants may be due to the perception that the survey has a strong bias toward the manufacturing industry, as the survey focusses primarily on material retention strategies. This was confirmed by some participants who dropped out and later contacted the survey team. Another reason for the high dropout rate could be that the survey targeted Europe, and countries outside of Europe were not explicitly listed in the demographic section, which may have made some participants feel excluded.

4.3.3 Sample Description

As the questionnaire design allowed participants to skip questions, it was necessary to evaluate the 548 responses for their level of completeness. Only responses that provided answers to all ten questions about whether or not the Rs are practised were included. As a result, 512 responses met the inclusion criteria and were used in the final analyses. The details of the final sample are seen in Table 4.3.

Most of the participants held leadership roles, and more than 44% identified themselves as CEOs or business owners. The manufacturing sector had the highest representation among the respondents, with 42% indicating that they are involved in the production of durable goods. The vast majority, 71%, indicated that they operate mainly in the Business-to-Business (B2B) market, and 61% of the companies can be classified as Small and Medium-sized Enterprises (SMEs) with fewer than 250 employees. On the other hand, 28% of the companies are considered large with an employee count between 250 up to 10,000 employees, and 10% are considered super large with more than 10,000 employees.

The distribution method proved to be successful in achieving a wide outreach, as evidenced by the participation of companies from 32 European countries. A geographical overview is presented in Fig. 4.1.

However, it is important to highlight that the majority of participating companies are based in Western Europe. Swiss companies have the highest percentage of participation at 17%, followed closely by German and Italian companies at 12%. The number of companies located in central Europe is considerably lower, where Poland recorded the highest participation rate at 1.8%. For a detailed breakdown of

Table 4.3 Characteristics of the survey participants

Dimension	Sub-dimension	n	Percentage
Position*	CEO/Business owner	223	43.6%
	Business development manager	73	14.3%
	Environmental/Sustainability/Governance manager	67	13.1%
	Other managerial positions	50	9.8%
	Other C-Suite Officers	39	7.6%
	Technical staff	29	5.7%
	Other	27	5.3%
	Administration and finance	3	0.6%
Economic activity	Manufacturer of durable products	214	41.8%
	Manufacturer of non-durable products	80	15.6%
	Service	80	15.6%
	Manufacturer of parts or components	49	9.6%
	Other	42	8.2%
	Retail/Wholesale	32	6.3%
	Repair and installation of machinery and equipment	10	2.0%
	Manufacturer of convenience products	5	1.0%
Employee count*	<10	115	22.5%
	10 –49	106	20.7%
	50 –249	93	18.2%
	250 –999	72	14.1%
	1,000 –10,000	72	14.2%
	>10,000	53	10.4%
Market of operation	B2B	362	70.7%
	B2B and B2C	105	20.5%
	B2C	36	7.0%
	Other	9	1.8%
Total (N)		512	

*For each of these questions, one participant chose not to answer

the company headquarters' locations and the number of participating entities, please refer to Table 4.4.

Fig. 4.1 Choropleth map of company headquarters distribution by country

Table 4.4 Location of company headquarters

Country	n	%	Country	n	%
Switzerland	88	17.2%	Slovakia	4	0.8%
Germany	62	12.1%	Greece	4	0.8%
Italy	61	11.9%	Ireland	4	0.8%
Netherlands	57	11.1%	Lithuania	2	0.4%
Sweden	39	7.6%	Serbia	2	0.4%
United Kingdom	36	7.0%	Portugal	2	0.4%
Spain	22	4.3%	Ukraine	2	0.4%
Finland	19	3.7%	Latvia	2	0.4%

(continued)

Table 4.4 (continued)

Country	n	%	Country	n	%
Austria	17	3.3%	Bulgaria	2	0.4%
France	17	3.3%	Hungary	2	0.4%
Belgium	12	2.3%	Estonia	2	0.4%
Denmark	10	2.0%	Romania	1	0.2%
Poland	9	1.8%	Luxembourg	1	0.2%
Other	8	1.6%	Croatia	1	0.2%
Norway	8	1.6%	Cyprus	1	0.2%
Czechia	7	1.4%	Liechtenstein	1	0.2%
Total (N)				512	100%

4.4 Survey Results and Discussion

The survey yielded an impressive number of completed questionnaires, allowing us to explore the key areas of interest.

4.4.1 The Impact of Circular Economy on Business Performance Factors

Participants were initially asked to provide information on the extent of CE implementation in their company to assess the impact on BPFs. Recognising that not all participants might be familiar with CE or have incorporated the concept, they were presented with five options for answering, as shown in Fig. 4.2. These options ranged from 'Do not know' and 'Not heard of CE before', indicating limited awareness, to more specific choices such as 'Separated from core business', 'Not implemented', and 'Core business', each reflecting a different degree of CE implementation within their organisations.

Among the 512 participants, 216 chose the option that CE was integrated into their core business operations, while 135 stated that it was separate from their core activities. The fundamental distinction between these two approaches lies in their integration within the company's overarching structure and strategy. In the first approach, CE principles become integral to all aspects of the organisation, while in the second approach, they are present as isolated initiatives outside the core business operations.

On the other hand, 139 participants stated that CE is not implemented in their company, while 9 participants were uncertain about their company's adoption of CE principles.

Only a very small group of 13 participants (2.5%) chose the option 'Not heard of CE before'. This indicates a considerable increase in the industry's awareness of CE compared to 2019 when almost 36% of participants in a survey conducted by Kumar et al. [14] had not heard of the concept before.

To evaluate the influence of CE strategies on BPFs, we focus on the 351 participants who indicated that CE is either incorporated into their core business or implemented separately. These participants were asked a follow-up question about the impact of their CE strategies on eight BPFs. An in-depth description of the value associated with these BPFs is provided below.

Material and Manufacturing Costs

As the global economic landscape expands and commercial activities increasingly transcend national boundaries, it is imperative for European companies to develop strategies that enhance their competitiveness in relation to both intra-European rivals and global market participants. Factors such as material, manufacturing and labour costs directly influence production expenses, rendering the pursuit of cost-effective material sourcing and manufacturing methods even more crucial for European organisations. This is particularly relevant for companies located in high-cost locations, as the optimisation of cost structures can increase their competitiveness.

Resilience against resource related risks

Resilience has become an important aspect for businesses, especially in light of disruptions caused by events such as the COVID-19 pandemic [40]. The magazine Fortune [41] reported that 75% of companies have had their supply chain disrupted due to the coronavirus. The automotive industry in Europe was one of the hardest hit, as the shutdown of Chinese factories disrupted their supply chain [42]. Moreover, resource scarcity is becoming an increasing concern. It will impact companies in many ways, such as the increase in raw material costs. Therefore, companies should focus on building resilience against these types of risks in order to remain competitive.

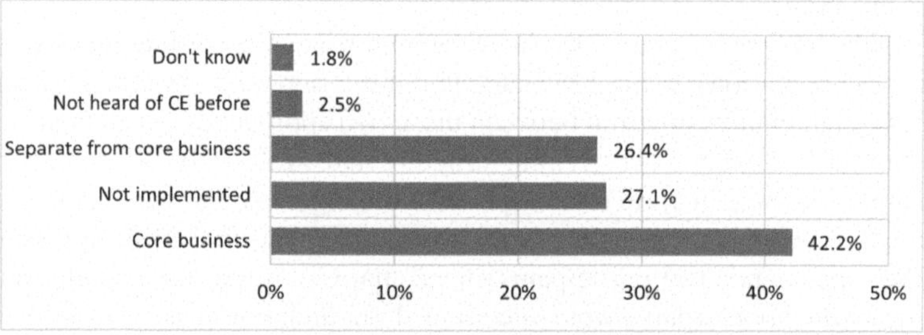

Fig. 4.2 Implementation of circular economy in companies (n = 512)

Profit

Profitability is a crucial aspect for businesses, as it reflects their financial performance and success. It is the amount of money that remains after all expenses have been paid and represents a key measure of a company's performance. A company that is profitable is able to sustain its operations, grow its business, and invest in future opportunities. Additionally, profitability also serves as a measure of the return on investment for shareholders and is also a vital aspect to attract new ones.

Technological advancements

A good performance of this BPF allows companies to maintain a competitive advantage and secure market benefits. It can also enhance efficiency and productivity, cut costs, and create new income sources. In the context of CE, technological advancements are also crucial for improving recycling and repurposing methods that allow material recovery, which in turn helps close the Product Lifecycle (PLC) loop and decrease the need for virgin material resources. Companies that achieve this (e.g., DESSO [43]) have also reported becoming more competitive and more robust in terms of resource-related risks.

Sales

Sales refers to the revenue generated from the sale of a product or service and directly relates to the profit. Hence, they are both financial performance measures and should positively correlate. A company with strong sales performance can increase not only its revenue but also its market share and profitability. In addition, a company that consistently meets or exceeds its sales targets can be seen as a market leader and can generate positive perceptions from investors and partners.

Reduction of waste

Waste reduction is an important aspect of business performance for companies because it can lead to significant savings in production costs. Companies that manage to reduce internal waste also benefit from reduced material consumption. In other words, they will need to purchase less and the amount of energy needed to process them can be decreased. This can lead to lower production costs, which again can increase profitability and competitiveness. Reducing waste also helps companies meet their environmental and sustainability goals, which is becoming increasingly important for customers and other stakeholders.

Company image

This BPF refers to customers', employees', and other stakeholders' perception of a company. A positive company image can help to increase its business performance in several ways. Firstly, it can attract new customers and retain existing ones. Customers are more likely to choose products or services from a company they perceive as trustworthy, reliable, and socially responsible. Additionally, customers with a positive image of a company are more likely to recommend the company to others, leading to increased sales and revenue. Secondly, a positive company image can also help to attract and retain talented employees. And thirdly, it can improve relationships with investors as they are more likely to invest in a company that they perceive as being well-managed, financially stable, and socially responsible. This can result in more accessible access to new funding opportunities and puts the company in a stronger position to negotiate terms. A positive image can also result in increased resilience in the face of a crisis. Companies with a strong reputation are more likely to be able to navigate difficult situations, such as economic downturns or public relations crises, without significant damage to their bottom line. Overall, a positive company image is of paramount importance to the overall business performance and growth of a company.

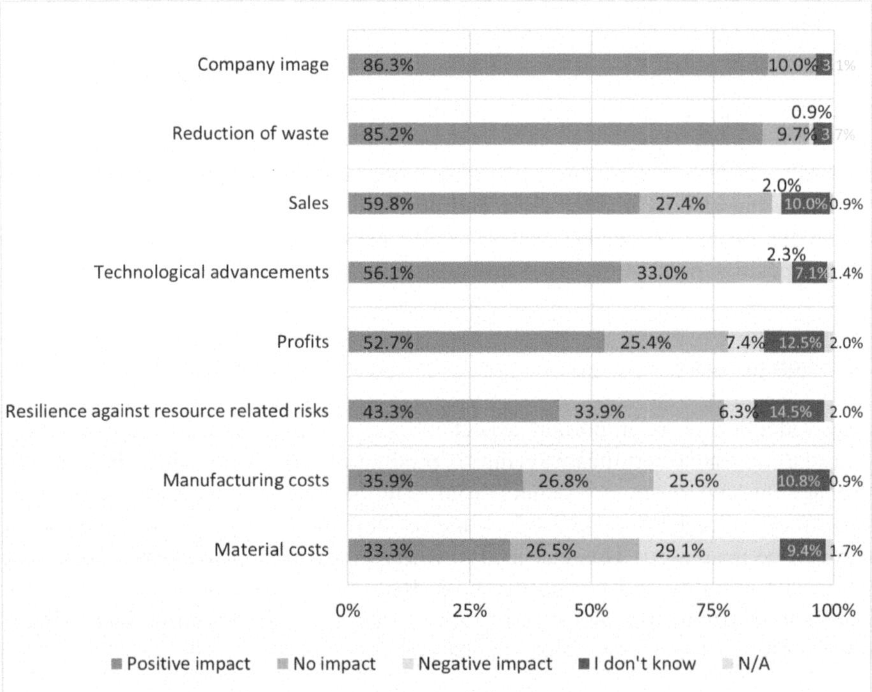

Fig. 4.3 Impact of circular economy strategies on BPFs (n = 351)

Overall Results of Circular Economy Impact on Business Performance Factors

Given the results seen in Fig. 4.3, incorporating CE into business operations presents the opportunity for a positive impact on all BPFs, which in turn can affect the overall effectiveness of a company. Remarkably, a significant number of participants have witnessed enhancements in their *company's image* (86.3%) and *waste reduction* (85.2%) as a result of adopting CE strategies.

For *sales*, 59.8% of respondents reported a positive impact and 27.4% experienced no impact. In the case of *technological advancements*, 56.1% of participants observed improvements, 33.0% noted no change, and 2.3% faced negative consequences. When considering *profits*, 53.7% of respondents indicated a positive influence, whereas 25.4% reported no change.

On the other hand, *manufacturing costs* and *material costs* seem to exhibit greater sensitivity to the implementation of CE principles. In contrast to the 0 to 7.4% range of negative impacts reported for all other factors, 25.6% and 29.1% of respondents encountered adverse impacts in these particular areas, respectively. Nonetheless, given that a significant number of companies observe positive effects on their *sales* and *profits*, while merely a small fraction experience negative impact on sales (2%) and profits (7.4%), it is evident that the advantages of adopting CE principles generally outweigh the potential rise in expenses for the majority of companies.

Concerning the impact of CE on *resilience* against resource-related risks, 43.3% of participants reported it to be improved. This suggests that adopting CE practices can enhance a company's ability to adapt and withstand potential disruptions in resource availability or market fluctuations. Additionally, it highlights the possible advantage of implementing CE principles to create a more stable and sustainable business model in the face of resource-related challenges. However, not all companies experienced such benefits. With 33.9% of respondents reporting no impact and 6.3% witnessing a negative impact, it is evident that the effectiveness of CE practices in bolstering resilience can vary depending on factors such as the specific industry or the extent of CE integration.

To further explore this, resilience along with the other BPFs will be examined individually to determine if the results remain consistent across different industries and implementation levels. Given the limited number of participants in the MCP industry group (only four), this industry will be excluded from the inter-industry comparison, along with participants who selected the 'other' industry category.

Assessing Business Performance Factors Across Industries and Circular Economy Implementation Levels

Material Costs

The analysis of various industries reveals distinct patterns in the impact of CE principles on material costs (see Fig. 4.4). MNPs have the highest negative impact rate at 40.8%, making them stand out compared to other industries. Along with RW,

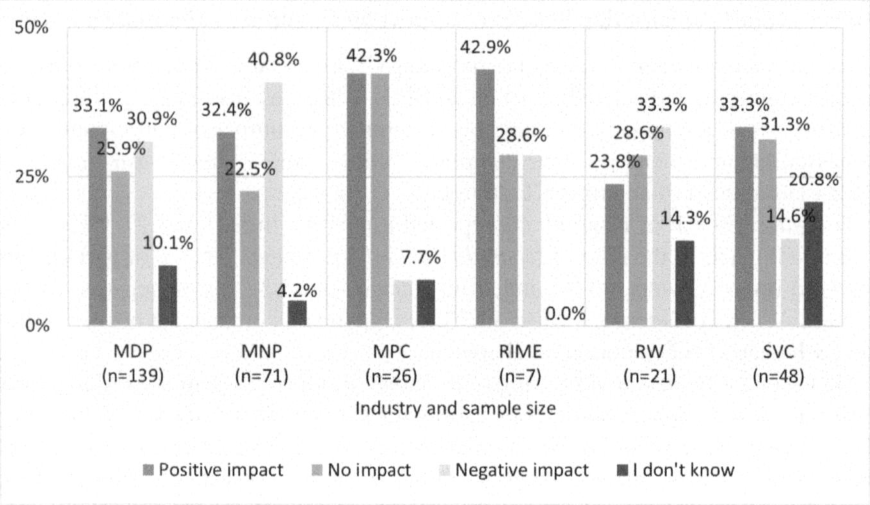

Fig. 4.4 Inter-industry comparison of CE impacts on material costs

these two sectors are the only ones where the majority of participants report negative impacts.

On the other hand, the RIME industry appears to be more positively affected, although the results should be interpreted with caution due to the sample size. Meanwhile, MPCs also report better performance, with 42.3% of respondents indicating positive effects; however, an equal percentage does not see an impact. Notably, this industry has the lowest percentage of participants reporting negative impacts, with 7.7%.

With one-third of MDPs and SVC reporting positive impacts through CE implementation, the results are similar to those of MNPs in this regard. However, the percentage of companies experiencing adverse effects varies between the three industries. MDPs face a higher negative impact rate at 30.9%, while SVCs reports a much lower percentage of 14.5%.

Hence, the impact of CE principles on material costs varies significantly across industries. While some sectors, such as RIME and MPC, have a higher tendency to experience positive effects, the MNP and RW sectors show a higher rate of negative impacts.

Differences in impact on material cost between the integration level (Fig. 4.5) are less pronounced than in the inter-industry comparison. In the core business group, which fully integrates CE practices into their primary business operations, 35.2% of companies report a positive impact on material costs. This indicates a substantial portion of companies experience beneficial outcomes, most likely due to more efficient material use and a reduction in waste. On the other hand, for companies that have integrated CE practices separately from their core business operations, a slightly lower percentage of 31.8% report positive impacts. Although it is only a

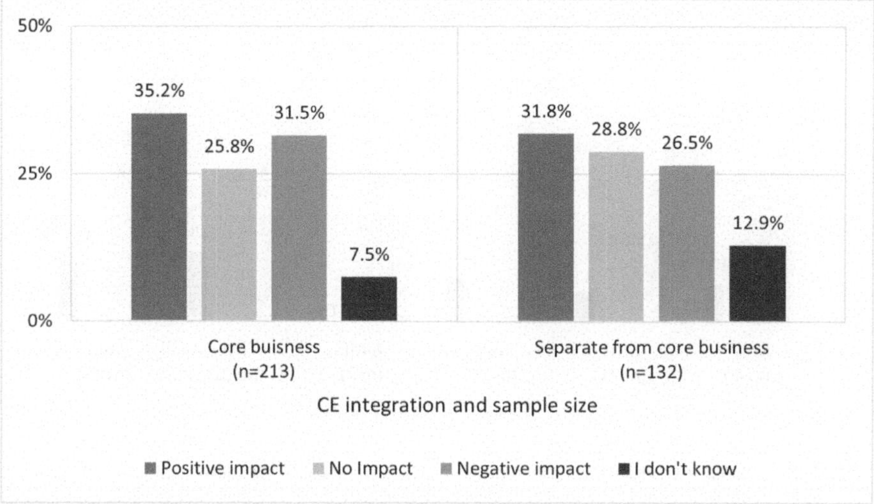

Fig. 4.5 Comparative analysis of CE impacts on material costs by integration level

minor difference, it may be related to the resources allocated to CE practices when they are not part of the central business strategy.

A noteworthy observation is the higher percentage of negative impacts within the core business group (31.5% compared to 26.5%). The negative impacts experienced might be due to the higher cost of using higher quality and more durable materials that can endure multiple use cycles compared to traditional materials. Although integrating CE into the core business results in a higher likelihood of achieving positive outcomes, it is important to consider that this approach also carries a higher risk of experiencing negative impacts. The balance between risk and reward needs to qbe considered when integrating CE practices.

Manufacturing Costs

The results of the study examining the influence of CE on manufacturing costs revealed a distinct lack of consensus among industries (Fig. 4.6). Although the majority of the groups showed mixed opinions, two particular groups emerged as notable exceptions.

MPC stands out with 53.8% of its participants observing a positive impact on their manufacturing costs. This group is unique in that it is the only one that indicates a clear and positive effect. In contrast, 71.4% of the companies in the RIME group saw their manufacturing costs remain unaffected and no negative impacts were reported.

In comparison, the MNP group shows positive outcomes for 36.5% of participants, ranking second in the inter-industry comparison. However, it also had the highest proportion of participants reporting negative effects (39.4%). Similarly, the MDP

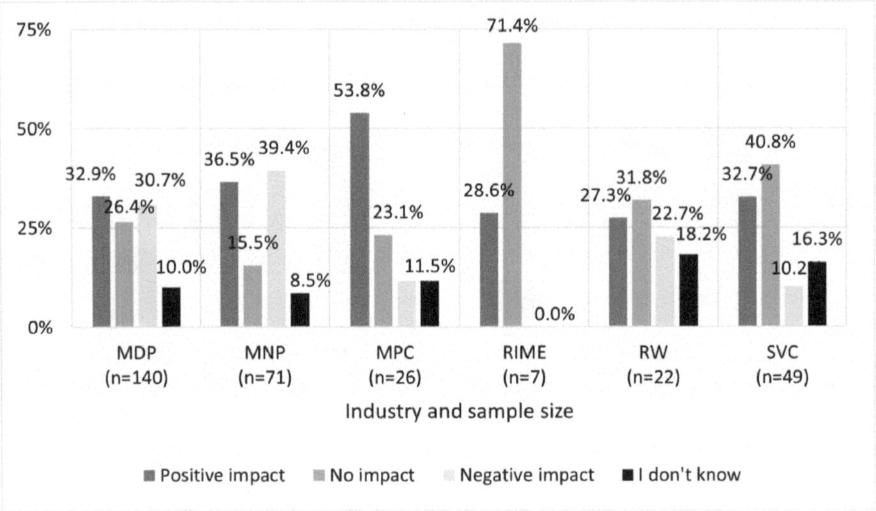

Fig. 4.6 Inter-industry comparison of CE impacts on manufacturing costs

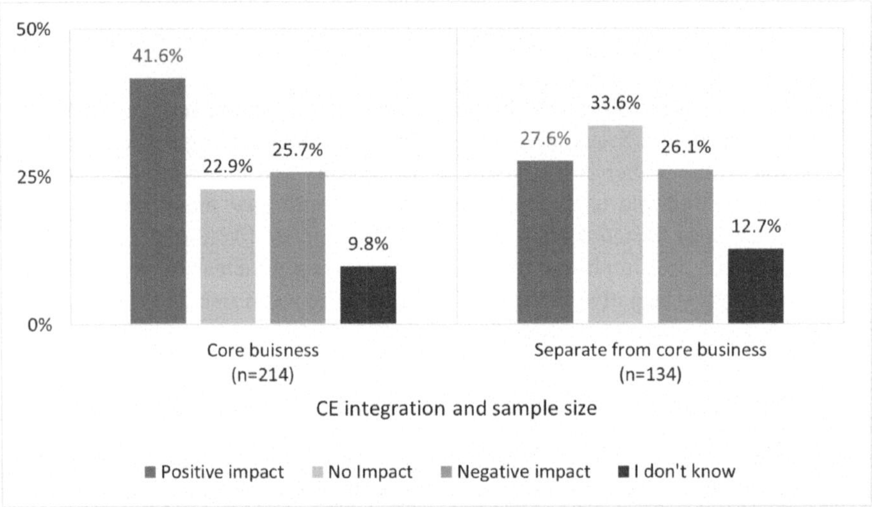

Fig. 4.7 Comparative analysis of CE impacts on manufacturing costs by integration level

and RW groups exhibited mixed results, with similar percentages of positive and negative impacts observed across both groups. The SVC industry is in between, with most participants experiencing either a positive or no impact on their manufacturing costs.

These findings emphasise the varied influence of CE on manufacturing costs in different industries. Companies looking to adopt CE practices should carefully consider their unique circumstances and tailor their strategies to achieve the most significant potential benefits.

The impact of CE on manufacturing costs differs depending on whether it is integrated into the core business or operated as a separate business (Fig. 4.7). In the core business group, 41.6% report a positive impact, whereas the second group it is 27.6%. These findings indicate that integrating CE practices into the core business has a higher probability of positively impacting manufacturing costs. However, it is worth noting that around a quarter in both groups experienced negative impacts. This cases can be due to several factors. For instance, CE often requires more complex and sophisticated manufacturing processes that can add to the cost of production. In addition, the implementation of new technology and infrastructure needed for CE practices can also come at a cost.

Resilience against resource related risks

The effect of CE on resilience is illustrated in Fig. 4.8. Based on the data, there is a positive impact trend, particularly in the MNP and SVC group, with more than 50% of the participants reporting positive impact. Whereas the resilience of companies that specialize in RIME remains largely unchanged (71.4%). Interestingly, three of the groups do not show negative impacts on resilience, showcasing the potential benefits of accepting CE. However, MNP appears to be an outlier, with 12.7% of participants reporting negative impacts. This discrepancy could be due to challenges

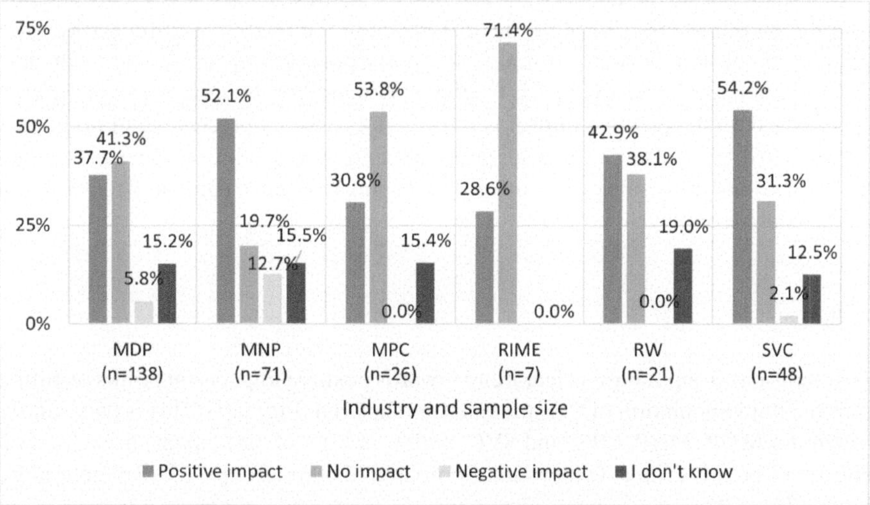

Fig. 4.8 Inter-industry comparison of CE impacts on company's resilience against resource-related risks

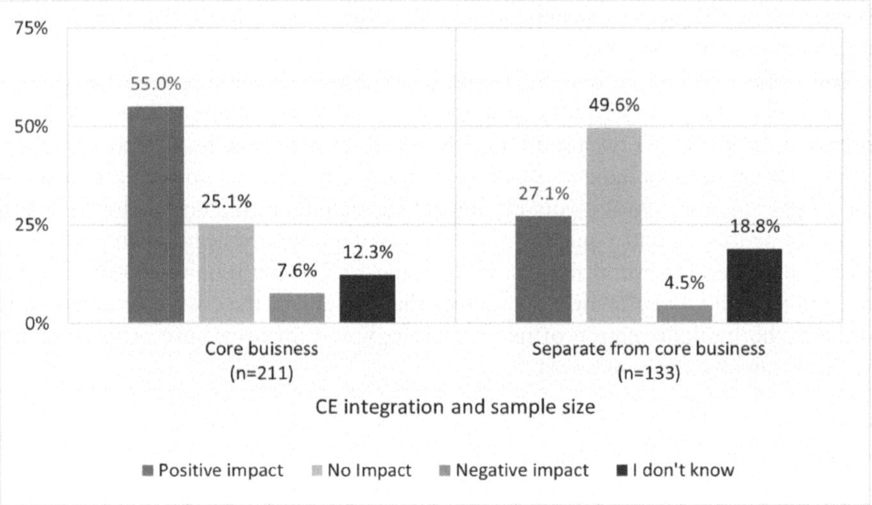

Fig. 4.9 Comparative analysis of CE impacts on company's resilience against resource-related risks by integration level

related to shorter product lifespans, less recyclable materials, and rapid consumer consumption.

In general, it can be concluded that the implementation of CE has the potential to positively impact the resilience of companies in all groups.

When comparing the impact of integration CE in the core business versus separate from the core business (Fig. 4.9), the difference in the results in resilience is more apparent. In the core business group, 55% report a positive impact on their resilience, while 25.1% experience no impact. On the other hand, for the separate from the core business group, the positive impact is only 27.1%, and the percentage of no impact increases to 49.6%. These results suggest that integrating CE into core business practice has a higher probability of positively impacting resilience compared to operating it as a separate entity.

Profit

Most industry groups show a clear trend towards positive impacts on profits resulting from the implementation of CE strategies, as shown in Fig. 4.10. This is particularly evident for MDP, MNP, MPC and SVC, with over 50% of participants in each group reporting a positive impact on profits. Compared to other groups, MNP benefit more frequently, as these participants also report the lowest rate of negative impact.

The RIME group stands out as the only one where the majority of participants (50%) reported no impact on their profits, whereas 33.3% reported a positive impact. The difference between this group and others might be that these companies already

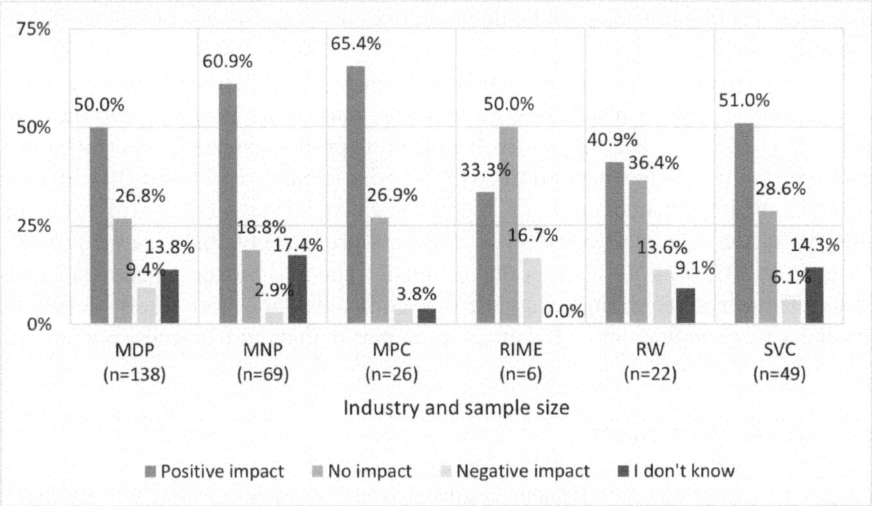

Fig. 4.10 Inter-industry comparison of CE impacts on company profits

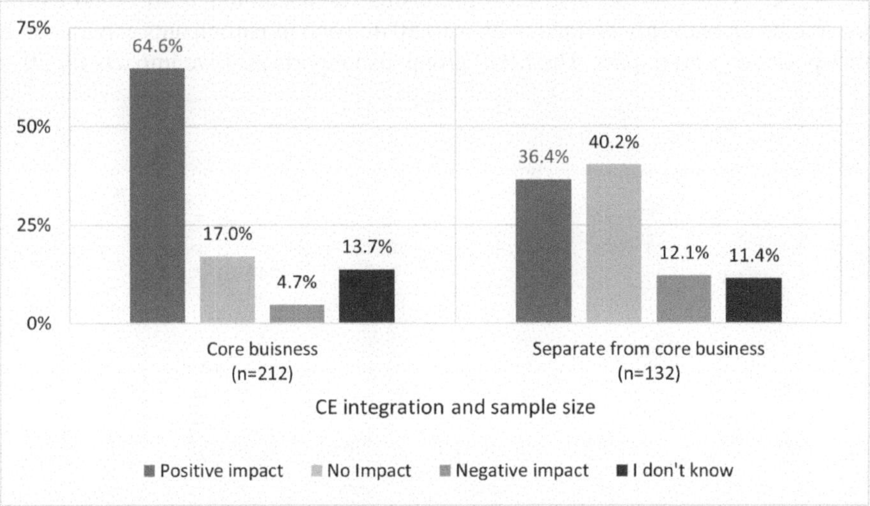

Fig. 4.11 Comparative analysis of CE impacts on profits by integration level

specialise in use cycle extending and renewing strategies and would not see a significant change in their profits if they had not deliberately taken steps to expand their business through other CE strategies.

Interestingly, the percentage of positive impact on profit is vastly higher than the one recorded for manufacturing and material costs. This suggests that some companies' profit gains may not be due to better cost performance but rather due to increased sales.

Compared to the inter-industry comparison, the level of CE integration (Fig. 4.11) shows a much different result. According to the data, integrating CE into the core business has a better chance of positively impacting profits compared to performing it separately. In the core business group, 64.6% of participants report a positive impact, 17% report no impact, and only 4.7% report a negative impact on their profits. On the contrary, for the separate from the core business group, only 36.4% saw a positive impact on profits, while 40.2% reported no impact, and 12.1% experienced a negative impact. These results strongly suggest that companies are more likely to benefit financially from implementing CE practices as part of their core business operations.

Technological advancements

In Fig. 4.12 technological advances resulting from the integration of CE are highlighted. The companies in the MNP and SVC group appear to be the most successful in this area, with more 68% and 63% of the participants, respectively, reporting positive impacts. Other sectors also show promising results, although MPC and RIME appear to be more evenly divided, with a 50:50 division of participants experiencing either positive or no impact. The MDP group also reports positive impacts for 50%

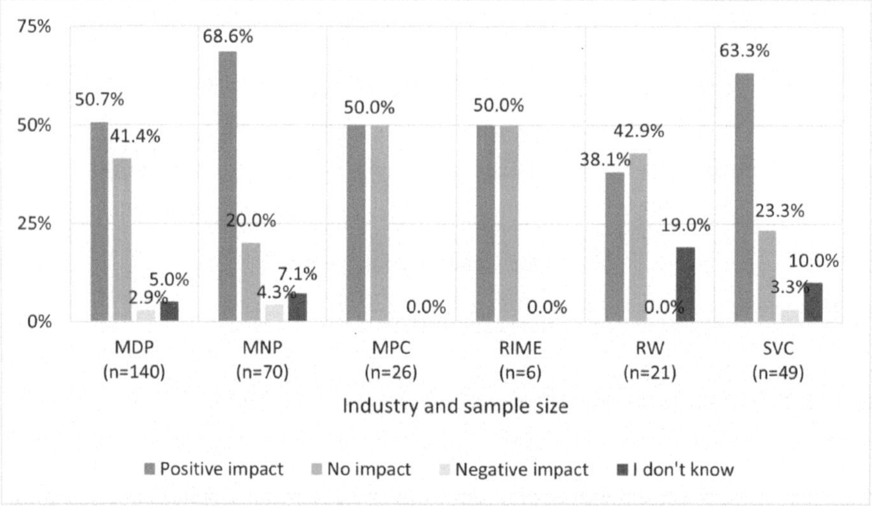

Fig. 4.12 Inter-industry comparison of CE impacts on technological advancements

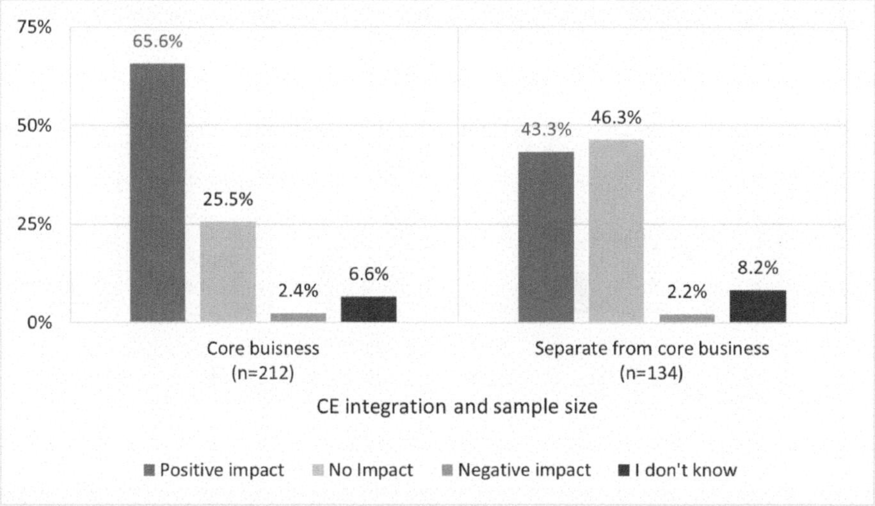

Fig. 4.13 Comparative analysis of CE impacts on technological advancements by integration level

of its participants. Overall, these findings suggest that the integration of CE has the potential to drive technological advancements across various industry sectors.

CE integration into core business presents significant differences in its impact on technological advancements, as shown in Fig. 4.13. The core business group surpasses the separate from the core business group with 65% of participants reporting a positive impact on technological advancements compared to 43.3%. Furthermore, the number of participants who report no impact on the separate from the core business group is nearly double that of the core business group. In particular, negative impacts were observed similarly in both groups. These findings suggest that fully integrating CE into core business operations can increase the likelihood of positively impacting technological advances. This may be attributed to the fact that the incorporation of CE into core business strategies allows a more comprehensive approach to the implementation of new technologies and processes that can support technological advances.

Sales

According to the results of the survey in Fig. 4.14, the CE initiatives have a significant positive impact on sales in various industries. Participants in each industry group reported a positive impact on sales, with percentages exceeding 53.8% in all cases.

There are some notable similarities and differences between the groups. The MDP, RIME, RW, and SVC groups all display a relatively similar response distribution,

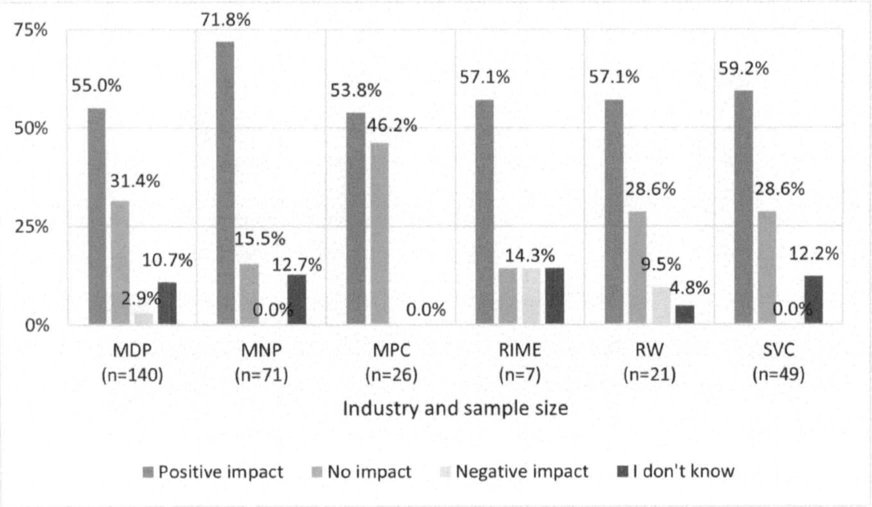

Fig. 4.14 Inter-industry comparison of CE impacts on sales

with positive impact percentages ranging from 55% to 59.2%. The MNP group stands out with the highest potential, as 71.8% of participants reported a positive impact on sales, while only 15.5% indicated no impact and none reported a negative impact. This suggests that CE initiatives are particularly effective in the MNP industry.

On the other hand, the MPC group exhibits a different pattern with 53.8% positive impact and a substantial 46.2% of participants who did not report any impact. However, similar to the MNP group, no negative impact was observed. This suggests that while the positive impact is lower in the MPC group, it is unlikely that initiatives will negatively impact sales when implemented CE.

Lastly, the RIME and RW groups experienced a mix of positive and negative impacts, 14.3% of the RIME participants and 9.5% of the RW participants reporting negative effects. Despite these adverse effects, the overall trend remains positive in both industries.

Upon analysis of the impact of CE on sales in different industries, it became evident that the majority of participants reported positive effects on their sales. However, a closer look reveals a striking difference between the impacts experienced by companies that have integrated CE into their core business versus those that have kept it separate (Fig. 4.15).

When CE is integrated into the core business, 73.4% of participants report a positive impact on sales, with only 15% indicating no impact, and a mere 2.3% experiencing negative consequences. On the other hand, when CE is treated separately from the core business, the impact on sales is notably different. In this case, only 39.6% of participants report a positive impact, while a significant 47.8% indicate no impact, and 1.5% experience negative effects.

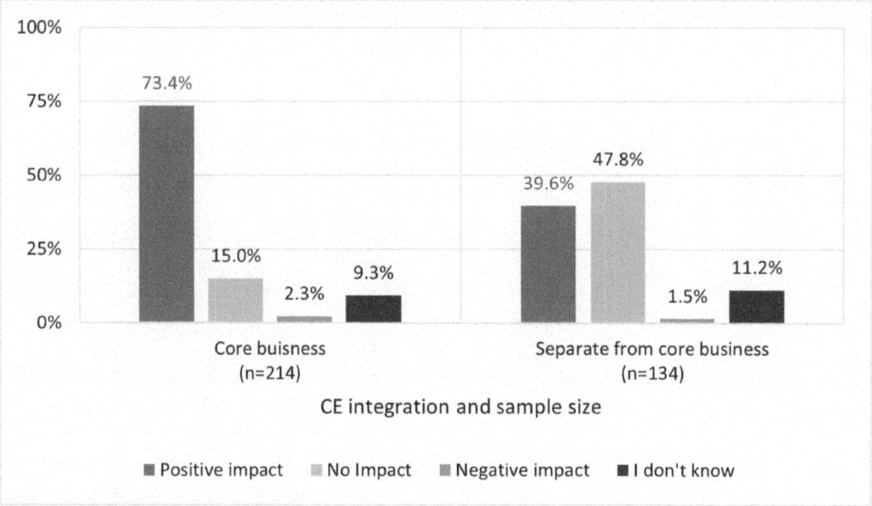

Fig. 4.15 Comparative analysis of CE impacts on sales by integration level

This disparity implies that keeping CE initiatives separate from the core business may hinder their potential to drive sales and create value for the company. The lack of integration could lead to missed opportunities for cross-functional collaboration, synergies, and more comprehensive adoption of CE principles throughout the organisation.

Reduction of Waste

According to the survey results (Fig. 4.16), the MNP group has the highest percentage of participants who reported a positive impact on waste reduction at 90.1%, closely followed by the MDP group at 87.1%. The MPC, RIME, and RW groups also exhibit significant positive impacts, ranging from 81.8% to 85.7%. Although still showing a positive effect, the SVC group has a slightly lower percentage of participants experiencing reduced waste at 75.5%. These results indicate that CE principles are generally successful in promoting waste reduction in various industries.

Upon examining the impact of CE on waste reduction at different implementation levels, as shown in Fig. 4.17, both approaches show a predominantly strong trend toward positive impacts. Companies that integrate CE into their core business report an impressive 90.2% positive impact on waste reduction, with only 7% experiencing no impact and a minimal 0.9% negative impact. On the contrary, organisations that treat CE as separate from their core business see a slightly lower percentage of positive impact at 78.4%, with 14.2% reporting no impact and 0.7% negative impact.

These findings suggest that integrating CE into the core business is more effective in reducing waste than keeping it separate from the primary operations. The results

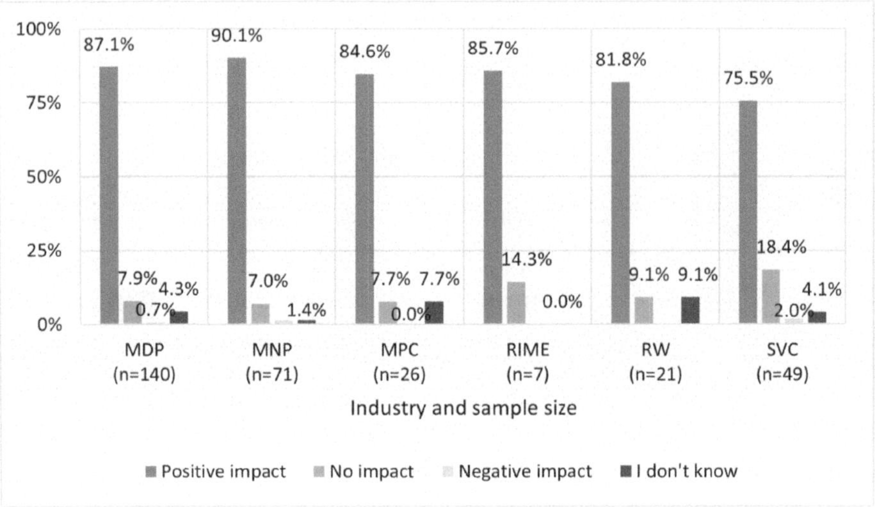

Fig. 4.16 Inter-industry comparison of CE impacts on the reduction of waste

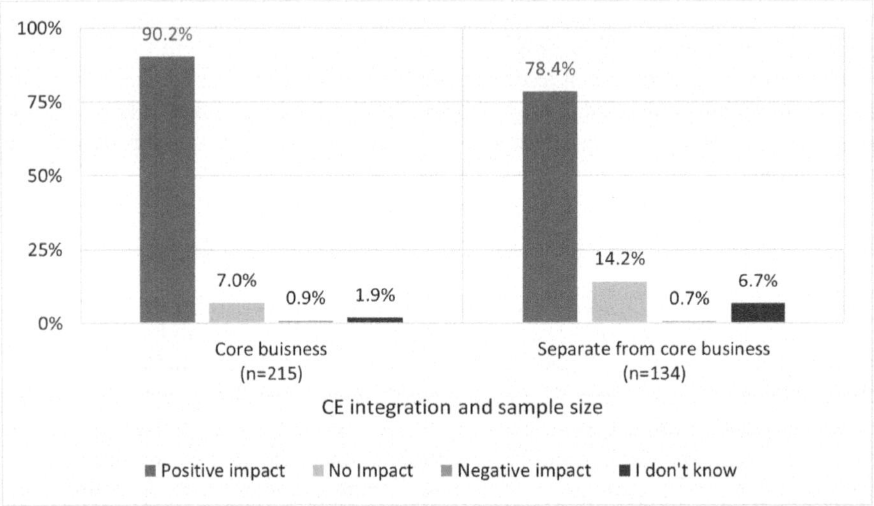

Fig. 4.17 Comparative analysis of CE impacts on reduction of waste by integration level

also highlight the importance of adopting CE principles as a central element of a company's strategy to achieve more significant waste reduction and maximise the environmental benefits of these initiatives.

Company Image

The results of the survey (Fig. 4.18) are truly staggering, with an astonishing 94% of participants from the MNP industry reporting that their company has improved their image due to CE initiatives. This is closely followed by the MDP, RIME, and RW groups with percentages ranging from 85.7% to 86.4%. The MPC group reports a positive impact of 84.6%, while the SVC group reports a slightly lower percentage at 73.5%. These findings strongly suggest that implementing CE initiatives can increase the reputation and image of a company, regardless of its industry. Unlike other impact comparisons, such as manufacturing or material costs, the numbers in different industries show similar positive outcomes. Therefore, companies from various industries can benefit from CE initiatives in terms of enhancing their reputation.

Interestingly, the survey findings indicate that the extent of CE implementation plays a crucial role in shaping a company's image (see Fig. 4.19). When CE principles are integrated into a company's core business, a significantly higher percentage of participants, 93.0%, report a positive impact on the company's image, compared to 76.9% when these principles are kept separate from core operations. This implies that wholeheartedly adopting the CE and weaving it into a company's primary functions can result in more considerable reputation enhancement.

The results underscore that CE initiatives incorporated into a company's core business are more likely to strengthen the reputation and brand image compared to when CE is considered an independent aspect. This serves as a compelling endorsement of the effectiveness of CE initiatives in both minimising waste and enhancing the reputation of a company. It is evident that CE presents a win-win situation for both the environment and a company's bottom line.

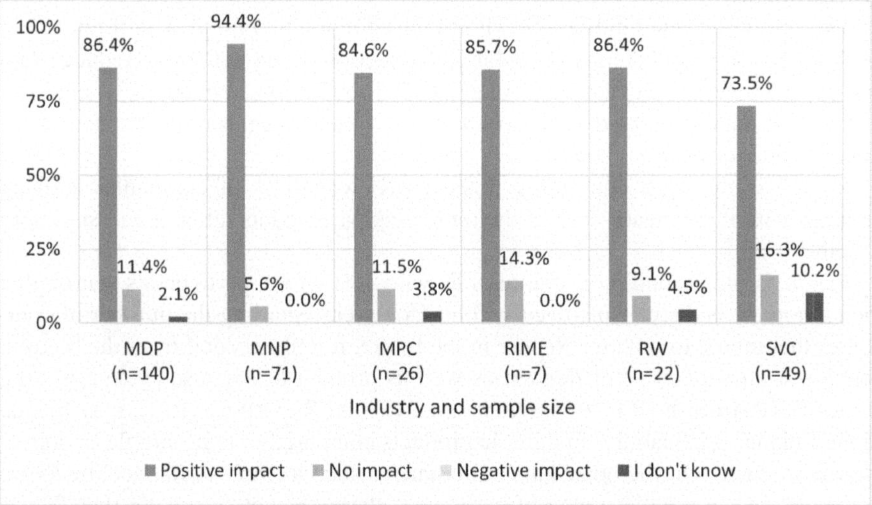

Fig. 4.18 Inter-industry comparison of CE impacts on company image

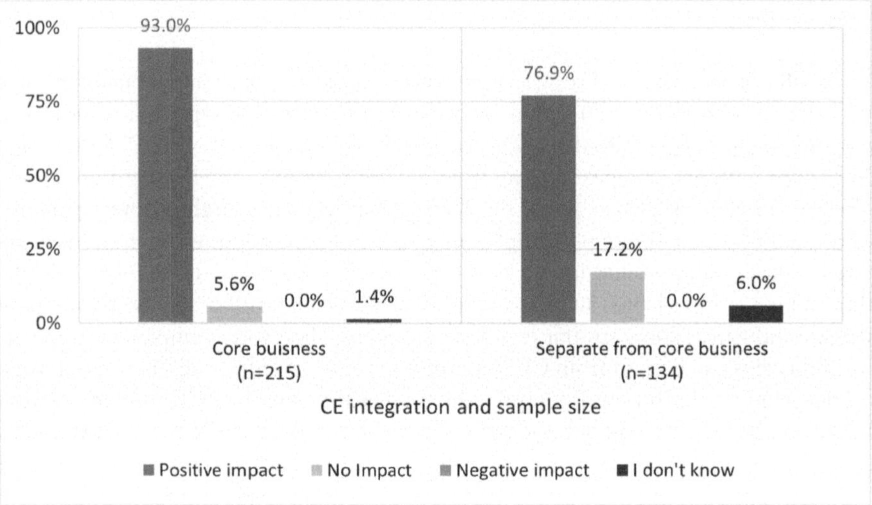

Fig. 4.19 Comparative analysis of CE impacts on company image by integration level

4.4.2 Industry Trends in Retention Strategies

The 512 companies will be studied to determine the prevalence and attractiveness of Rs. The objective is to distinguish between the most and least implemented Rs and to determine if contemporary sustainable business practices prioritize strategies that maintain product structure and function over those that mainly focus on preserving material value.

The graph presented in Fig. 4.20 shows the adoption and frequency of interest in 10 R options across all companies. Material reduction strategy (R0 and Reduce (R1)) and product use cycle extending strategies (Reuse (R2) and Repair (R3)) emerged as the most widely adopted and also stand out as the strategies most companies are interested in adopting in the future.

Reduce is the most commonly practised R, with 62% of companies actively engaged with it and nearly half of the remaining participants intending to start with it in the future.

Repair is the second most common R, with 48% of the participants stating that their company is directly involved with it. However, regarding the number of companies that intend to practice *Repair* in the future, it ranks second from the bottom. The lower attractiveness of *Repair*, as well as product or part use cycle renewing strategies (Refurbish (R4), Remanufacture (R5) and Repurpose (R6)), may be due to their higher applicability to durable products compared to non-durable or convenience products. On the other hand, alternatives such as *Refuse*, *Reduce*, or *Reuse* are applicable in a wider range of cases. It is therefore not surprising that *Refuse* ranks as the third most practised R with 43% of companies actively engaged with it.

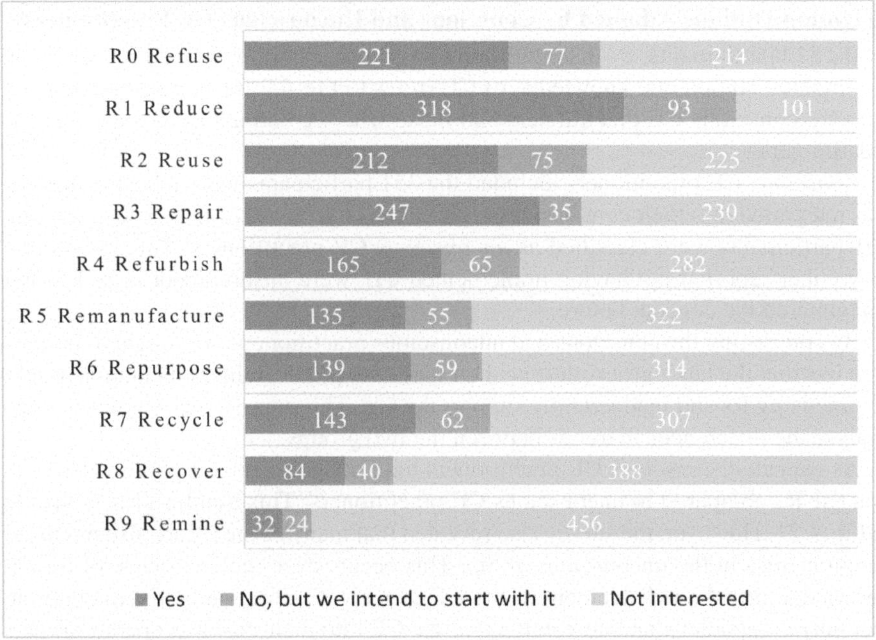

Fig. 4.20 The adoption and interest frequency of Rs among all participants (n = 512)

At the same time, it holds the second position in terms of interest, as 18% plan to implement this approach in the future. Whereas *Reuse* is the fourth most practised R, with 41%, and in terms of interest, it ranks third.

There is a marked decrease in the number of companies practising R0 to R3 compared to R4 to R9. The least frequently practised Rs involve material processing strategies (Recycle (R7), Recover (R8), and Remine (R9)). *Remine* (6%) and *Recover* (16%) not only rank as the least practised R but also garner the lowest interest among companies that have not yet adopted them. In contrast, *Recycle* (28%) is practised more widely. This trend may be attributed to the fact that *Recycle* allows the internal processing of production waste, offering a more direct and manageable approach for companies. On the contrary, *Recover* and *Remine* generally demand more specialised processes that require significant investment in specialised equipment and infrastructure. This can make them less attractive to companies that prioritise cost-effectiveness. Essentially, this indicates a propensity among companies to focus on reducing material consumption rather than investing in methods to reclaim materials after they have left the company's premises.

Retention Options Adopted by Conscious and Unconscious CE Practitioners
Of the 512 participants, we distinguished two groups according to their responses on the implementation and knowledge of CE (refer to Fig. 4.2 for more details): those who are conscious CE practitioners and those who we defined as unconscious CE practitioners.

Conscious CE practitioners included the 351 participants who indicated that CE was integrated into their core business operations or separate from it. The remaining 161 participants were classified as *unconscious* CE practitioners. This means that they either reported not having implemented CE, were unsure about it, or had not encountered the concept before.

By contrasting the conscious and unconscious practitioners, we sought to investigate whether the latter group, despite their lack of explicit CE implementation, might be practising Rs and inadvertently support the CE. This approach also allowed us to analyse the differences in trends between the two groups.

As expected, conscious CE practitioners have a significantly higher tendency to practise Rs compared to unconscious CE practitioners. This trend is clearly visible in Fig. 4.21. However, the survey also revealed that many of the Rs are also practised by companies in the unconscious group. This occurs even in the absence of formal knowledge or acknowledgement of the CE concept. Consequently, the assumption that only companies that knowingly integrate CE into their business operations participate in R practices can be refuted. In fact, Rs appear to be more widespread than other research would suggest.

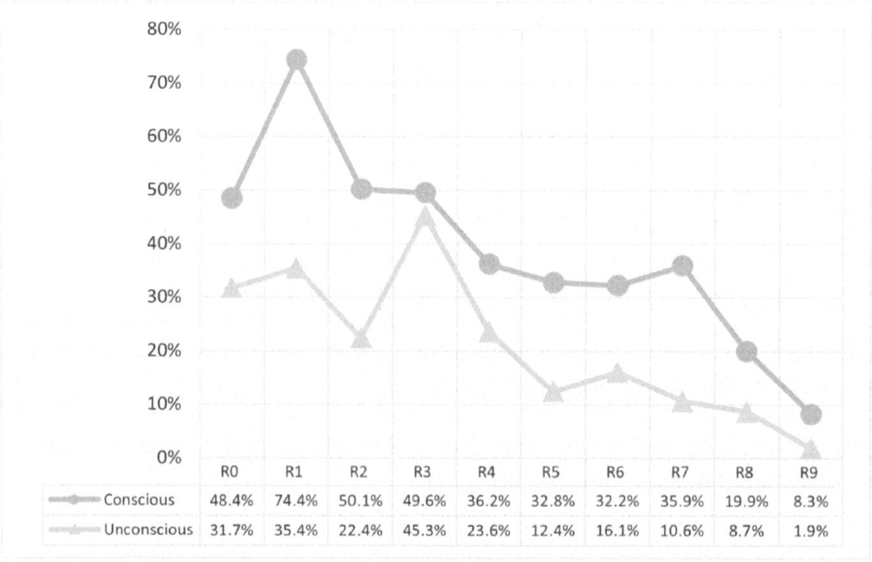

Fig. 4.21 Rs practised by conscious (n = 351) and unconscious (n = 161) CE practitioners

A significant difference between the two groups lies in the Rs that are prioritised. *Conscious* practitioners tend to focus heavily on *Reduce* as a strategy, highlighting the importance of waste reduction and resource efficiency. *Refuse*, *Reuse* and *Repair* are also widely adopted, with around half of the conscious practitioners practising these Rs.

On the contrary, *unconscious* practitioners exhibit a strong preference for the *Repair* aspect, devoting nearly as much attention to this R as their conscious counterparts. This suggests that *Repair* is a popular strategy among organisations, regardless of their level of engagement and knowledge of CE.

Both conscious and unconscious practitioners show a noticeable decrease in interest concerning product or part use cycle renewal (R4–R6) and material processing (R7–R9) strategies. Among these practitioners, those who are conscious tend to prefer *Recycling*, despite the fact that it does not retain any of the product's value except for the material content. This can be explained by the fact that *Recycling* is crucial in maintaining a balance between consumption and material conservation, and as a result supports other approaches, particularly *Reduce*. Hence, the higher adoption rate of *Recycling* can be related to the increased implementation of the *Reduce* approach. This connection can be attributed to the synergistic relationship between the two strategies, which can lead to improved economic and environmental advantages.

On the other hand, unconscious practitioners tend to practice product or part use cycle renewing strategies over material processing, of which *Remanufacture* is the least practised. The reasons may be that *Refurbish* and *Repurpose* often involve relatively simple processes, such as cleaning, repairing, or replacing specific parts, and modifying or updating the product for a new use. These processes typically require fewer resources and lower investment compared to *Remanufacture*, which involves complete disassembly and rebuilding of the product to meet the same specifications as a new one.

In conclusion, the survey results demonstrate that companies are engaging with CE strategies without explicitly recognising or labelling them as such. But the involvement of companies that recognise CE is much higher than that of companies that do not. This highlights the importance of raising awareness about CE and its benefits, as well as providing guidance for companies to better understand and implement CE strategies systematically.

Future Retention Strategies Plans for Conscious and Unconscious CE Practitioners

Figure 4.22 shows the rate of companies that intend to adopt a specific R in the future. In particular, the R plans of unconscious practitioners CE appear to align with the trends already established by conscious practitioners, as seen earlier in Fig. 4.21. This alignment indicates a potential trend in which unconscious practitioners are increasingly following the footsteps of their conscious counterparts.

Unconscious participants generally show a higher interest in R0 to R2, with a particular high focus on R1. On the contrary, conscious practitioners exhibit a lower and relatively consistent planned adoption rate in all of these Rs. Since 74% of

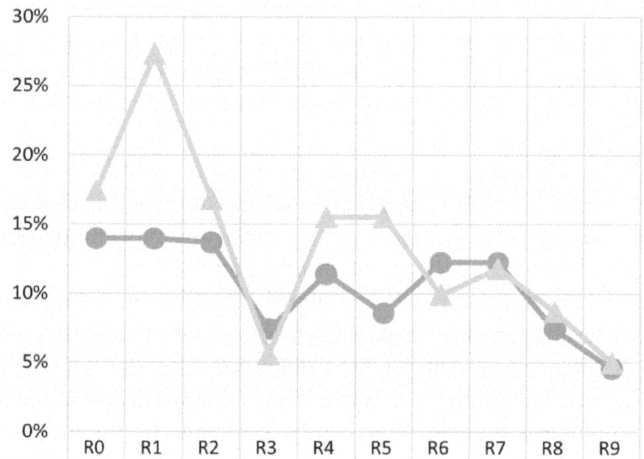

	R0	R1	R2	R3	R4	R5	R6	R7	R8	R9
Conscious (intend)	14.0%	14.0%	13.7%	7.4%	11.4%	8.5%	12.3%	12.3%	7.4%	4.6%
Unconscious (intend)	17.4%	27.3%	16.8%	5.6%	15.5%	15.5%	9.9%	11.8%	8.7%	5.0%

Fig. 4.22 Future R plans for conscious (n = 351) and unconscious (n = 161) CE practitioners

conscious and 35% of unconscious CE practitioners already practice *Reduce*, it is expected that this strategy will remain the most widely adopted R.

Both groups show limited interest in Repair (R3), suggesting that many companies capable of implementing this strategy may already be practising it. A notable difference between the groups is seen in their approach to Repurpose (R6), with conscious practitioners expressing increased interest, indicating its growing traction in the industry. Furthermore, R4 and R5 exhibit higher future implementation rates among companies than material processing strategies. With respect to material processing strategies, *Recycling* is preferred, while *Recover* and *Remine* seem less appealing to most.

In alignment with Rs adopted by conscious and unconscious CE practitioners, companies continue to prioritise Rs that contribute to waste reduction and minimise material requirements in the production process. In contrast, retention options critical for material supply (R7 to R9) appear less appealing, indicating a trend that favours immediate benefits for waste reduction over long-term supply considerations.

4.4.3 Prevalence and Relevance of Drivers for Adopted Retention Strategies

As observed, there is a variation in the adoption rates of Rs and the number of companies with plans to implement them in the future. Understanding the key drivers that have motivated companies to implement these practices is crucial to further promote and support the transition from a linear economy to a circular economy.

By aggregating and analysing the collected data, we have identified the drivers most frequently chosen in the multiple choice questions in all 10 R. Based on 5,259 selections, the results offer a comprehensive overview of the main factors that have guided the companies in their adoption of CE principles, as presented in Fig. 4.23.

The results highlight that *environmental and resource management drivers* are the most influential. This is evident from the high selection percentages for D.E1 (15.8%) and D.E2 (15.6%), indicating that companies recognise the importance of reducing resource consumption and decreasing the carbon footprint of their products. This

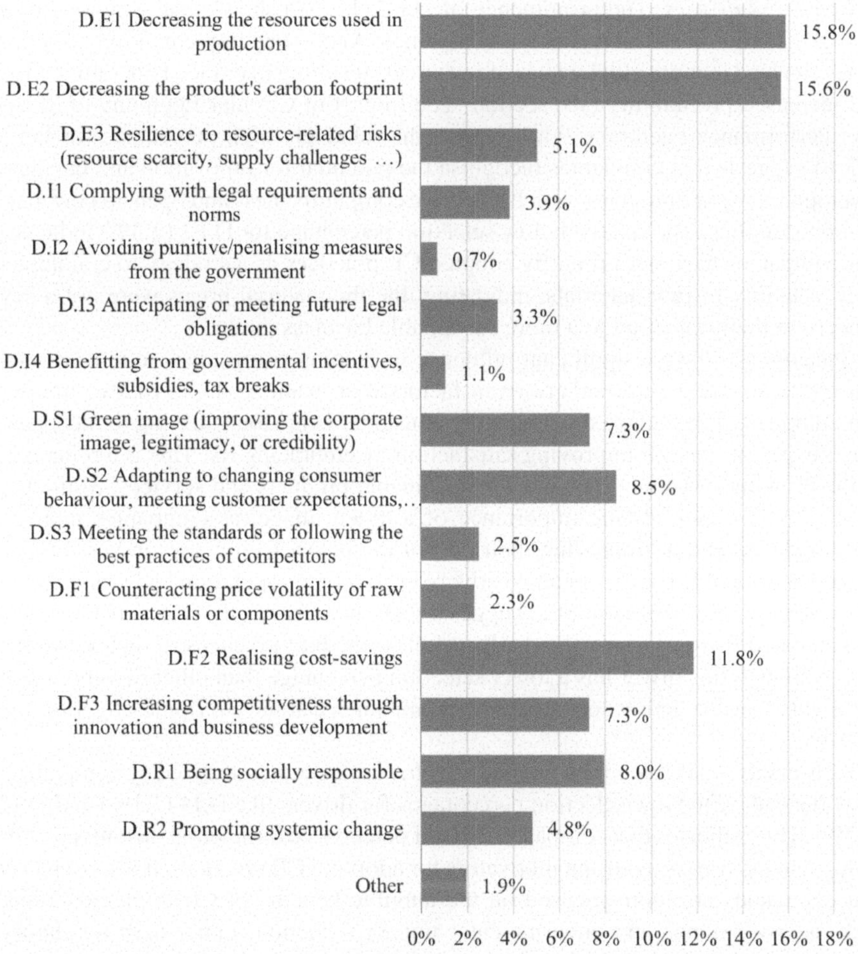

Fig. 4.23 Weighted collective results for drivers ordered according to clusters (n = 5,259)

focus on sustainability likely stems from companies' understanding of the need to adapt to the changing global landscape, where resource scarcity and environmental degradation are becoming pressing issues. As companies strive to maintain long-term growth, adopting CE practices that reduce their environmental impact and optimise resource use is essential. In addition, companies are increasingly under pressure from *stakeholders*, such as investors and consumers, to demonstrate their commitment to sustainable practices. Although D.E3 (5.1%) has been selected considerably less often, it still highlights that multiple companies have recognised that they can build *resilience* to resource-related risks by adopting Rs. It also shows that companies are aware of the vulnerabilities in their supply chains and actively seek to mitigate potential disruptions caused by resource constraints.

Furthermore, the survey results revealed that companies also emphasise *financial and business drivers*. The prominence of D.F2 (11.8%) shows that achieving cost savings is a significant motivator for companies. And in fact implementing CE principles can lead to more efficient use of resources, resulting in reduced operating costs and increased profitability (see section "The Impact of Circular Economy on Business Performance Factors"). Furthermore, the relatively high percentage for D.F3 (7.3%) suggests that companies recognise the potential for innovation and business development by adopting Rs, which can drive competitiveness and open up new market opportunities. The relatively low selection percentage for D.F1 (2.3%) indicates that companies have not primarily adopted CE practices as a strategy to counteract price volatility of raw materials, indicating that the material prices were not a key concern in their transition to a more sustainable business model.

Stakeholders have a significant influence in shaping companies' actions, as evidenced by strategic and market-driven factors. For example, D.S2 (8.5%) implies that companies have adapted to changing consumer behaviour, striving to meet customer expectations and improving satisfaction by embracing Rs. This is a reflection of the growing consumer demand for sustainable products and services. Similarly, D.S1 (7.3%) highlights the importance of a green image, as companies increasingly understand that improving their *corporate image*, legitimacy, and credibility through sustainable practices can positively impact their reputation and brand value. D.S3, with a 2.5% selection rate, suggests that only a minority of companies have implemented Rs to meet the standards or follow the best practices of their competitors. Although this driver has a lower selection percentage than other drivers, it still highlights that the actions of pioneering companies can have a ripple effect on the industry.

Conversely, *institutional and regulatory drivers* are less prioritised by companies, as evidenced by the low selection percentages for drivers like D.I4 (1.1%) and D.I2 (0.7%). This indicates that companies did not perceive governmental incentives, subsidies, and tax breaks as strong motivators for adopting CE practices. It also suggests that companies are more focused on the tangible benefits of CE implementation, such as cost savings and environmental impact reduction, rather than regulatory compliance.

Responsibility drivers provide insight into how companies view their role in fostering systemic change and their commitment to social responsibility. The 8.0%

selection rate for D.R1 reveals that numerous companies acknowledge their social responsibility as a reason for implementing CE practices. However, the 4.8% selection rate for D.R2 indicates that only a minority consider the promotion of systemic change as a main reason for adopting circular practices.

The Importance of Drivers in Relation to Individual Retention Strategies

Further analysis was performed to investigate the relationship between each driver and the Rs that companies have already implemented. The results are visualised as a heat map in Fig. 4.24 that illustrates the percentage of each driver selected for each R. Shades of green represent the most popular choice, while shades of red represent the least popular choice of drivers for each R. A comparison of all drivers for each Rs is also illustrated in Fig. 4.25.

The results show that some drivers exhibited a greater range of variability in their influence on people's decisions to engage with a particular R, compared to other factors.

D.E1 Decreasing resources used in production. This driver stands out as one of the most significant for companies to actively engage with any of the Rs.
The available data indicate that specifically R2, R5, R6, and R7 are being practised to achieve this goal. Among these, R7 stands out as a particularly strong example.

D.E2 Decreasing the product's carbon footprint. R1, R7, R8, and R9 are most influenced by this driver, which implies that these practices are indeed effective in reducing a company's carbon emissions. For R3 and R4, the responsiveness is slightly lower, implying that carbon reduction is not the primary benefit of implementing them compared to the other Rs.

Driver	R0	R1	R2	R3	R4	R5	R6	R7	R8	R9
D.E1	13.0%	11.1%	18.2%	12.5%	16.6%	17.6%	18.0%	19.9%	15.1%	16.3%
D.E2	16.1%	17.4%	16.7%	9.4%	11.8%	13.9%	16.1%	17.4%	19.0%	18.5%
D.E3	5.2%	4.4%	4.6%	4.4%	5.0%	5.0%	4.1%	6.5%	6.9%	4.3%
D.I1	5.0%	4.6%	2.4%	3.8%	2.2%	2.9%	1.0%	5.6%	5.6%	5.4%
D.I2	1.8%	0.9%	0.3%	0.3%	0.7%	0.3%	0.5%	0.9%	0.4%	1.1%
D.I3	3.2%	4.9%	2.7%	2.2%	2.0%	2.6%	2.9%	4.3%	3.0%	5.4%
D.I4	0.8%	1.1%	0.2%	0.3%	1.5%	0.5%	0.5%	0.9%	1.7%	3.3%
D.S1	8.6%	10.0%	6.5%	6.0%	5.7%	5.2%	9.0%	8.4%	7.3%	6.5%
D.S2	6.6%	8.3%	7.6%	12.9%	11.6%	9.7%	9.0%	8.4%	7.3%	3.3%
D.S3	2.0%	2.3%	1.6%	5.1%	4.4%	3.1%	2.7%	1.4%	2.6%	0.0%
D.F1	2.0%	2.0%	3.3%	3.4%	2.6%	3.7%	2.2%	1.1%	0.9%	2.2%
D.F2	10.9%	8.1%	13.9%	14.1%	16.6%	15.2%	12.9%	7.0%	10.8%	8.7%
D.F3	8.3%	10.4%	6.6%	8.5%	7.7%	6.8%	7.1%	5.0%	7.3%	5.4%
D.R1	10.1%	8.3%	7.9%	8.8%	6.1%	7.1%	8.0%	7.4%	6.0%	9.8%
D.R2	5.6%	5.1%	5.5%	3.4%	3.3%	4.7%	4.9%	4.3%	3.9%	7.6%
Other	0.7%	1.0%	2.1%	5.0%	2.2%	1.6%	1.0%	1.6%	2.2%	2.2%
n =	713	1216	633	682	457	381	410	443	232	92

Fig. 4.24 Driver popularity heatmap for R0 to R9 where 'n' denotes the aggregate count of driver selections for a given R

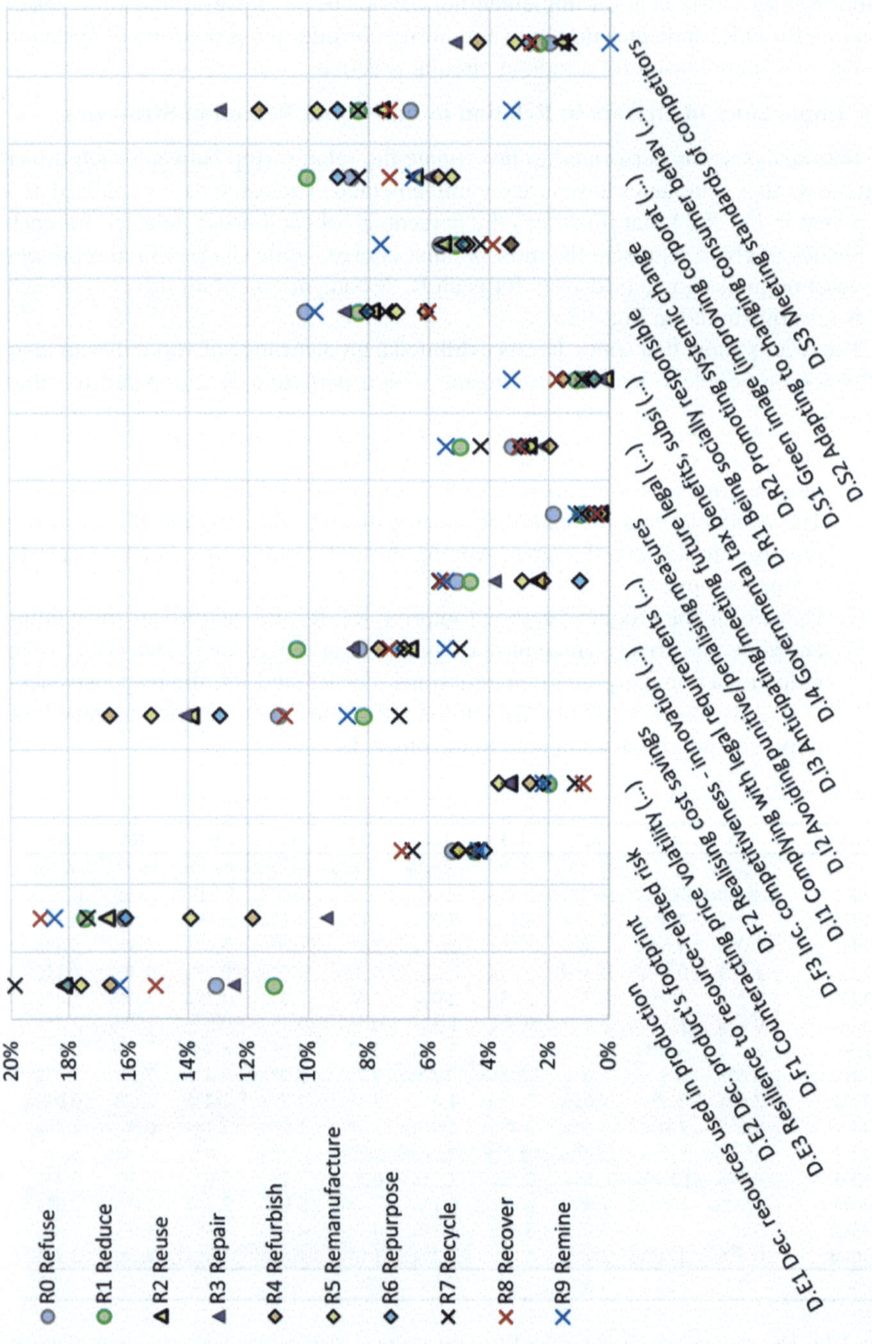

Fig. 4.25 Comparison of participants choices for Rs across 15 different drivers

D.E3 Resilience to resource-related risks. This driver demonstrates a low response across all Rs. However, companies that practise R7 or R8 exhibit a higher affinity for this driver, indicating that these two practices are more effective in mitigating resource-related risks than other Rs.

D.I1 Complying with legal requirements and norms. Legal compliance appears to have a limited influence on initiating specific Rs, as all percentages are below 5.6%. The highest rate is observed in material processing strategies, namely R7, R8 and R9. The low rate for recycling is unexpected, considering the stringent regulations in place within the European Union (EU). Nevertheless, the findings indicate that companies focus mainly on these strategies as well as R1 and R2, to fulfil regulatory mandates.

D.I2 Avoiding punitive/penalising measures from the government. This driver is the least influential of all Rs. Therefore, it appears that the punitive measures in place have been only marginally effective in the past to push companies to adopt Rs. Despite this, the response rate of 1.8% for R0, which is twice that of other strategies, suggests that punitive measures are primarily focused on regulating the use of specific products or materials, rather than directing companies to implement any of the other strategies.

D.I3 Anticipating or meeting future legal obligations. R1, R7, and R9 exhibit a stronger response to this driver, emphasising that legal obligations are more prevalent for the type of material used in production, as well as how material is recovered at the end of a product's life. In contrast, R3 and R4 are less driven by such concerns.

D.I4 Benefiting from governmental incentives, subsidies, tax breaks. R9 shows the strongest response to this driver, highlighting that companies implementing this practice are more likely to leverage governmental support to enhance their CE initiatives. On the contrary, all other Rs are substantially less influenced by this driver, suggesting that these practices are not as dependent on or specifically targeted by government incentives and support.

D.S1 Green image. The relatively higher percentages across all Rs compared to most other drivers suggest that companies are genuinely focused on cultivating a green corporate image. R1 stands out as the particular strategy companies have implemented to promote this image. In comparison, R4 and R5 show to be less driven by this goal.

D.S2 Adapting to customers' changing expectations and increasing satisfaction. Including adapting to consumer behaviour. R3 and R4 are clearly more influenced by this driver, while R9 shows the least responsiveness, suggesting that products that can be repaired or refurbished are particularly valued by consumers. Other Rs range between 6.6% and 9.7%, suggesting that there is a market for circular products that companies can take advantage of, as these products are appreciated and desired by consumers. This emphasises the importance of thoughtful product design, to ensure that these strategies can be successfully implemented.

D.S3 Meeting the standards or following best practices of competitors. The percentages for this driver remain relatively low across all Rs. However, R3 stands out with a 5.1% response, suggesting that product repair is becoming increasingly important. In contrast, R9 has a 0.0% response, indicating that it is not driven by stakeholder pressure at all.

D.F1 Counteracting price volatility of raw materials or components. With all percentages remaining below 3.7%, price volatility does not appear to be a primary concern for most companies. However, implementing R2, R3, or R5 displays a stronger response to this driver, suggesting that these strategies are more favoured for addressing price-related challenges. On the contrary, R7 and R8 are less commonly adopted for this purpose.

D.F2 Realising cost savings. Most companies view this driver as the main reason for their adoption of R2, R3, R4, or R5. This suggests that these practices are effective in achieving cost reductions. On the other hand, R1, R7, and R9 display a weaker response, implying that realising cost savings with these strategies may be more challenging compared to other Rs.

D.F3 Increasing competitiveness through innovation and business development. Companies practising R1 have chosen this driver at a notably higher percentage as the rationale for its adoption compared to other Rs. In contrast, R7 and R9 are the least practised strategies for this specific reason, suggesting that there is less competition in these areas. This presents an opportunity for businesses to explore and differentiate themselves in the market by focusing on R7 and R9, which can lead to untapped growth and a unique competitive advantage in the evolving landscape of sustainable business practices.

D.R1 Being socially responsible. Based on the results, companies demonstrate their social responsibility primarily by practising R0 or R9. On the contrary, R4 and R8 are adopted less frequently for this purpose. However, compared to the other drivers, D.R1 occupies a middle position, indicating that companies are increasingly focusing on social responsibility and striving to achieve it through various Rs.

D.R2 Promoting systemic change. R9 clearly is the primary R that companies practice to achieve this goal. Although other strategies do fall behind, the grand picture of drivers still reveals that systemic change is something companies strive for.

Ultimately, Rs are driven by varying factors. *Environmental and institutional drivers* play a more significant role in material processing strategies, while drivers from the *stakeholder* and *financial* cluster predominantly influence product use cycle extending strategies and product or part use cycle renewing strategies. Conversely, *responsibility-driven* factors tend to be the key motivators for companies actively engaging in material reduction strategies and R9. By understanding the relationships between various drivers and R-strategies, companies can identify and implement the most suitable approach that aligns with their individual objectives.

4.4.4 *Prevalence and Importance of Enablers for Adopted Retention Strategies*

The aggregated and weighted results of what participants considered the most important enablers are shown in Fig. 4.26. Based on 1,669 selections, it is apparent that two enablers are considered particularly important for successfully practising CE. Specifically, E.V1 (26.0%) and E.T3 (20.4%) were chosen most frequently, indicating that most companies practising Rs regard establishing close relationships, collaborations or partnerships with other value chain parties, and having in-house technical and technological know-how as the most crucial prerequisite for success. This highlights the importance of interconnectedness and expertise within the business ecosystem when practising Rs. On the other hand, it is interesting to see that E.V2 (8.2%) enhanced information sharing and transparency along the value chain is considered less often as the necessary prerequisite for becoming effective in implementing CE strategies.

The survey results also indicate that companies place importance on technology development and collaboration when choosing the most important enabler for CE practices. The prominence of E.T2 (13.6%) and E.T1 (12.2%) demonstrates that tech-

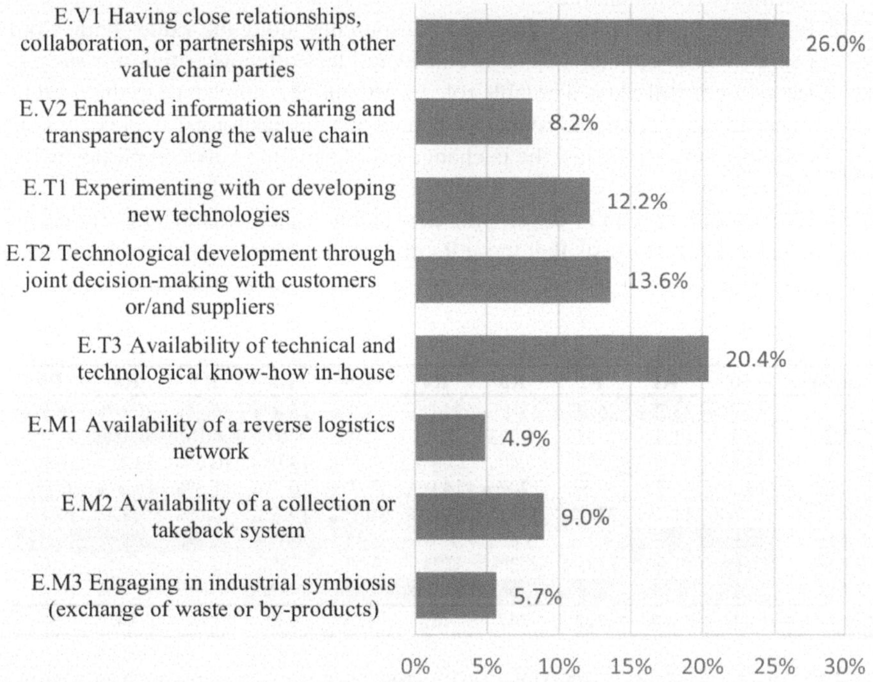

Fig. 4.26 Weighted collective results for enablers across all Rs

nological development through joint decision-making with customers or suppliers, and experimenting with or developing new technologies are perceived as essential factors. This finding emphasises the need for companies to invest in innovation and work closely with stakeholders to foster sustainable growth.

The variation in selection percentages between E.M1 (4.9%), E.M3 (5.7%), and E.M2 (9.0%) suggests that companies are less concerned about the availability of a logistics network or engaging in industrial symbiosis with other companies. Instead, they emphasise the implementation of strategies for efficient product collection and retrieval to ensure a steady supply of material.

The Importance of Enablers in Relation to Individual Retention Strategies

The percentage of selection for each enabler, chosen as the most important prerequisite for the implementation of each R, is represented using a heatmap format in Fig. 4.27 using a heatmap format, while Fig. 4.28 offers a visual comparison of these results.

E.V1 Having close relationships, collaboration, or partnerships with other value chain parties. This enabler is the most important prerequisite for nearly all Rs, highlighting the significance of collaboration in a CE. However, E.V1 is particularly important for R1 and R7. The higher percentage for these two Rs can be due to their interconnection – *reducing* virgin material by substituting it with *recycled* material.

E.V2 Enhanced information sharing and transparency along the value chain. Compared to other enablers, E.V2 is considered less often as critical for success. However, it still plays a notable role for R0 and R1, emphasising the need for transparency in material reduction strategys, whereas for use cycle extending and renewing strategies, the exchange of information between parties is less of a critical factor.

E.T1 Experimenting with or developing new technologies. It is most prevalent for R0 and R9. It suggests that these Rs involve innovative approaches, showing that technological advancements are crucial in these areas and not readily

Enabler	R0	R1	R2	R3	R4	R5	R6	R7	R8	R9
E.V1	23.4%	33.2%	26.3%	25.1%	24.2%	18.8%	25.4%	36.9%	20.0%	26.7%
E.V2	15.1%	11.4%	8.6%	6.2%	7.5%	7.5%	8.0%	5.0%	6.3%	6.7%
E.T1	20.6%	13.0%	7.2%	6.6%	5.6%	11.3%	13.0%	10.6%	13.8%	20.0%
E.T2	14.2%	15.5%	10.5%	12.8%	14.9%	17.3%	16.7%	11.3%	16.3%	6.7%
E.T3	15.6%	11.7%	22.0%	34.6%	34.8%	30.1%	19.6%	10.6%	8.8%	16.7%
E.M1	2.3%	2.8%	9.1%	5.8%	5.6%	5.3%	4.3%	4.3%	6.3%	3.3%
E.M2	4.1%	5.7%	11.0%	6.2%	6.2%	9.0%	7.2%	14.2%	16.3%	10.0%
E.M3	4.6%	6.6%	5.3%	2.9%	1.2%	0.8%	5.8%	7.1%	12.5%	10.0%
n =	221	318	212	247	165	135	139	143	84	32

Fig. 4.27 Enabler popularity heatmap for R0 to R9 where 'n' denotes the aggregate count of enabler selections for a given R

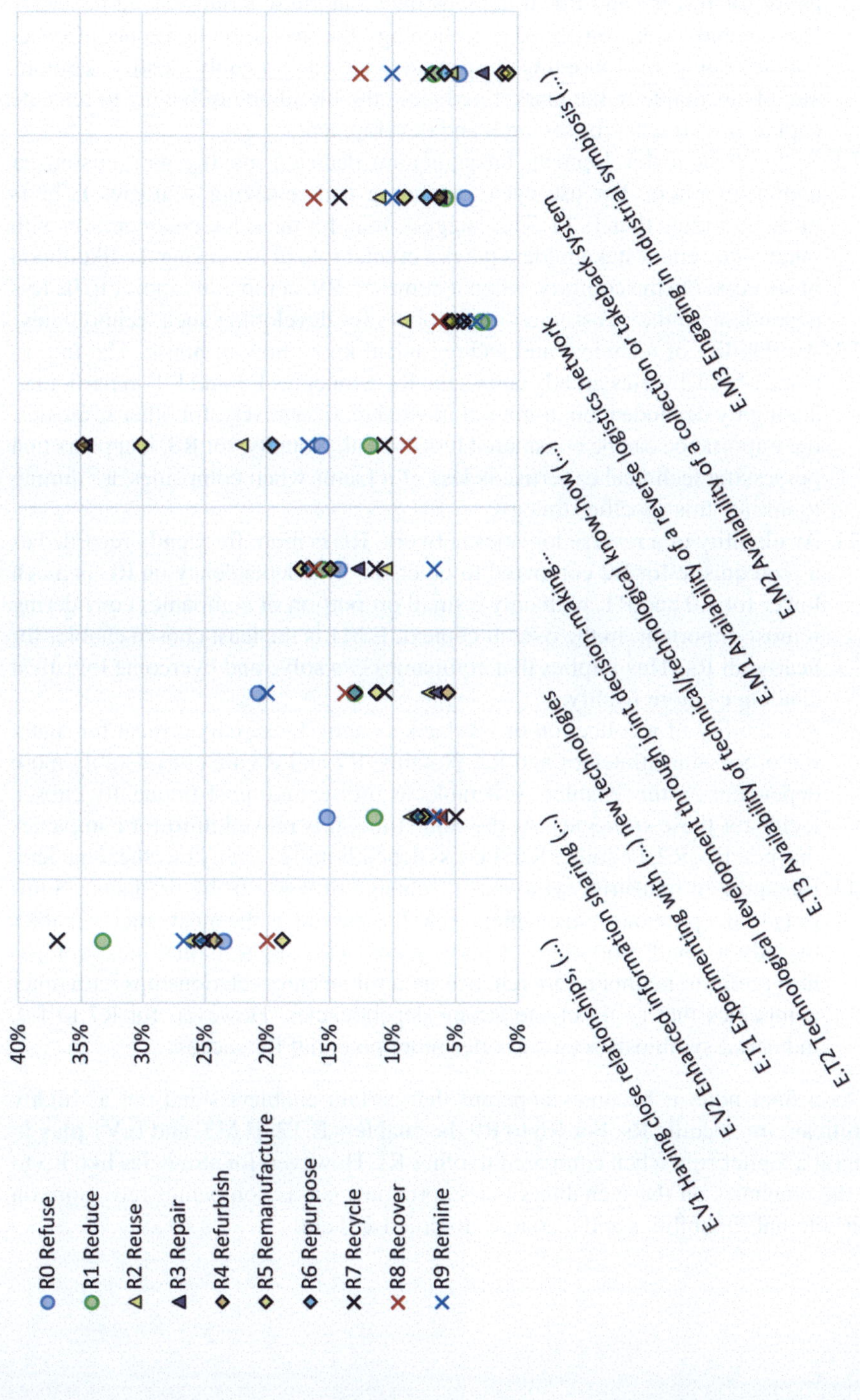

Fig. 4.28 Comparison of participant choices for Rs across eight different enablers

available. The importance of E.T1 is comparatively lower for other Rs, especially for R2, R3 and R4. It demonstrates that these strategies are generally less dependent on cutting-edge technology that has to be developed. Consequently, companies adopting these approaches can probably identify solutions already available in the market, reducing the likelihood of having to allocate capital investments for research and development.

E.T2 Technological development through joint decision-making with customers and/or suppliers. For use cycle extending and renewing strategies E.T2 is more important than E.T1. This suggests that, for these Rs, collaboration with other value chain stakeholders plays a crucial role in increasing the likelihood of success. On the contrary, when it comes to R9, companies appear to be less dependent on their customers or suppliers for developing such technologies.

E.T3 Availability of technical and technological know-how in-house. The importance of E.T3 varies greatly across the Rs. Strategies R3 and R4, in particular, are highly dependent on in-house knowledge. Conversely, for other strategies, the importance can be considered less critical. Notably, for R8, it appears that possessing technical expertise is less of a factor when companies are aiming to initiate this specific strategy.

E.M1 Availability of a reverse logistics network. RL is more frequently regarded as a prerequisite for R2 compared to other Rs. The dependency on RL is much lower for R0 and R1, with only a small proportion of companies considering it most important. In the overall context, E.M1 is the least chosen enabler for nearly all Rs. This implies that companies can solve and overcome logistical challenges more readily.

E.M2 Availability of a collection or takeback system. Especially critical for material processing strategies and R2. Notably, R7 and R8 are considerably more dependent on this enabler, as it ranks as the second most frequently chosen factor for these strategies. At the same time, it is ranked third for companies that practice R2. For other Rs, success depends more heavily on other enablers.

E.M3 Engaging in industrial symbiosis (exchange of waste or by-products). Similarly to the previous two enablers, E.M1 is viewed as the most crucial enabler by only a small proportion of participants. This suggests that most companies prefer to maintain self-reliance rather than enter relationships with other companies that could create strong dependencies. However, for R7 to R9, industrial symbiosis seems to offer more potential for success.

As a final note, it becomes apparent that certain enablers stand out as highly significant for specific Rs. For R6 to R9 the enablers E.T2, E.M3, and E.V1 play in general a higher role when compared to other Rs. However, for strategies like R3 to R5 the emphasis on these enablers is less pronounced, as companies rely more on their internal capabilities and resources to practise them.

4.4.5 Rationale for Adopting Retention Strategies in the Future

This examination delves into the factors that affect a company's decision to adopt CE practices. It focuses on the primary reason, that is, the main motivation factor *(M.F)* for participants who expressed that they intend to adopt a specific R in the future.

As answer choices, the original 15 drivers were consolidated into a shorter list of eight factors (Table 4.5). This was done to simplify the survey, making it more accessible and less overwhelming to respondents. It can also lead to higher response rates and better data quality since a shorter list reduces the cognitive load on respondents, allowing them to concentrate better on each motivation factor M.F and provide more accurate responses. Additionally, by presenting an altered set of answer options, respondents are less often faced with repetitive choices. This variety can help maintain their interest and motivation throughout the survey, reducing the likelihood of survey fatigue or disengagement.

In particular, the results are based on a single choice question, as opposed to the multiple choice question about drivers for already adopted Rs (Sect. 4.4.3). Figure 4.29 presents the collected answers for all Rs. The results reveal some parallels with the drivers behind Rs that are already being practised by companies.

M.F1 Environmental reasons emerge as the most influential factor, with 37.8% of respondents considering it their primary motivation. This demonstrates a strong awareness and concern for the environment, emphasising the importance of addressing climate change and reducing environmental impact.

M.F2 Business development is the second most cited reason, accounting for 18.7% of responses. It strongly contrasts the earlier driver analysis, where even multiple selections were possible. This discrepancy highlights the increasing recognition of the potential for growth and competitiveness that CE strategies can offer businesses in the long term.

M.F3 Growing customer demand is cited the third most by 11.7% of participants, reflecting an awareness of changing consumer preferences and the need for businesses to adapt to these shifts in order to remain competitive and meet customer expectations.

M.F4 Economic reasons such as economic crises, increasing resource prices, and scarcity, are selected by 11.0% of respondents. This suggests that businesses are increasingly recognising the potential economic benefits and risk mitigation that can be achieved through sustainable practices.

M.F5 Green image is considered the primary motivation for 7.0% of participants, indicating that the value of a positive environmental reputation remains a motivating factor for some businesses. Still, the lower percentage, when compared to M.F1 to M.F4 also implies that companies are increasingly focusing on a broader set of concerns that encompass various aspects of sustainability and address the multiple dimensions of their operations. It reflects an evolving understanding of the interconnected nature of sustainability and the need

Table 4.5 Mapping the 15 original drivers to 8 consolidated motivation factors

Motivational factor	Related drivers	Reasoning
M.F1 Environmental reasons (climate emergency, environmental impact,...)	D.E1 Decreasing the resources used in production D.E2 Decreasing the product's carbon footprint D.R2 Promoting systemic change	Reducing carbon footprint and resource usage directly impacts the environment Systemic change addresses environmental concerns
M.F2 Business development	D.F3 Increasing competitiveness through innovation and business development	Innovation and business development contribute to a company's overall competitiveness and growth
M.F3 Growing customer demand	D.S2 Adapting to changing consumer behaviour, meeting customer expectations, or increasing customer satisfaction	Adapting to changing consumer behaviour meets growing demand for sustainable products
M.F4 Economic reasons (economic crises, increasing resource prices and scarcity, etc.)	D.F1 Counteracting price volatility of raw materials or components D.E3 Resilience to resource-related risks (resource scarcity, supply challenges …)	Counteracting price volatility addresses economic challenges related to resource prices Building resilience to resource-related risks responds to economic pressures
M.F5 Green image	D.R1 Being socially responsible D.S1 Green image (improving the corporate image, legitimacy, or credibility)	Social responsibility contributes to a green image
M.F6 Production cost reduction	D.F2 Realising cost-savings	Cost-savings and production cost reduction both focus on decreasing expenses to improve profitability
M.F7 Institutional reasons (global pacts, strategies, government programmes, subsidies, tax benefits, etc.)	D.I4 Benefiting from governmental incentives, subsidies, tax breaks D.I3 Anticipating or meeting future legal obligations D.I2 Avoiding punitive/penalising measures from the government D.I1 Complying with legal requirements and norms	Governmental incentives, subsidies, and tax breaks encourage circular economy practices Anticipating/meeting legal obligations and avoiding punitive measures are part of a company's response to institutional pressures Compliance with legal requirements is a response to institutional regulations
M.F8 Access to new technological developments	D.S3 Meeting the standards or following the best practices of competitors	Adopting new technologies in order to meet industry standards and follow best practices set by competitors

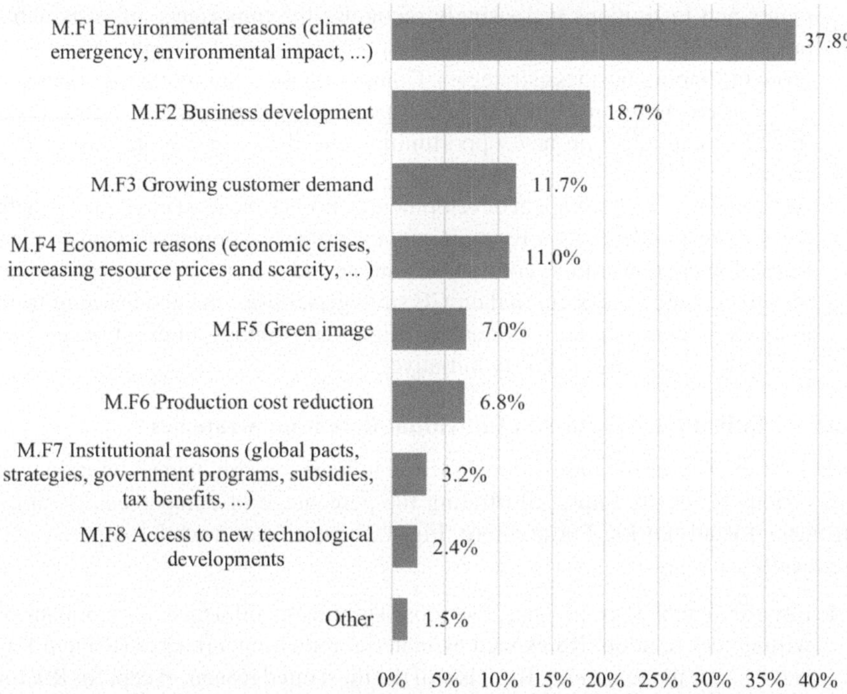

Fig. 4.29 Weighted collective results for motivational factors why entities intend to start Rs (n = 583 choices)

for a more systemic approach to address the complex challenges faced by businesses today.

M.F6 Production cost reduction accounts for 6.8% of responses, indicating that while cost savings and efficiency improvements are important to companies, they do not appear to be the primary driving force behind the adoption of CE practices. This suggests that businesses recognise that even if production costs are not significantly reduced, there are other benefits associated with embracing CE principles, such as customer retention and satisfaction. The relatively lower emphasis on cost reduction as a primary motivator could also signal that companies are starting to consider a wider range of factors when evaluating possible benefits. These can include environmental and social aspects, alongside economic considerations, which collectively contribute to a more resilient and sustainable business model.

M.F7 Institutional reasons such as global pacts, strategies, government programmes, subsidies, and tax benefits, are cited by 3.2% of participants. Although not the most prevalent factor, it is important not to underestimate the potential influence of institutional driver, as they can provide the necessary support and framework to encourage businesses to transition towards a CE. As governments and institutions increasingly recognise the importance of sustainable practices, it is possible that the role of policy and regulatory incentives will grow in shaping business strategies. Companies that stay informed about and adapt to changing institutional landscapes may find themselves better positioned to capitalise on new opportunities and thrive in a more sustainable future.

M.F8 Access to new technological developments ranks as the least cited factor, with 2.4% of respondents identifying it as their primary motivation. However, as technological innovations continue to emerge and mature, and support companies to become successful in their CE strategies, observing and learning from industry leaders could offer valuable insights and opportunities for businesses to improve their sustainability initiatives.

Primary Motivational Factors to Individual Retention Strategies

Figure 4.30 showcases the motivational factors influencing companies' intentions to adopt various Rs in the future, illustrating the percentage of companies driven by each factor for all ten Rs. Furthermore, Fig. 4.31 provides a visual comparison of these results.

M.F1 Environmental reasons have the most significant influence on companies' willingness to adopt R8 as well as material reduction strategys (R0 and R1) and R6. For the remaining Rs, it is still the most cited reason, except for R9, for which M.F2 excels and M.F1 is second. However, it shows that companies' willingness to adopt Rs is mainly to mitigate environmental concerns, as they are looking to reduce the usage of virgin materials and waste generation and retain the value of a product through up- and down-cycling.

M.F2 Business development is the second most cited reason for companies' willingness to adopt Rs in their operations, and it is the most mentioned factor for adopting R9. R4 is another strategy that stands out, and 26.2% of the participants prefer to implement it for business development purposes compared to other Rs. However, it is important to note that for the remaining strategies, M.F2's influence significantly decreases. For instance, in the case of

Motivation	R0	R1	R2	R3	R4	R5	R6	R7	R8	R9
M.F1	46.8%	44.1%	36.0%	25.7%	32.3%	36.4%	44.1%	41.9%	50.0%	20.8%
M.F2	11.7%	12.9%	17.3%	20.0%	26.2%	20.0%	18.6%	17.7%	13.2%	29.2%
M.F3	5.2%	7.5%	13.3%	20.0%	15.4%	9.1%	5.1%	9.7%	10.5%	20.8%
M.F4	7.8%	11.8%	16.0%	14.3%	12.3%	9.1%	10.2%	8.1%	7.9%	12.5%
M.F5	6.5%	3.2%	6.7%	11.4%	6.2%	10.9%	8.5%	9.7%	2.6%	4.2%
M.F6	6.5%	11.8%	4.0%	2.9%	4.6%	7.3%	3.4%	9.7%	5.3%	12.5%
M.F7	9.1%	5.4%	1.3%	2.9%	1.5%	0.0%	1.7%	0.0%	10.5%	0.0%
M.F8	2.6%	3.2%	2.7%	0.0%	1.5%	5.5%	5.1%	3.2%	0.0%	0.0%
Other	3.9%	0.0%	2.7%	2.9%	0.0%	1.8%	3.4%	0.0%	0.0%	0.0%
n =	77	93	75	35	65	55	59	62	38	24

Fig. 4.30 Motivational factors popularity heatmap for R0 to R9 adoption in the future

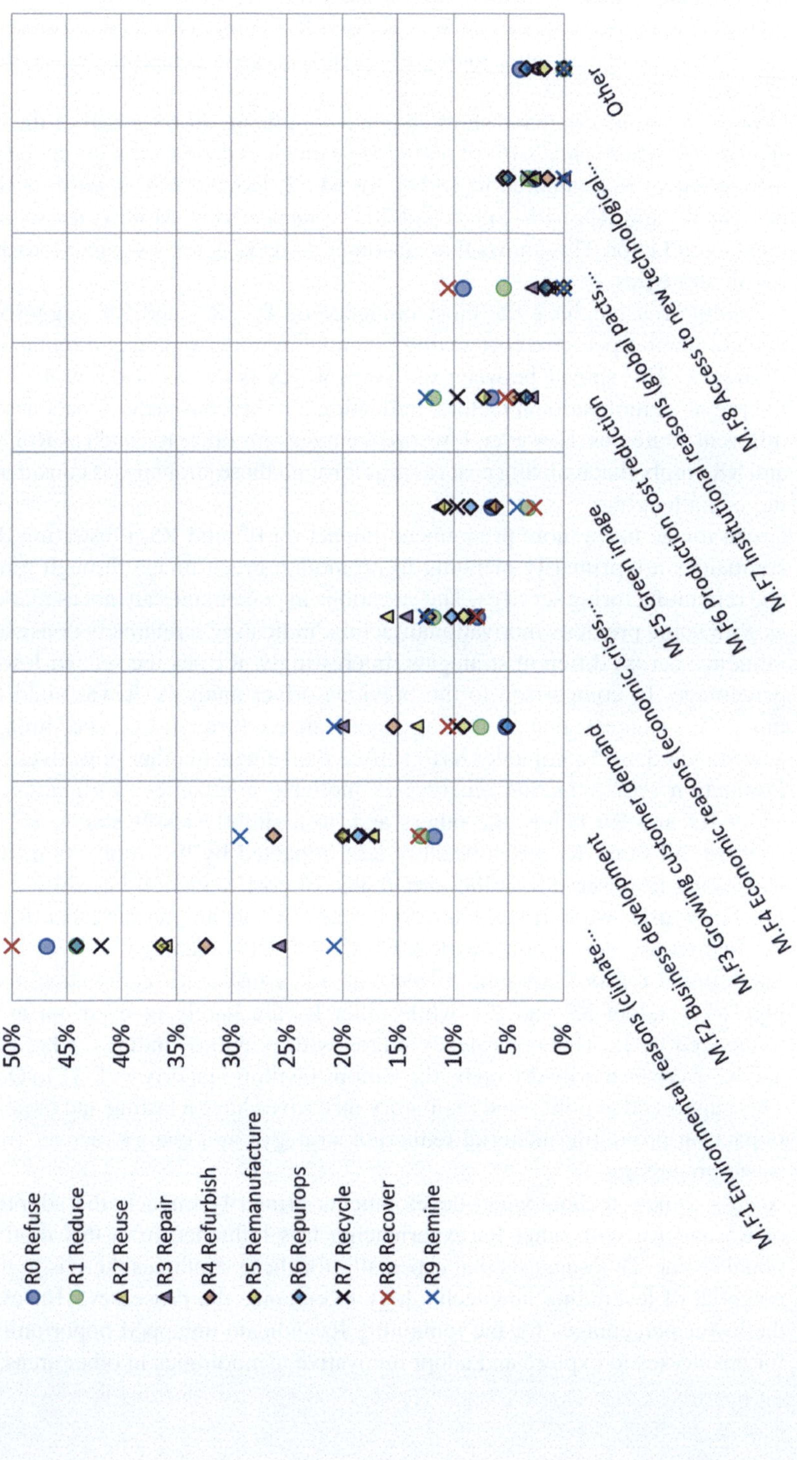

Fig. 4.31 Comparison of eight motivational factors driving companies' intention to implement Rs in the future

R0, M.F2 accounts for just 11.7%, while M.F1 has a much higher impact at 46.8%. This suggests that although business development is a key motivating factor for specific strategies such as R9 and R4, companies tend to prioritise environmental factors much more prominently when considering the adoption of other Rs.

M.F3 Growing customer demand shows again a significant drop for half of the Rs, such as R6, where only 5.1% of participants cite it as their reason for intending to implement it, compared to 18.6% for M.F2. However, it is worth noting that for R9 and R3, with 20.8% and 20.0% respectively, M.F3 is the second most cited factor. This shows that customer demand is not a strong motivator for all strategies.

M.F4 Economic reasons have the most influence on R2, R3, and R9, suggesting that companies perceive cost-saving potential in reuse, repair, and remaining strategies. The spread between the percentages is not as wide as it is for the previous motivational factors, indicating a more consistent impact across different strategies. However, lower percentages for other Rs, such as R0, R7, and R8, imply that realising cost savings through these strategies is considered more challenging.

M.F5 Green image has a more pronounced impact on R3 and R5, illustrating that companies are primarily planning to promote a green image through repair and remanufacturing services. The variations in percentages are not as marked as with some previous motivational factors, indicating a relatively consistent influence across different strategies. Interestingly, R1 has the second-lowest percentage. In comparison to the previous driver analysis, it was cited the most. This suggests that promoting a green image through R0 is becoming a new standard, and companies have shifted their focus to other objectives.

M.F6 Production cost reduction emerges to motivate companies' willingness to adopt R1 and R9 before the others and, to a slightly lesser extent, R7. In contrast, all other Rs are noticeably less impacted by this motivation. It is interesting to observe how the importance of cost reduction has shifted for R3. Previously, when framed as cost savings, it was among the most driven Rs. In contrast, now it holds with 2.9%, the lowest percentage.

M.F7 Institutional reasons continue to serve as a key driver for companies' willingness to adopt R0 and R8, while other Rs are barely or even not at all influenced by it. This consistency aligns with previous findings, where R0 and R8 were primarily driven by the various institutional drivers D.I1 to D.I4. This suggests that policy and regulatory incentives have a lasting and targeted impact on promoting material reduction strategys and energy recover from waste processing.

M.F8 Access to new technological developments primarily impacts the adoption of R5 and R6 with other Rs experiencing less influence from this motivational factor. This suggests that especially for these strategies there is higher potential of leveraging new technology to enhance the processes. However, the lower percentages for the remaining Rs indicate untapped opportunities for businesses to explore and adopt innovative technologies in other areas.

Analysis of the motivational factors that drive companies' willingness to adopt specific Rs revealed that environmental reasons are the most influential for almost all Rs. This reflects the importance of addressing environmental concerns in today's business landscape.

It is also evident that the drivers behind the already implemented Rs and the motivations behind the intention to implement the same strategies have changed. This highlights the dynamic nature of motivations and the evolving priorities of companies as they engage with and adopt these strategies. Understanding these shifts is crucial for companies to adapt and align their objectives effectively with the changing landscape of the CE.

4.5 Conclusion and Insights

4.5.1 Circular Economy Enhances Business Performance Factors

The impact of CE initiatives on BPFs was assessed by a comparison between industries and whether CE is integrated into the core business or separate from the core business. The analysis showed that the majority of companies across sectors experienced positive effects, with the non-durable goods manufacturing sector ranking 1st in technological advancements, sales, waste reduction, and company image improvements. Companies that integrated CE into their core operations demonstrated to be more often positively impacted compared to companies that practice CE separate from the core business. This is evident specifically in manufacturing costs, resilience, profits and sales. In contrast, the impact on waste reduction and company image is positively impacted for the greater majority regardless of integration level.

The data revealed that the enhancement of the company image and the reduction of waste were the most positively affected BPFs. Consequently, companies that have not yet adopted CE principles should consider doing so to achieve these long-term benefits, which can result in increased sales and profits. A circular approach to resource management can also improve resilience against resource-related risks and thereby strengthen competitive advantage. We also saw that the integration of CE promotes technological advancements between industries, ultimately stimulating sustainable growth and strengthening overall industry performance.

Recognising that cost factors were negatively influenced for roughly a third of participants, it is indispensable to examine various performance factors to assess the effectiveness of CE practices within a broader context. While certain aspects may not immediately meet expectations, it is vital to recognise that other BPFs can counterbalance these shortcomings. In doing so, we emphasise the importance of conducting a comprehensive evaluation, which will ultimately reveal that long-term benefits outweigh short-term challenges in pursuing sustainable success.

4.5.2 *Some Value Retention Options Show a Limited Prevalence*

Our evaluation of R strategies among 512 companies shows that material reduction and product use cycle extension techniques are predominantly adopted, with the highest interest for future implementation. Most companies actively practice *Reduce*, *Repair*, *Refuse*, and *Reuse*, whereas there is a considerable drop in frequency for R4–R9 strategies.

Remine and *Recover*, which focus on material processing and often require substantial investments in equipment and infrastructure, are the least commonly practised Rs. In return, their lower interest and implementation rates compared to other strategies present a unique opportunity for companies to differentiate.

Our research reveals that both conscious and unconscious practitioners engage in R practices. In particular, while conscious practitioners have a higher focus on material reduction strategys, unconscious practitioners tend to focus on strategies related to renewing the product use cycle. These findings also suggest that Rs are more widespread than previously believed.

Both groups have adopted a variety of retention options, with conscious practitioners prioritising the *Reduce* approach for waste reduction and resource efficiency. In contrast, unconscious practitioners display a strong preference for *Repair*, which means that it is a popular strategy among organisations, regardless of their knowledge or participation in CE.

A critical distinction between the two groups is their adoption of material processing and use cycle renewal strategies. Conscious practitioners prefer *Recycling*, which balances consumption and material conservation, and often aligns with the *Reduce* approach to contribute equally to economic and environmental benefits. On the other hand, unconscious practitioners tend more towards use cycle renewal strategies, with *Repurpose* being the most adopted one of this group.

Our analysis also indicates that there is a potential convergence of practices between unconscious and conscious CE practitioners. Both groups show preferences for strategies that emphasise waste reduction and minimising material requirements in the early stages of production, particularly focusing on *Reduce*. The growing interest in *Repurpose* among conscious practitioners highlights its increasing prominence in the industry. Although there are differences in the adoption of other strategies such as *Repair*, *Recycling*, *Recover*, and *Remine*, the general trend indicates that unconscious CE practitioners gradually align with the practices of their conscious counterparts. This implies that awareness and implementation of Rs continue to spread. Companies that have not yet considered adopting strategies R should re-evaluate their position to avoid being left behind in the advancing field of sustainable business practices.

4.5.3 Future Adoption of Value Retention Options Driven by Few Motivators

Based on the findings of this examination, it is clear that environmental concerns are the primary motivator for companies to adopt retention strategies in the future. Businesses are becoming increasingly aware of the impact of their operations on the environment and are looking to mitigate this impact by reducing waste generation and virgin material usage while reallocating used resources through up- and down-cycling.

However, it is also essential to note that the factors vary depending on the selected retention option, similar to the drivers from the past. For example, business development is the most cited reason to adopt retention options R9 and R4, but not for R1 and R0. However, the growing customer demand has significant impact for R9 and R3 and is relatively immaterial for R6 and R0.

Economic reasons, such as cost savings and risk mitigation, are clearly not the top priority, but are more often the motivator behind R2, R3, and R9 and much less for R0 and R8. Whereas reduction in production costs is a primary motivator for R9, R1 and R7. Hence, businesses seem to be increasingly recognising the multifaceted advantages of incorporating circularity into their operations, extending beyond mere cost savings to encompass broader sustainability objectives and long-term value creation.

Although technological innovations continue to emerge and mature and contribute to becoming more efficient in practising Rs, access to technology is seldom the motivator for any Rs. However, observing and learning from industry leaders is important and could offer valuable insight and opportunities for businesses to improve their own initiatives.

4.6 Recommendations for Business

4.6.1 Select Retention Options Based on Your Business Drivers

The analysis shows that environmental and resource management considerations are the primary drivers, followed by financial and business, strategic and market-driven factors, and finally institutional and regulatory drivers. Nevertheless, companies should adopt retention options that align with their individual objectives, industry, and target market.

For companies focusing on environmental and resource management improvements, prioritising strategies such as R2, R4 and R7 will help reduce resource consumption and minimise their carbon footprint.

On the other hand, when financial and business goals like cost savings and increased competitiveness through innovation and business development are the primary objectives, companies should consider implementing strategies like R3, R4, and R5, which are more influenced by these factors.

For businesses that emphasise strategic and market-driven factors, such as adapting to changing consumer behaviour and promoting a green corporate image, strategies like R3 and R4 are particularly effective in meeting customer expectations and enhancing brand reputation.

Although institutional and regulatory drivers are less influential, they should not be ignored. Companies can showcase their commitment to social responsibility and institutional drivers by adopting strategies like R0 and R9. Additionally, businesses should stay informed about relevant legal requirements and norms, and explore opportunities to capitalise on government incentives, subsidies, and tax breaks to support their transition.

It is evident that various circumstances can influence which R is the most suitable for achieving a company's goals. Companies, therefore, need to carefully assess their primary objectives and tailor their approach accordingly.

4.6.2 Leverage Your Success Through Enablers

To effectively implement CE practices, companies should prioritise building strong relationships, collaborations or partnerships with stakeholders in the value chain and cultivating in-house technical and technological capabilities. Realising these key factors will enable organisations to navigate the transition to a CE more successfully. However, it is essential to recognise that the importance of various enablers differs depending on the specific R pursued. Therefore, companies should carefully assess the relevance of each enabler in the context of their chosen R to maximise the chances of success.

Companies focusing on material reduction and processing strategies should concentrate heavily on fostering collaborations and partnerships within their value chain. On the other hand, businesses that implement use cycle extending and renewing strategies, such as R3 and R4, should invest more in developing their in-house technical expertise.

Furthermore, companies should not overlook the importance of enhanced information sharing and transparency, especially when implementing material reduction strategys. Experimenting with or developing new technologies is essential for R0 and R9, while technological development through joint decision-making with customers or suppliers is more critical for use cycle extending and renewing strategies.

Lastly, companies should consider the availability of a RL network, collection or take-back systems, and engagement in industrial symbiosis when selecting the appropriate Rs. Although these enablers may not be as critical for certain strategies, they specifically offer greater potential for success when implementing R7 to R9.

References

1. Web of Science. (2022). Search throughout "All database" using as Keyword "Circular Economy" from 1989 to 2022. https://www.webofscience.com/wos/woscc/analyze-results/8565a6aa-8d21-441b-81e1-e72cccb90ddf-68e87c00.
2. OECD. (2018). *Business models for the circular economy: Opportunities and challenges from a policy perspective*. Paris: OECD Publishing.
3. Masi, D., Kumar, V., Garza-Reyes, J. A. & Godsell, J. (2018). Towards a more circular economy: Exploring the awareness, practices, and barriers from a focal firm perspective. *Production Planning & Control, 29*, 539–550. ISSN: 0953-7287, 1366-5871.
4. Sá, M. M., et al. (2022). Integration of the circular economy paradigm in companies from the Northwest of the Iberian Peninsula. *Sustainability, 14*, 7940.
5. Reike, D., Vermeulen, W. J. & Witjes, S. (2018). The circular economy: New or refurbished as CE 3.0? – exploring controversies in the conceptualization of the circular economy through a focus on history and resource value retention options. *Resources, Conservation and Recycling, 135*, 246–264. ISSN 0921-3449.
6. Govindan, K. & Hasanagic, M. (2018). A systematic review on drivers, barriers, and practices towards circular economy: A supply chain perspective. *International Journal of Production Research, 56*, 278–311. ISSN: 0020-7543, 1366-588X.
7. Camilleri, M. A. (2020). European environment policy for the circular economy: Implications for business and industry stakeholders. *Sustainable Development, 28*, 1804–1812. ISSN: 0968-0802, 1099-1719.
8. Gusmerotti, N. M., Testa, F., Corsini, F., Pretner, G. & Iraldo, F. (2019). Drivers and approaches to the circular economy in manufacturing firms. *Journal of Cleaner Production, 230*, 314–327. ISSN: 0959-6526.
9. Ranta, V., Aarikka-Stenroos, L., Ritala, P. & Mäkinen, S. J. (2018). Exploring institutional drivers and barriers of the circular economy: A cross-regional comparison of China, the US, and Europe. *Resources, Conservation and Recycling, 135*, 70–82. ISSN: 0921-3449.
10. Vermunt, D., Negro, S., Verweij, P., Kuppens, D. & Hekkert, M. (2019). Exploring barriers to implementing different circular business models. *Journal of Cleaner Production, 222*, 891–902. ISSN: 0959-6526.
11. Calzolari, T., Genovese, A. & Brint, A. (2021). The adoption of circular economy practices in supply chains-An assessment of European multi-national enterprises. *Journal of Cleaner Production, 312*, 127616. ISSN: 0959-6526.
12. Dieckmann, E., Sheldrick, L., Tennant, M., Myers, R. & Cheeseman, C. (2020). Analysis of barriers to transitioning from a linear to a circular economy for end of life materials: A case study for waste feathers. *Sustainability, 12*. ISSN: 2071-1050.
13. Fonseca, L., Domingues, J., Pereira, M., Martins, F. & Zimon, D. (2018). Assessment of circular economy within Portuguese organizations. *Sustainability, 10*, 2521. ISSN: 2071-1050.
14. Kumar, V., Sezersan, I., Garza-Reyes, J. A., Gonzalez, E. D. & Moh'd Anwer AL-Shboul. (2019). Circular economy in the manufacturing sector: Benefits, opportunities and barriers. *Management Decision, 57,* 1067–1086. ISSN: 00251747.
15. Rizos, V. et al. (2016). Implementation of circular economy business models by small and medium-sized enterprises (SMEs): Barriers and enablers. *Sustainability, 8,* 1212. ISSN: 2071-1050.

16. Van Langen, S. K., & Passaro, R. (2021). The Dutch green deals policy and its applicability to circular economy policies. *Sustainability, 13*, 11683.
17. Guldmann, E. & Huulgaard, R. (2020). Barriers to circular business model innovation: A multiple-case study. *Journal of Cleaner Production, 243*. ISSN: 0959-6526.
18. Henry, M., Bauwens, T., Hekkert, M., & Kirchherr, J. (2020). A typology of circular start-ups: Analysis of 128 circular business models. *Journal of Cleaner Production, 245,* 118528. ISSN: ISSN 0959-6526
19. Langen, S. et al. (2021). Promoting circular economy transition: A study about perceptions and awareness by different stakeholders groups. *Journal of Cleaner Production, 316*, ISSN: 0959-6526.
20. Niinimäki, K. et al. (2022). The environmental price of fast fashion. *Nature Reviews Earth & Environment, 1*, 189–200. ISSN: 2662-138X.
21. Manninen, K. et al. (2018). Do circular economy business models capture intended environmental value propositions? *Journal of Cleaner Production, 171*, 413–422. ISSN: 09596526.
22. Vatamanescu, E., Dabija, D., Gazzola, P., Cegarro-Navarro, J. & Buzzi, T. (2021). Before and after the outbreak of Covid-19: Linking fashion companies' corporate social responsibility approach to consumers' demand for sustainable products. *Journal of Cleaner Production, 321*, ISSN: 0959-6526.
23. Peiro, L. T., Polverini, D., Ardente, F. & Mathieux, F. (2020). Advances towards circular economy policies in the EU: The new Ecodesign regulation of enterprise servers. *Resources, Conservation and Recycling, 154*, ISSN: 0921-3449
24. Van Eijk, F. (2015). *Barriers & Drivers towards a Circular Economy – Literature Review* A-140315-R-Final (Acceleratio).
25. European Commission & Directorate-General for Environment. (2020). *Circular Economy Action Plan – For a cleaner and more competitive Europe*. Publications Office of the European Union. https://data.europa.eu/doi/10.2779/05068.
26. European Commission. (2020). *A new Circular Economy Action Plan For a cleaner and more competitive Europe* Issue: COM/2020/98.
27. Bansal, P. & Roth, K. (2000). Why companies go green: A model of ecological responsiveness. *Academy of Management Journal, 43*, 717–736. ISSN: 0001-4273, 1948-0989.
28. Morlet, A. et al. (2017). A new textiles economy: Redesigning fashion's future. Ellen MacArthur Foundation.
29. Deutz, P., Baxter, H., Gibbs, D., Mayes, W., & Gomes, H. (2017). Resource recovery and remediation of highly alkaline residues: A political industrial ecology approach to building a circular economy. *Geoforum; Journal of physical, human, and regional geosciences, 85*, 336–344. ISSN: 0016-7185.
30. Tura, N. et al. (2019). Unlocking circular business: A framework of barriers and drivers. *Journal of Cleaner Production, 212*, 90–98. ISSN: 0959-6526.
31. Ratner, S., Gomonov, K., Lazanyuk, I. & Revinova, S. (2021). Barriers and drivers for circular economy 2.0 on the firm level: Russian case. *Sustainability, 13*, 11080. https://doi.org/10.3390/su131911080
32. Ilić, M. & Nikolić, M. (2016). Drivers for development of circular economy – A case study of Serbia. *Habitat International, 56*, 191–200. ISSN: 01973975.
33. European Investment Bank. (2021). *Circular Economy - Overview 2021.*
34. Kirchherr, J. et al. (2018). Barriers to the circular economy: evidence from the European Union (EU). Ecological Economics, *150*, 264–272. ISSN: 0921-8009.
35. Hart, J., Adams, K., Giesekam, J., Tingley, D. D. & Pomponi, F. (2019). Barriers and drivers in a circular economy: The case of the built environment. *Procedia CIRP, 80*, 619–624. ISSN: 2212-8271.
36. Galvão, G. D. A., Evans, S., Ferrer, P. S. S. & de Carvalho, M. M. (2022). Circular business model: Breaking down barriers towards sustainable development. *Business Strategy and the Environment, 31,* 1504–1524. ISSN: 0964-4733.
37. UNIPARK. (n .d.). *Online Survey Software: Surveys Made Easy with Unipark* in collab. with TIVIAN. https://www.unipark.com/en/ (2022).

38. Dillman, D. A., Smyth, J. D. & Christian, L. M. (2014). Internet, phone, mail, and mixed-mode surveys: The tailored design method 4th edn. Hoboken: Wiley. ISBN: 978-1-118-45614-9.
39. Fowler, F. J. (2014). *Survey research methods* (Fifth edition). SAGE. ISBN: 978-1-4522-5900-0.
40. Luban, K. & Hänggi, R. (eds). (2022). *Erfolgreiche Unternehmensführung durch Resilienzmanagement*. Heidelberg, Berlin: Springer. ISBN 978-3-662-64022-7.
41. Lance Lambert. (2020). 75% of companies report coronavirus has disrupted their supply chains. *Fortune*. Section: Finance Economy. Retrieved January 1, 2023 from https://fortune.com/2020/03/11/75-of-companies-report-coronavirus-has-disrupted-their-supply-chains/
42. European Parliament et al. (2021). Impacts of the COVID-19 pandemic on EU industries – Study. https://data.europa.eu/doi/10.2861/052356. European Parliament.
43. Desso. (2015). *As a Cradle to Cradle® company, we go beyond what most people understand by sustainability*. Retrieved April 14, 2022 from https://vloeren.projecten.tarkett.nl/nl_NL/node/onze-duurzaamheidsstrategie-omgezet-in-acties-1831.

Part II
Analysis of Successful Circular Economy Cases

Chapter 5
Introduction to Part II

André Podleisek

Abstract The broad knowledge gained in Part I is now deepened and connected to practical insights with use cases explored by means of interviews. The methodology for each retention option is to describe them, find a previously not discussed use case, analyse the use case with the help of the drivers and enablers from Part I and derive learnings about the impact and outcome of the case. This approach enables the reader to learn from the combination of concepts, practice examples, and outcomes analysed. Repurpose and Remine follow a slightly different approach due to limitations in available use cases and interview partners. This will form the basis for Part III and its focus on practical implication models.

Intention of Case Study Analyses

In Part I, a broad understanding of Circular Economy (CE) and its drivers and enablers was created based on literature (see Chap. 3) and further expanded by research on application in the European industry (see Chap. 4). The initial hypothesis that there are common factors for successful CE implementation was proven with a comprehensive literature review and a large-scale industry survey in 512 European industrial entities. Furthermore, success factors on how to successfully practice CE were identified. Part II is about deepening this knowledge with the help of case studies. In discussions with industry, we found that the description of Rs was often too generic to see how it could be used in practice. A more detailed description of the major aspects of each Retention Option (R) and how it exists in the real world was needed. With respect to the drivers and enablers discussed above and their identified level of application for each R, a statistical relevance was demonstrated. However, it was not yet clear *how* a company would experience the drivers and make use of the enablers. In addition, we wanted to evaluate the actual impact in terms of sustainability and economics. This additional knowledge from this deep dive is then used in Part 3 to generalise our findings, conclude on implications, and create a model for companies introducing CE into their business.

A. Podleisek (✉)
Wettswil am Albis, Switzerland
e-mail: andre.podleisek@ost.ch

A. Podleisek et al. (eds.), *Circular Economy: The Next Level of Company Success*,
https://doi.org/10.1007/978-3-031-66406-9_5

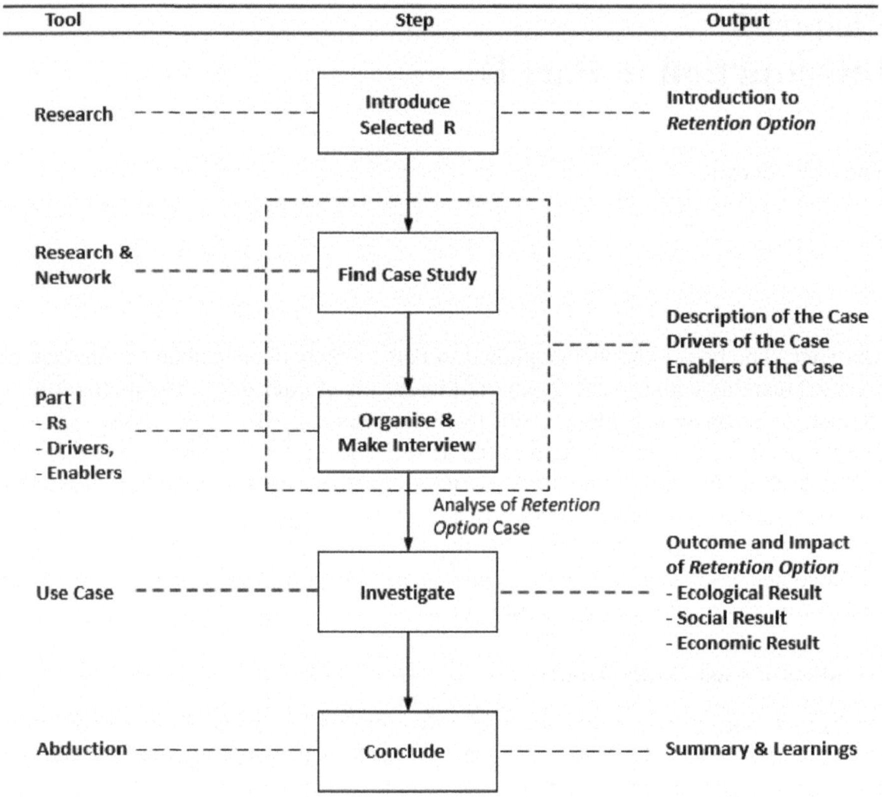

Fig. 5.1 Methodology case studies Part II

Methodology of Case Study Analysis

The methodology of the case study analyses is displayed in Fig. 5.1. The steps, used tools and the outputs are visualised.

First, additional literature research was conducted on each R. The result of this research is a comprehensive description and introduction to the retention option and its application. To complement this outline of an R, a suitable use case was searched for study. Prerequisites for the use case were:

- real-life, i.e. existing application of the R, not just a concept,
- new, i.e. not one of the very successful examples already widely discussed in the literature (e.g., Interface, Caterpillar),
- accessible, i.e. available for interviews.

For *Repurpose* and *Remine* it showed to be difficult to meet all three criteria, and preference was given to real-life examples over the other two. One explanation for why these difficulties materialised for these two R is that *Repurpose* by definition requires a new application for components, which is usually a different market – a move not much in favour for companies already active in a defined segment. In Sect. 12.1 an explanation is given why *Recycling* is often more suitable. Similarly, *Remine* is not yet widespread, as shown by the lowest score in practice and plans for this R in Sect. 4.4.2 (Figs. 4.21 and 4.22). We assume that this is also because, for *Remine*, significant investments in plants and machines are necessary and the legal issues to redefine waste as a resource have not yet been solved.

For each R, a set of use cases was presented and discussed internally and the most suitable use case was selected for the book. For this use case, an in-depth analysis of the company and its business model was performed, including details on the implementation of R. This was complemented by in-person interviews with key personalities. Applying the definitions and learning from Chaps. 3 and 4 resulted in a detailed description of the case itself, how the R was implemented, and its drivers and enablers. Since we believe CE to be a means rather than a result in itself, we added an investigation on the practical outcomes and impacts of the R, especially with regard to ecological, social, and economic results. In some cases, the authors were not able to populate all three results with the same level of detail. Finally, abductive reasoning lead to a summary and learnings from this use case for the application of its R.

All case studies together with the insights from Part I resulted in a more comprehensive understanding of the practical application of retention options, their requirements, and results. Through this, the practical implications of CE were better understood, forming the basis for an in-depth analysis and application in Part III.

Chapter 6
R0 Refuse and R1 Reduce

Nico Bleisch and Katharina Luban

Abstract Companies can engage in Circular Economy by designing products and processes that are more resource-efficient and less wasteful, and by encouraging customers to use products in a more sustainable way. Our findings show that the application of the retention options *Refuse* and *Reduce* do not only open up ecological opportunities, but also provides economical opportunities. The following chapter explains these observations and explores drivers and enablers for the successful implementation of these retention options.

6.1 Introduction to *Refuse* and *Reduce*

Refuse and *Reduce* are the first two Retention Option (Rs) mentioned in the United Nations Environmental Programme (UNEP) Framework. Their goal is to minimise or avoid waste and resource consumption. This can be achieved by designing products and processes that are more resource efficient and less wasteful, and by facilitating and encouraging customers to use products in a more sustainable way [1].

The two Rs are considered together in this chapter because they highlight a similar area in the value chain: The *Refuse* option means that companies should strive to *eliminate* waste generation and resource consumption in the first place. The *Reduce* option, on the other hand, means that companies should aim to *reduce* the amount of waste generated and the resources consumed throughout the lifecycle of their products. This can be achieved through measures such as optimising production processes, reducing packaging, using less virgin materials within products, or providing more efficient functions during the lifetime of the products [2–4].

The activities to realise the retention options *Refuse* and *Reduce* take effect before the product reaches the customer. Therefore, the framework developed in Sect. 3.1.3

N. Bleisch
Mels, Switzerland

K. Luban (✉)
Sevelen, Switzerland
e-mail: katharina.luban@ost.ch

A. Podleisek et al. (eds.), *Circular Economy: The Next Level of Company Success*,
https://doi.org/10.1007/978-3-031-66406-9_6

classifies these Rs as 'short loops'. Short loops build the fundamental basis for Circular Economy (CE) and have a preparatory character because they occur before any other R [4]. The most direct way to CE is obviously when fewer resources are used or when no resources are consumed or wasted at all in the first place. Therefore, short loop Rs are considered the most effective strategy to reduce the input of virgin material and waste. To maximise their potential, short loop Rs should be addressed during the product design phase in Product Lifecycle Management (PLM) (see also Sect. 3.1.5) to promote efficient and innovative material use and prevent waste production in products and manufacturing processes.

The scope of *Refuse* and *Reduce* is very broad and can range from small product or process improvements to revolutionary product and process innovations. It is quite obvious and trivial to enforce this CE approach through legal regulations: Rules and laws can require the use of alternative and sustainable materials, they can ban certain materials, or they can encourage lower material consumption by imposing taxes on these materials.

A more complex approach leaves more room for innovation by focussing on sustainability aspects throughout the entire lifecycle of a product. Brezet et al. point out that adapting product design strategies in a way that all aspects of the lifecycle are considered is a key to achieve CE [5]. This concept of considering sustainability aspects in product development is called Ecodesign, also known as Sustainable Design or Green Design.

The focus of Ecodesign strategies lies on the creation of new, added value products or services that can stimulate the economic competitiveness of companies and industries through cost savings, improved brand reputation, and increased access to new markets. At the same time, these products contribute to more sustainable forms of consumption and lifestyle scenarios. The concept of Ecodesign leads to a lower overall environmental impact over a lifetime: reduction of waste and pollution caused by products and promotion of their responsible production and consumption [6].

With this, Ecodesign strategies also meet the Sustainable Development Goals (SDGs), which call for improved solutions throughout the Product Lifecycle (PLC), promoting the use of fewer and environmentally friendly materials and increasing the ecological efficiency of products, as well as other social innovations and user-orientated strategies [7].

Even if the Ecodesign framework is already several decades old, it provides a holistic, comprehensive, and practical view which is still valid [5]. It can guide designers and manufacturers to create products that can contribute to reducing waste and energy consumption and can encourage the creation of new business models that rely on sustainable principles. Therefore, the Ecodesign is seen as an all-encompassing framework to analyse and realise initiatives concerning the retention options *Refuse* and *Reduce*.

Mestre and Cooper have formulated eight product design strategies to ensure a holistic view [7]. In the following, their framework shown in Fig. 6.1 will be explained and illustrated with examples. It is important to mention that some of the mentioned product design strategies are very similar, they influence each other, and cannot always be clearly separated. In addition, some examples may contain more than

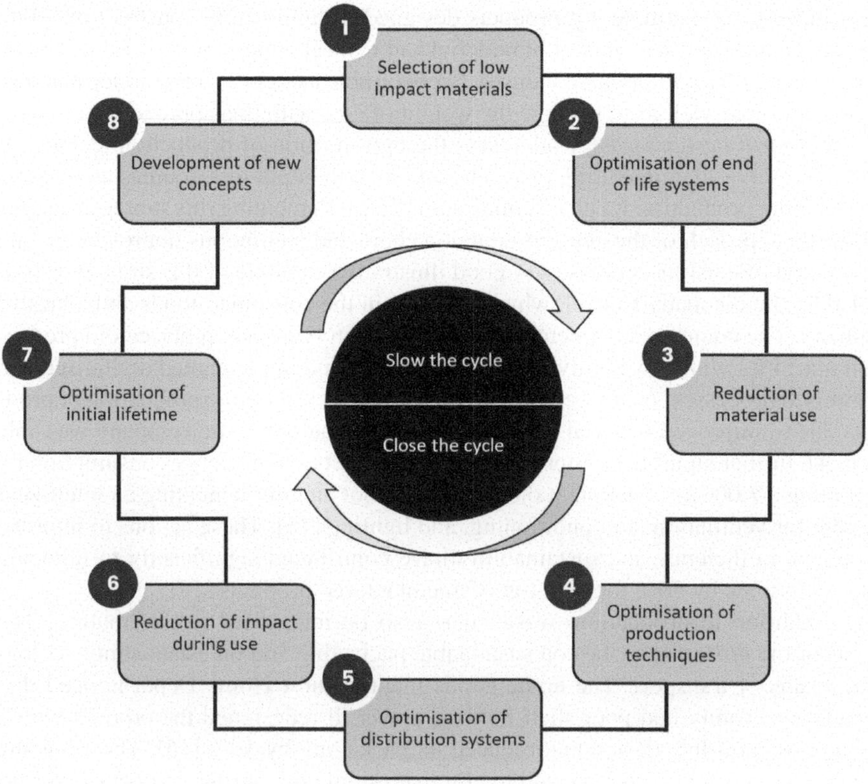

Fig. 6.1 Overview of the product design strategies (own representation based on [7])

one strategy. However, the explanation will show and clearly define the relationship between design strategies and the Rs *Refuse* and *Reduce*.

The *first two strategies* aim at *selecting low impact materials* and *optimising of End-of-Life (EOL) systems*. The focus is on choosing recyclable materials that require little energy for reprocessing and reuse. These could be biodegradable or photodegradable materials. For example, the company Starbucks has completely changed the material of its coffee cups. They have eliminated foamed polystyrene and together with their suppliers they developed a fully recyclable and compostable cup solution [8, 9]. Another well-known example is the biodegradable plastic bag, which is intended to provide an environmentally friendly alternative to the conventional single-use plastic bag [10].

In addition to the choice of a suitable material (i.e., avoiding non-ecological materials), the *third design strategy* focuses on the reduction of material use and the reduction of resource consumption. In some countries, such as Switzerland or Sweden, there are efficiency and reduction targets, for example, regarding the withdrawal of water and targets for the use of certain materials [11]. A reduction in material can be achieved, for example, by products designed in lightweight construction. With this concept, a minimal amount of material and a lower weight of the product can be achieved [12]. The automotive industry has been able to significantly reduce material consumption as well as to manage the weight of cars with this approach.

The *fourth design strategy* addresses the optimisation of production techniques with the aim of manufacturing products with as little input of resources as possible and keeping production waste to a minimum [7]. Implementing this strategy requires a holistic approach of the entire production chain and continuous improvement and adaptation of existing processes. A good illustrative example of this strategy is provided by the company Henkel, which operates in the consumer goods and adhesive industry. The company completely changed the factory layout, replaced old production machines with new highly efficient machines, and implemented digitalised and optimised processes. As a result, fewer resources are used for the production of products due to improved material flow and higher productivity. The company was able to make further significant improvements and reductions in energy consumption by installing a 7,000 m^2 solar plant on the factory roof and implementing an intelligent system for ventilation, air conditioning, and lighting [13]. These actions to improve production efficiency and sustainability have contributed significantly to reducing CO_2 emissions by 50% for each ton of manufactured products [14].

In addition to production, waste must also be minimised in distribution. The focus of the *fifth strategy* lies on sustainable packaging and on establishing ecological means of transport. The home goods manufacturer Home Depot noticed that a bathroom vanity had poor shelf holding power. It redesigned the product with a folding origami-like design that reduced its packaging by 30% [15]. The company Apeel Sciences has even managed to eliminate single-use shrink wrap plastic packaging for fresh fruits and vegetables. The innovation consists of its edible coating on a vegetable base, which is applied to fresh products and mimics and improves the natural protective mechanisms of fruits and vegetables. This innovation has two main advantages: On the one hand, it prevents premature decay of food by reducing water loss and oxidation. On the other hand, it eliminates the need for synthetic packaging [16].

Another important component of product design is addressed with the *sixth and seventh strategies*, which focus on the reduction of negative impact during use and the optimisation of initial lifetime. Impact during use refers to how a product affects its surroundings, users, and environment, including aspects such as energy consumption, emissions, and noise. Optimisation strategies for the initial lifetime aim to achieve high reliability and durability, easy maintenance, and standardisation of the products. A good example is the company Georg Fischer which offers a plastic piping system for refrigeration plants called 'COOL-FIT 4.0'. The special plastic design of the system eliminates the risk of corrosion, smooth inner surfaces prevent deposits and making the system 100% maintenance-free. Furthermore, the system is also 65% lighter than a metal piping system, can be installed 50% faster due to standardised connections, and achieves 30% higher energy efficiency during use due to a high-efficiency foam [17]. It is needless to say that these advantages also have a positive impact on cost reduction.

The *last strategy* addresses the *development of completely new concepts*. The focus lies on dematerialisation through the integration of functions into products or the digitalisation of products up to complete new product innovations or concepts. An adequate example for the integration of further functions into an existing product is the multifunctional pressure cooker from the company Instant pot, which can be used as a slow cooker, rice cooker and steamer and can replace several other kitchen appliances [18]. An adequate example for dematerialisation with a new business model is provided by Mobility. The company was able to assert itself successfully on the market with a new concept of shared mobility. Private customers and companies have access to 2,860 vehicles at 1,530 strategic locations in Switzerland. Users can rent the desired vehicle for a certain period of time and return it at the same or another location. With shared mobility, resources can be saved because the number of cars can be reduced and the rate of utilisation per car can be increased [19].

In addition to the mentioned product design strategies, the Ecodesign framework follows two product design concepts, which are 'slow the cycle' (or in this book, medium loop R) and 'close the cycle' (or long loop R). These two concepts are the underlying concepts of product design strategies. The slow-the-cycle concept focusses on the integration of easy-to-implement product design strategies, which follow the idea of extending the lifespan of products and keeping them in use as long as possible by means of a high durability. The goal is to reduce waste and the need for new resources [20]. The closing-the-cycle concept is an extension of the slow-the-cycle concept with material processing strategy. The concept aims at integrating as many product design strategies as possible. It is important to note that there may be tensions between product design strategies (e.g., between durability and material reduction) that need to be resolved [21].

There is no standard solution for the development of a product design that can be fully integrated into CE. It takes time and know-how to develop viable solutions. However, companies should apply the slow-the-cycle concept in any case. This will contribute to the development of sustainable solutions and is a good starting point for further developments towards the close-the-cycle concept and full integration of the product in the CE. But even if companies are willing to apply the close-the-cycle concept, the responsibility is not only with producers. Equally responsible are consumers, who should reject products that use hazardous substances, non-recyclable materials, inefficient processes, and they should reject products that are not of high quality and durability [3].

From the various examples described, it is clear that the eight product design strategies and the two cycle concepts provide a good indication of how to realise the retention options *Refuse* and *Reduce*. However, the Ecodesign framework itself only forms the basis for a successful implementation of Rs. Therefore, it is necessary to examine in detail the drivers and enablers that are considered relevant for successful implementation. This will be done by first describing and then analysing a specific use case.

6.2 Analysis of a *Refuse* and *Reduce* Case

6.2.1 Description of the Case

This chapter describes a successfully implemented use case with a focus on the Rs *Refuse* and *Reduce*. The use case is taken from a global industrial company with more than 8,000 employees in 31 countries, with a base in Switzerland. The use case company is the leading provider of flow solutions for safe transport of fluids. The product range encompasses industry-leading high-quality, leak-free piping solutions for numerous demanding end-market segments. The strong focus on products and innovation is reflected in its global sales and manufacturing presence and its portfolio that includes fittings, valves, pipes, automation, fabrication, and jointing technologies.

For a detailed description and elaboration of the use case, interviews were conducted with A. O., Corporate Sustainability Manager, and D. R., Global Business Development Manager. The company has successfully established new standards in the field of Pressure Regulating Valves (PRVs). The use of PRVs enables a precise and optimised pressure management, ensures a balance within the water supply network and provides a large contribution to eliminate water losses in distribution networks. In cooperation with a spin-off of the Oxford University a PRV with an innovative design was developed. The PRV will be referred to as NewValve in this chapter.

The main difference between NewValve and other conventional PRVs is the choice of material and a completely new product design. In contrast to conventional PRVs, which are entirely made of metal, the main components of the NewValve, like the housing, the piston and valve core, are made of plastic.

During the development phase, many difficulties and trade-offs had to be solved. From a technical point of view, difficulties with respect to the change of material from metal to thermoplastic Poly-Oxy-Methylene (POM) had to be overcome (strategies 1 and 2). In addition to the positive aspects of POM, such as high durability, low weight, and recyclability, there are also negative aspects. To achieve consistently high reliability and complete tightness, a suitable product design had to be found (A. O., personal communication, November 25, 2022).

During the product development phase, a strong focus was also placed on high-quality, simplicity, and efficient processes (strategy 4). The simplicity of the product is characterised by only three primary parts and one movable component that is axially aligned with the flow direction and allows full functionality with precision flow stability (strategy 3), (D. R., personal communication, November 25, 2022).

As can be seen from the described structure and function of NewValve, the definition of *Refuse* is fulfilled, because the new developed solution offered the same function in a completely different design and material, but refused using metals for the housing. The predecessor product became obsolete.

In addition to the known function, NewValve provides the possibility to connect a control devices which enables the monitoring and remote control of the PRV (strategy

6 and 7). This enables simple operability, high reliability, and short reaction time in the event of problems. Based on the recorded data, it is possible to create transparency about the water supply system. Transparency enables a better understanding of the water supply system and further optimisations and adjustments can be made to the water flow rate.

In addition, the new product design fulfils the *Reduce* approach by reducing the number of individual parts and reducing the weight through the change of the material and compact design. The new product design has a significant advantage not only for the company, but also for the customer. Due to its reliable flow regulation, NewValve can significantly reduce pipe bursts and water wastage. Furthermore, the innovative design of NewValve enables fast installation and maintenance, which results in significantly lower costs for the customer [22].

6.2.2 Drivers of the Case

Undoubtedly, NewValve is a successfully launched product that was developed to refuse and reduce material and resource consumption. The use case shows the existing advantages as well as difficulties that arise with the development of a new product with the CE approach. But how to define a promising field to deploy *Refuse* and *Reduce* as concepts of choice? In the following chapter, this question will be answered by analysing drivers that were responsible for completing the case. Table 6.1 provides an overview of the individual drivers and their relationship to the use case together with an assessment of relevance, with 0 representing insignificant importance, + indicating minor relevance, ++ reflecting notable relevance, and +++ signifying crucial importance.

The evaluation of the drivers in connection with NewValve and the Rs *Refuse* and *Reduce* showed that all clusters with the exception of *Institutional* drivers are important. Companies are often forced by new laws to reduce resources or to manufacture a product with a different material because the current material will be banned. However, the decision to use plastic to manufacture NewValve was not based on legal requirements and standards, which therefore is not considered as a driver (D.I1), (D. R., personal communication, November 25, 2022). Also, no future legal obligations (D.I2, D.I3) or taxation benefits (D.I4) pushed the development of NewValve with its focus on sustainable product design.

According to D. R. (personal communication, November 25, 2022), the change in material from metal to POM was not due to the scarcity of materials and supply chain problems (D.E3). The material change is rather related to the cluster *Environment and Responsibility*. Strong drivers are the reduction of resources and energy (D.E1) as well as the reduction of the carbon footprint (D.E2). Resources are saved due to lower transportation emissions and less packaging by the compact design with less individual parts. In addition to that, resources are also saved in the use phase of the NewValve: Water losses can be reduced by preventing pipe bursts, and existing pumps need less energy because of uniform pressure. The development of a durable design

Table 6.1 Drivers of the *Refuse* and *Reduce* case

Cluster	Code	Driver	Context	Relevance
Environmental	D.E1	Resources	Reduction of material requirements due to the new design with few individual parts	+++
	D.E2	Footprint	Energy and CO_2 savings due to a good volume-weight ratio during transportation of the PRV, simple installation without lifting equipment and reduction of water losses	++
	D.E3	Risk	Reliable suppliers of raw material and foreign parts in the country of production or near abroad	0
Institutional	D.I1	Legal	The legal requirements and norms need to be fulfilled but the change of material from metal to plastic and the new product design is not driven by these factors	0
	D.I2	Punitive	No punitive measures available	0
	D.I3	Juridification	No foreseeable future obligations	0
	D.I4	Tax	New product design and place of production has no taxation benefits	0
Stakeholder	D.S1	Image	Product defines new standards in sustainability and quality and positively influences the image of the company	++
	D.S2	Customer	Growing customer awareness calls for sustainably produced high-quality products that meet customer expectations	+++
	D.S3	Practices	Existing standards are exceeded by better performance, quality, and value for the customer compared to competitors	++
Financial	D.F1	Volatility	The change of material and the new product design is not related to the cost and fluctuation of the material prices of metal or plastic	0
	D.F2	Savings	Lower transportation and packaging cost due to lightweight and compact design and lower production cost due to fewer and smaller individual parts	+
	D.F3	Competitiveness	Sustainable and innovative product enables a competitive advantage over the competitor through better performance	+++
	D.F4	Value	Create value through local sourcing	0
Responsibility	D.R1	CSR	Social responsibility towards the customer with a user-friendly and reliable product	+
	D.R2	Promoter	Product supports the company philosophy and the commitment for a sustainable system change	++

0 representing insignificant importance, + indicating minor relevance, ++ reflecting notable relevance, and +++ signifying crucial importance

with an enhanced function shows the commitment of the company to assume social and ethical responsibility with respect to the environment and customers (D.R1). NewValve complements the overall product portfolio and confirms the company's dedicated aim to be a promoter of a sustainable systemic change (D.R2).

This commitment to assume responsibility is based on clear customer expectations and awareness of negative environmental impacts (D.S2). The expectations of society for sustainable products that nevertheless meet the necessary quality requirements have increased sharply. The combination of sustainability and high quality has strongly influenced product development with the CE approach (D.S3). The change in awareness of sustainability of *stakeholders* influences the image of the company (D.S1). A positive evaluation through the development of sustainable and innovative products from the company ensures long-term success (A. O., personal communication, November 25, 2022).

In the cluster of *financial* drivers, competitiveness is considered a strong driver (D.F3). By reducing the amount of individual parts and using products more efficiently and sustainably, new solutions are created, helping a company to differentiate itself from competitors.

In addition, financial advantages can be achieved by lower transportation and packaging costs and the use of fewer and smaller individual parts in the production process (D.F2). Price fluctuations (D.F1) and additional value of local sourcing (D.F4) are no drivers in this use case. Metal and plastic price fluctuations are approximately the same and no additional benefit can be generated from waste because of quality regulations (D. R., personal communication, November 25, 2022).

6.2.3 Enablers of the Case

Drivers are relevant when choosing a promising field of action, and the successful implementation of the Rs is favoured by certain enablers as described in Sect. 4.4.4. Table 6.2 provides an overview of the individual enablers and their relationship to the use case together with an assessment of the relevance.

The evaluation of enablers in this use case shows that enablers in the cluster *Firm* are strong enablers for the Rs *Refuse* and *Reduce*. The company is strongly focused on the development of sustainable solutions with circular product development (E.F1). This claim is supported by the fact that sustainability is an integral part of the strategy 2025 with concrete and measurable targets in the reduction of such as CO_2 emissions and water consumption. Furthermore, the company needs sufficient capital for the development of sustainable products and the establishment of production processes (E.F2). The NewValve product development involved additional research and sustainability evaluation, which the company must be willing to pay. The co-development with the Oxford University was advantageous for the target-oriented development of NewValve (E.V1). The product design can be tested for performance and sustainability with different experiments, and it can positively

Table 6.2 Enablers of the *Refuse* and *Reduce* case

Cluster	Code	Enabler	Context	Relevance
Firm	E.F1	Culture	Sustainability framework focused on sustainable long-term and responsible corporate entrepreneurial action	+++
	E.F2	Capital	Continuous investment in further product development, new technologies and optimised, resource saving processes	++
Value chain	E.V1	Partnership	Targeted partnership in the development of the product design and functionality	+
	E.V2	Transparency	Sharing of information in the value chain is not present	0
Technological	E.T1	Research	Research and feasibility studies support sustainable product design ans production technology	+
	E.T2	Mutuality	No joint decisions were made with suppliers and customers during product development	0
	E.T3	R-technology	Company has a large product portfolio for pipe systems and know-how in production	+++
	E.T4	Outsourcing	All activities are done by the company itself and are not outsourced	0
Customer	E.C1	Engagement	Understand the existing problems of the customers and eliminate it trough the product solution	++
Material supply	E.M1	Logistics	A reverse logistic network does not exist and is not required for the product	0
	E.M2	Collection	No take-back system in place due to quality conditions and application area	0
	E.M3	Symbiosis	No targeted symbiosis for material suppliers	0

0 representing insignificant importance, + indicating minor relevance, ++ reflecting notable relevance, and +++ signifying crucial importance

influence the further development of the product (E.T1). In addition to the development of NewValve, no other information has been shared with other partners in the value chain (E.V2).

The company has a great deal of know-how in production and assembly and can incorporate this into the sustainable product development and production of NewValve (E.T3). The existing knowledge with this technology supports the development and probability of success. Outsourcing of activities (E.T2) and arrangements with suppliers (E.T4) are not necessary for NewValve. The company produces many other products made of plastic in large quantities and NewValve can be easily integrated into the existing system.

The Rs *Refuse* and *Reduce* are strongly supported by the inclusion of the customer (E.C1). By identifying current problems and wasted resources on the customer side, the current situation can be mapped and the weaknesses identified. The existing problems and the customer's required quality and sustainability goals can be incorporated into the product development (A. O., personal communication, November 25, 2022).

The quality requirements for the product and raw material are very high due to the application area and the resulting safety regulations. The quality requirements can currently only be achieved with virgin raw material. Additionally, NewValve is used mostly in chambers which can be located above or below ground with a lot of pollution and poor access. A reverse logistic or a take back system as well as symbiosis with material suppliers does not exist so far (D. R., personal communication, November 25, 2022).

6.3 Outcome and Impact of *Refuse* and *Reduce*

6.3.1 Ecological Results

The ecological impact of NewValve can be seen in the design of the product itself and through its use. With the innovative design and the choice of material, NewValve is nine times lighter and five times more compact compared to a standard metal PRV [23]. The compact lightweight design with few individual parts makes it possible to manufacture a product with fewer materials and recyclable materials, which have a lower carbon footprint than metal alternatives. Due to the lower weight and volume and the global use of NewValve, transport emissions can be saved (D. R., personal communication, November 25, 2022).

Even greater savings in resources can be achieved during the use of NewValve. Around the world, water supply companies are losing approximately 126 bill. cubic metres of water per year. Reducing water leakage by 5% alone in the United States can save up to 1 bill. litres of water, 313 mill. kWh of electricity and 225,000 tonnes of CO_2 annually [24]. Reduction of water pressure in the pipe systems with NewValve is an efficient way to preventively protect the entire infrastructure from pipe bursts. A pressure reduction of 25% in the water supply network reduces the frequency of pipe bursting by 75%. Reduction of water losses can also reduce the energy consumption of pumps. By reducing water losses, less water must be treated and pumped through the system [25]. Unfortunately, even with sophisticated pressure management systems, pipe bursts might occur. However, with the use of NewValve, water losses can be minimised by reducing the flow rate, and the pipe can be repaired faster and easier [23].

Additional resource savings can be achieved through the extended lifetime of NewValve. Polymers cannot rust and consequently have less abrasion than metals. In addition, the entire water supply network can be protected from unnecessary mechanical stress and increased abrasion by pressure management. The lifetime of the entire water supply network can be increased and resources can be saved due to longer operating times of the pipes.

A practical case of a water supplier in Italy, which supplies water to more than 2.5 mill. customers with a piping network of roughly 20,000 km illustrate the advantages of NewValve: The company had to deal with major problems with water losses of up to 30% on average. With the installation and operation of NewValve, piping bursts in the piping network were reduced by more than 63% [26]. The measurable successes achieved through NewValve are good sales arguments and can provide the company with economic benefits.

6.3.2 Social Results

The social outcomes in this use case are difficult to identify and quantify. However, three different perspectives can be adopted: the company, the water suppliers, and the water consumer. The successful development and market launch of NewValve confirms the company's commitment to developing sustainable products. The pleasant result of the product development has a positive effect on the company culture and the desire to integrate the CE approach into future product developments. However, it should also be said that the product in view of the size of the company and the large product portfolio only has a limited influence and impact on the overall company culture (D. R., personal communication, November 25, 2022).

The social outcome for the water supplier is facilitated by the design of the product and its function. The innovative design allows NewValve to be installed in existing water systems without any construction work. The compact design allows operation on the tightest shafts and under difficult conditions without the need for structural extension [23].

The possibility of connecting a control device on the NewValve with various monitoring sensors and control units provide up-to-date data on the water supply network and enable digital remote monitoring of the flow rate and supply pressure. The control parameters can be automatically adjusted with the recorded data via the actuators according to the customer's requirements. A unique feature of NewValve is the reliable and continuous control of low and high water pressures from 0.2 bar to 16 bar. This function ensures a simple operation of NewValve by the water supplier and creates transparency. Additionally, the large control span as well as the material properties allow the use in many different locations, even underwater. The customer can therefore use the same PRV for the entire water supply network [23].

The social impact for water consumers can be found in water quality and increased service level. Through leaks, unwanted foreign substances can enter the water system and might contaminate drinking water. The reduction of leakage through NewValve leads to a minimisation of impurities in the water and to a high quality of water for the consumers. In addition, water consumers can benefit from consistent water pressure and uninterrupted access to drinking water (D. R., personal communication, November 25, 2022).

6.3.3 Economic Results

The economic results of the company with respect to the NewValve cannot be conclusively assessed due to the short market launch of less than two years. As is common in the industry of system providers, the product must first establish itself and become an integral part of system solutions. Furthermore, customers should be informed about the new product and be convinced of its functionality and advantages. However, with concrete application examples, such as the Italian water supplier, the benefit and added value generated for the customer are confirmed.

The innovative product design of NewValve allows differentiation from the other competitors. Reduced volume and weight, as well as high reliability and durability, makes it possible to offer customers an improved and resource-saving product. The simple design and change of material allow for efficient and cost-effective production. Cost savings are achievable through a streamlined design featuring fewer components, enhanced transport efficiency with reduced packaging and product weight and improved product quality, which directly contribute to positive financial outcomes for the company. (A. O., personal communication, November 25, 2022).

The existing potential and the positive economic results for the water suppliers are rather large. The cost of global water losses per year is estimated at about 39 bills. USD [27]. Water suppliers can save a lot of money with just a small reduction in water losses. In addition to the costs of water losses, indirect material costs such as energy consumption costs from pump operation and fines for noncompliance with water loss requirements by regulators will be avoided. The great advantage is that reduction can be achieved with NewValve in the existing water system without expensive construction work in underground facilities [23].

The configurable product design of NewValve reduces the installation time by up to 40% compared to standard metal PRV. Furthermore, the maintenance requirements of NewValve are low due to the simplicity that facilitates complete maintenance in less than one hour. Low maintenance efforts, longevity, and fewer pipe bursts not only save costs, but will also give water providers more time for other activities. The water system can be examined for leaks and weaknesses and can be further improved by adjusting the parameters.

6.4 *Refuse* and *Reduce*: Summary and Learnings

The application of Refuse (R0) and Reduce (R1) are strongly linked to the design of the product. For this reason, the Rs must be an integral part of the development of the product and its production. The framework with the eight product design strategies can help developers to consider different perspectives for a sustainable product design and incorporate them into the product development.

Evaluation of drivers showed that the reduction of resources and carbon footprint, stakeholders, and corporate responsibility are driving forces to engage in the CE concepts of *Refuse* and *Reduce*. Companies are under increasing pressure to develop sustainable products with fewer resources that meet the required demands of customers. In addition, customers are demanding products that are characterised by simplicity, durability, and reliability. By addressing resource consumption and waste reduction in the product development phase, these requirements can be met.

Evaluation of enablers has shown that the supporting factors for achieving requirements are corporate culture, the availability of technological know-how, and the participation of customers. If the CE approach is strongly embedded in the corporate culture and the know-how and required technologies are available, it is easier to develop and produce high-quality products with fewer resources. However, a sustainable product design does not guarantee success in the market. The product must create additional value and a positive impact for customers.

In the end, the development of products within the framework of the CE approach must be financially worthwhile for the company. The retention options *Refuse* and *Reduce* favour the development and production of products that can deliver financial benefits for the company and for the customers: The cost of the product and the cost of production can be reduced by using alternative and sustainable materials, lower material consumption, and efficient production. With high-quality products that provide high reliability and lifetime, financial benefits can also be created for the customer. Costs can also be reduced by fewer failures and by avoiding replacement of the product after a short period of time.

As can be seen in the described use case, the innovative product design creates the desired benefits for both the company and the customers. It has been shown that with the development of a product with Refuse (R0) and Reduce (R1) in mind, the requirements of stakeholders and customers can be met, and the existing problems in the water supply network and the waste of resources due to water losses can be reduced. By switching the material from metals to plastic, the durability of the PRV has increased and operation under the most difficult environmental conditions is possible. The innovative product design provides additional benefits to the customer compared to other PRV. The low weight and volume, as well as the simple design with few individual parts, allow for a fast installation on existing water systems and easy and fast maintenance. The possibility to connect a control device for the recording of data on water consumption and water pressure allows further optimisation of the water distribution network as well as an automatic, remote-controlled regulation of

NewValve. The use case shows that active integration and implementation of Rs in product development help to develop high-quality and resource-saving products.

The CE approach and the implementation of Rs are not completed once the product is launched on the market. Further technical developments and digitalisation enable further continuous improvements of the products. By building and maintaining close customer relationships, remaining problems with the product and other customer wishes can be incorporated into future developments.

For a holistic implementation of the CE approach, the perspective must be extended beyond the development and production of the product. Reduced resources and the use of other materials can also be applied in other areas such as material procurement and logistics. Furthermore, the decision about the continuation of the use of the product after reaching service life can influence the design of the product. The broader perspective and decisions can influence other Rs positively but also negatively. If, for example, a product is reprocessed and used again after reaching the end of service life, it must be possible to disassemble the product into individual parts. In addition, a take-back system must be set up together with the customers so that the used product can be returned.

Finally, it can be said that Refuse (R0) and Reduce (R1) are crucial for the full application of a CE. The Rs contribute significantly to reducing the required resources, producing high-quality products, and strengthening the market position of a company.

References

1. UNEP. (2019). *Circularity*. UNEP - United Nations Environment Programme. http://www.unep.org/circularity.
2. Potting, J., Hekkert, M., Worrell, E., & Hanemaaijer, A. (2017). *Circular economy: Measuring innovation in the product chain PBL publication number: 2544*. The Hague: PBL Netherlands Environmental Assessment Agency.
3. Reike, D., Vermeulen, W. J. & Witjes, S. (2018). The circular economy: New or Refurbished as CE 3.0? – Exploring controversies in the conceptualization of the circular economy through a focus on history and resource value retention options. *Resources, Conservation and Recycling, 135,* 246–264. ISSN: 0921-3449.
4. Morseletto, P. (2020). Targets for a circular economy. *Resources, Conservation and Recycling, 153,* 104553. ISSN: 0921-3449.
5. Brezet, H., Van Hemel, C., Brezet, H. & Instituut, R. (1997). *Ecodesign: A promising approach to sustainable production and consumption* (1st edn, 346 pp). Paris: UNEP. ISBN: 978-92-807-1631-3.
6. Bhamra, T. & Lofthouse, V. (2007). *Design for sustainability: A practical approach* (184 pp). Gower; Ashgate Pub, Aldershot, England: Burlington, VT. OCLC: ocn173238972. ISBN: 978-0-566-08704-2.
7. Mestre, A. & Cooper, T. (2017). Circular product design. A multiple loops life cycle design approach for the circular economy. *The Design Journal, 20,* S1620–S1635. ISSN: 1460-6925, 1756-3062.
8. Boffey, D. (2018). EU declares war on plastic waste. *The Guardian*. Retrieved October 17, 2022 from https://www.theguardian.com/environment/2018/jan/16/eu-declares-war-on-plastic-waste-2030.

9. Warnick, J. (2022). *Starbucks solidifies pathway to a planet positive future*. Starbucks Stories. https://stories.starbucks.com/stories/2020/starbucks-solidifies-pathway-to-a-planet-positive-future/.
10. Barrett, A. (2022). *Best bioplastics companies*. https://bioplasticsnews.com/2022/01/05/top-10-best-bioplastics-companies-ranking-2021/.
11. Agency, E. E. (Ed.) (2016). *More from less - material resource efficiency in Europe: 2015 overview of policies, instruments and targets in 32 countries*. EEA Report No 10/2016 (151 pp). Luxembourg: Publications Office of the European Union. ISBN: 978-92-9213-737-3.
12. Otto, F., Nerdinger, W., Barthel, R., München, T. U. & der Moderne, P. (Eds.) (2005). *Frei Otto - das Gesamtwerk: leicht bauen, natürlich gestalten ; [anlässlich der Ausstellung "Frei Otto - Leicht Bauen, Natürlich Gestalten" im Architekturmuseum der TU München in der Pinakothek der Moderne vom 26. Mai bis 28. August 2005]* (391 pp). Meeting Name: Ausstellung Frei Otto - Leicht Bauen, Natürlich Gestalten. Basel, Berlin: Birkhäuser. ISBN: 978-3-7643-7233-0.
13. Henkel. (2022). *Saving energy in production*. https://www.henkel.com/spotlight/2020-10-06-saving-energy-in-production-1118478.
14. Henkel. (2022). *Sustainability Report 2021*. Henkel. Retrieved February 23, 2022 from https://www.henkel.com/sustainability/sustainability-report.
15. Fedewa, D., Swan, D., Teichner, W. & Wiseman, B. (2022). *Skinny design: Smaller is better*. McKinsey & Company.
16. Apeel. (2022). *Working with nature to make food last longer: Apeel*. Ellen Macarthur Foundation. Retrieved, October 7, 2022 from. https://ellenmacarthurfoundation.org/circular-examples/apeel.
17. Georg Fischer. (2022). *COOL-FIT 4.0*. GF piping systems. Retrieved, December 2, 2022 from https://www.gfps.com/de-de/products-solutions/systems/cool-fit-4.html.
18. Mhapankar, M. (2023). *Instant Pot Pro*. Instant pot. Retrieved, January 1, 2023 from https://www.instanthome.com/product/instant-pot/pro/6-quart-multi-use-pressure-cooker.
19. Mobility. (2022). *Mobility Jahresbericht 2021*. Retrieved January 1, 2023 from https://www.mobility.ch/fileadmin/files/documents/annual-reports/Mobility-Jahresbericht_2021.pdf.
20. Vezzoli, C., & Manzini, E. (2008). *Design for environmental sustainability*. London: Springer. ISBN: 978-1-84800-163-3.
21. Collab. with Colaco Maruta Mestre, A. & Brezet, J. (2014). *Cork design: A design action intervention approach towards sustainable product innovation*. OCLC: 945104196. Delft: Delft Academic Press. ISBN: 978-90-6562-365-2.
22. Georg Fischer. (2022). *Product Range PE Systems .pdf*. https://gf.showpad.com/share/hKr8zaTM5vpHH9N8q3wEq.
23. Georg Fischer. (2022). *Wasser im Gleichgewicht*. Retrieved January 3, 2022 from https://www.gfps.com/com/de/products-solutions/innovation/neoflow-pressure-reducing-valve/neoflow_gc.html.
24. Karnib, A. (2016). *A quantitative assessment framework for water, energy and food nexus*. Retrieved November 29, 2016 from http://www.fao.org/nr/water/aquastat/data/query/index.html?lang=en.
25. Georg Fischer. (2021). *Solving water loss for life*. Retrieved November 5, 2022 from https://www.gfps.com/content/dam/gfps/com/brochures-and-flyers/de/gfps-00068-brochure-non-revenue-water-de.pdf.
26. Georg Fischer. (2021). *Sustainability Report 2021*. Georg Fischer AG. Retrieved October 17, 2022 from https://sustainability-report.georgfischer.com/21/en/.
27. Liemberger, R. & Wyatt, A. (2019). Quantifying the global non-revenue water problem. *Water Supply, 19*. Publisher: IWA Publishing, 831–837. ISSN: 1606-9749.

Chapter 7
R2 Reuse

Roger Rinderer and Katharina Luban

Abstract The concept of *Reuse* has many challenges concerning the collaboration between partners, logistics and customers. In this chapter, we will analyse drivers, enablers as well as ecological, social and economic results of a successful *Reuse* case. Learnings will be derived, generalised, and transferred to other cases.

7.1 Introduction to *Reuse*

According to Reike's 10 retention options of Circular Economy (CE) [1], the option *Reuse* (R2) marks the first option that goes beyond the absolute avoidance or reduction of resource consumption (see Chap. 3 for details). *Reuse* describes a circle which, from a consumer perspective, implies 'buying second hand or finding a buyer for a product that was not or hardly in use, possibly after some cleaning or minor adaptations for quality restoration by the consumer' [1, p. 255].

Reuse marks the first real loop and refers to the Retention Option (R), where the highest form of the initial value of a product is retained. According to the literature, *Reuse* is a well-known concept, but it is difficult to identify a unique definition. This is because attributes such as the extend of reprocessing before the change of ownership are sometimes included and sometimes excluded.

According to Reike et al. [1], *Reuse* most commonly applies to a second ownership of a product, which hardly needs any adaptions and works 'as new' [2], 'with the same purpose' [3], 'without refurbishment' [4], or 'without repair' [1, 5]. We also find the notion of *Direct Reuse* [1, 6, 7], referring to controlled collection, inspection, testing, cleaning, and delivery of a rarely used product back to the market. Hence, *Direct Reuse* is defined by a second or more lifecycles of a product under a change of ownership, that hardly needs any reprocessing to work as new for the same purpose.

R. Rinderer
Flums, Switzerland

K. Luban (✉)
Sevelen, Switzerland
e-mail: katharina.luban@ost.ch

A. Podleisek et al. (eds.), *Circular Economy: The Next Level of Company Success*,
https://doi.org/10.1007/978-3-031-66406-9_7

Reuse in this chapter is defined in the sense of *Direct Reuse*, referring to the reuse of products, where no additional production processes are allowed, but small aesthetic retouchings, such as cleaning or minor repair, can be applied. In case of adding value enhancement steps before reintroducing the product into the market, we talk about the retention options of *Repair*, *Refurbishment*, *Remanufacture* and *Repurpose*.

A good example to illustrate the different business models of (direct) *Reuse* and *Repair*, is the outdoor clothing company Patagonia. Patagonia's 'Worn Wear' programme promotes the direct reuse and repair of clothing items rather than buying new ones. Regarding the *Reuse* strategy, Patagonia encourages customers through their online platform to buy and sell used clothes, which are still in good condition. In case a piece is damaged, Patagonia either offers a repair services or provides the resources and advice to customers if they want to repair their own clothes at home [8].

A good example of the distinction between *Reuse* and *Refurbish* is the business case of the Swedish furniture company Ikea. Ikea has a programme called 'Buy Back', which encourages customers to sell their used Ikea furniture back to the company. The company then has the option to resell the products either directly (*Reuse*), after mending (*Repair*) or after improving certain parts of the products (*Refurbish*). The difference between these options lies within the share of reprocessing before reintroducing the product into the market.

An important facilitator for all *Reuse* options is the market place. In addition to flea markets and second-hand shops, online platforms such as eBay or Amazon have gained major importance. Whether from a Business-to-Business (B2B), Business-to-Consumer (B2C), or Consumer-to-Consumer (C2C) relationship, online platforms offer the opportunity to buy and sell used or pre-owned products. eBay, with 134 mill. active buyers and a total of 1.7 bill. listed products in 2022, is estimated to sell 80% new and 20% reused or refurbished products [9].

7.2 Analysis of a *Reuse* Case

7.2.1 Description of the Case

To fully understand the *Reuse* strategy within CE, a particularly successful use case will be discussed in the following text. The *Reuse* of food-packaging material is a current theme. Therefore, the case of a startup called kooky (kooky) was chosen as the case for analysis.

Kooky provides an alternative solution to disposable tableware and offers a reusable cup system for takeaway cups with an own restitution infrastructure. The first system was launched in 2021, and by 2022 kooky was able to attract over 20,000 users in only one and a half years. Since then, kooky has grown significantly and currently employs around 40 people.

The fact that kooky was launched as a startup does not influence the learnings as it is still imaginable that a company would consider introducing *Reuse* as an option for their existing products.

To reveal the significant factors that have contributed to the successful execution of the *Reuse*-based business model, this chapter provides insight into the central drivers and enablers of CE as well as results that were achieved through implementation. During the analysis, interviews were conducted with T. B., CEO, and M. H., General Manager Switzerland of kooky. According to T. B., the startup pursues the vision to have a lasting impact on the future of reusability and enable waste-free consumption by providing a leading circular infrastructure (personal communication, November 28, 2022).

Infrastructure

The infrastructure of the kooky business model consists of an initial set of kooky cups, a restitution station, and a smartphone application (see Fig. 7.1). The kooky cup consists of three different items: The cup itself, a lid, and a chip for tracking and data transmission. The cup and lid are made of toxin-free recyclable Polypropylene (PP) produced by an established reusable cup manufacturer in Germany. Currently, the size of the cup is limited to 2–3 dl, but as stated by M. H., the goal is to provide additional sizes up to 5.5 dl for hot and cold beverages. There is also interest in using a mineral oil-free bio-compostable version of the cup in the future. The chip integrated into the cup is based on Near Field Communication (NFC) technology and ensures that every cup is traceable. The restitution stations are made of steel to prevent vandalism and are powered by replaceable battery packs.

The third element of the infrastructure is the smartphone application. This application is the mastermind behind the kooky concept and controls the whole usage process and the monetary flow for each kooky user. The concept of how the usage of the kooky concept works will be described in the following by means of a customer journey when buying a takeaway coffee.

Concept of Reusable Cups

The journey of a user who wants to avoid disposable coffee cups begins with finding a kooky partner shop on a map within the kooky app. At the shop, the user buys a takeaway coffee and has to pay 1 CHF as deposit for a kooky cup instead of getting a standard disposable cup for free. After the purchase process, the user assigns the cup to the personal account by scanning a QR code printed on the cup with the app. This personalising is necessary to be able to return the cup at one of the restitution stations scattered throughout the city.

Having finished the drink, the user can use the kooky application to find one of the return stations. After returning the cup, the deposit of 1 CHF will be automatically credited to the account. At the moment, the user can either transfer the credit to its bank account or donate the credit to the World Wide Fund for Nature (WWF). According to T. B., there will be further options in the future such as different loyalty programs or using the credit to buy another cup (personal communication, November 28, 2022).

Fig. 7.1 Infrastructure of kooky [10]

After the user has returned the cup, kooky takes care of the Reverse Logistics (RL) of the cups. A kooky courier collects all the cups by e-bikes and delivers them for cleaning. The cleaning of the cups is done in cooperation with partner companies such as canteens, which already have a professional and resource-efficient cleaning infrastructure. After cleaning, kooky takes care of the logistics to transport clean cups to partner coffee shops again. (T. B. and M. H., personal communication, November 28, 2022).

Establishing the Case

Establishing a successful CE case is a step-by-step approach, as was - in this case - the establishment of the startup kooky.

At first, kooky started a four-month pilot project together with the largest Swiss public transport company, Swiss Federal Railways (SBB). Zurich main station was predestined for a trial because it is frequented by many commuters and the infrastructure consisted of several shops with takeaway services. The success of the pilot project led to a unique multi-year partnership throughout Europe with SBB. In June 2022, kooky introduced the concept at additional Swiss train stations in Bern and St. Gallen. In addition to the partnership with SBB, kooky was also able to introduce its concept to other public locations, including four hospitals and three universities.

In July 2022, kooky established a partnership with Valora Holding AG (Valora). Valora is one of the largest suppliers in the food and beverage sector at high-frequency locations in Switzerland. Is has 2,700 sales points, including chains such as avec, k kiosk, Brezelkonig, CaffeSpettacolo and SuperGuud [11]. These shops are located in train stations, as well as in highly frequented urban areas.

The next hallmark was the expansion of kooky into the city centre of Bern in October 2022. Bern is pursuing a cleanness initiative, in which participating local gastronomy and retail businesses contribute to reducing waste in public places and increasing awareness of waste in the community [12]. By providing a part of the solution for this initiative, kooky has also come into discussions with retail giants such as Coop, Migros and Lidl.

One year later, another milestone was reached. Kooky won the German gastronomy provider YORMA's as a customer and was able to expand its business to Germany into the city of Munich. This international expansion was triggered by discussions about a new reusable packaging law in Germany, which requires restaurant owners to offer reusable tableware as an alternative to disposable packaging for takeaway products. This results in additional work for the restaurant owners, as the cups and dishes have to be cleaned according to the high hygiene standards in Germany. The fact that many restaurants do not have the ability to professionally clean reusable tableware according to these standards opens up an enormous potential for the solution that kooky provides.

Currently, kooky is negotiating to expand its concept not only to further locations in Germany but also to Austria. Regarding the fact that the kooky cups provide a sustainable solution to standard disposable cups, it is clear that the potential of growth is enormous and the story has just begun (T. B. and M. H., personal communication, November 28, 2022).

7.2.2 Drivers of the Case

Progress, development, and innovation do not take place on their own. We call the forces behind these advances 'drivers'. Driver can be created intentionally or can originate from circumstances.

To learn from the kooky case on how to implement a *Reuse* option successfully, the drivers of this case were carefully identified and analysed. The discussion considers all drivers that make the *Reuse* case successful. Thus, it is irrelevant that the case was realised by a startup, since the drivers refer to the case itself and not to the legal entity that realised it.

The analysis was carried out using the general framework of drivers of CE, as proposed in Chap. 3. The general drivers will be classified as relevant or not relevant to the case. Those drivers with the most significant impact on the success of the kooky *Reuse* case will be pointed out. The results are shown in Table 7.1 and will be explained in the following.

Within the group of drivers that arise from *Environmental* issues, the main driver is clearly the reduction in resource consumption (D.E1). The whole idea of kooky is based on the ambition of reducing the resources needed for disposable cups in takeaway restaurants and cafés. The idea of 'less resource consumption' is also considered in the overall processes necessary to realise the case, for example, in the logistics (using e-bikes instead of combustion engines). Kooky also emphasises the resource-efficient cleaning of returnable cups by using existing and certified washing lines in canteens. This canteen cleaning infrastructure is used only in short periods of the day, when they process large numbers of tableware and cutlery in

Table 7.1 Drivers of the *Reuse* case

Cluster	Code	Driver	Context	Relevance
Environmental	D.E1	Resources	Reduction of resources used to produce disposable cups and a resource-conserving cleaning process	+++
	D.E2	Footprint	Reduction of the logistic footprint by using e-mobility	++
	D.E3	Risk	The case does not address any existing resource-related risks	0
Institutional	D.I1	Legal	Comply with legal requirements regarding reusable packaging in the European Union	++
	D.I2	Punitive	Avoid punitive measures on disposable packaging	++
	D.I3	Juridification	Anticipate and meet future legal obligations regarding on the traceability of cups and on reusable packaging	++
	D.I4	Tax	Access to supportive funds from the government	+
Stakeholder	D.S1	Image	Improve corporate image by providing a sustainable solution for takeaway cups	+
	D.S2	Customer	Adapt to new customer expectations by providing a sustainable solution for disposable cups	+
	D.S3	Practices	Follow the best practices of competitors is not represented in the first place	0
Financial	D.F1	Volatility	Price volatility of raw material is not necessary, because the materials used have no high values	0
	D.F2	Savings	Providing cost saving opportunities through a comprehensive introduction of the kooky concept	+
	D.F3	Competitiveness	Innovation according to the idea of the restitution infrastructure increases competitiveness	0
	D.F4	Value	Create value from waste and production side streams is not a driver of kooky in the first place	0
Responsibility	D.R1	CSR	Social and ethical responsibility in connection with awareness sensitisation	+++
	D.R2	Promotor	Become a promoter of systemic change in connection with awareness sensitisation	+++

0 representing insignificant importance, + indicating minor relevance, ++ reflecting notable relevance, and +++ signifying crucial importance

short times. Thus, except during peak hours, the canteens have free capacity, which is used to clean the kooky cups. Additionally, kooky also aims to further reduce their material consumption by using the recycled material of the kooky cups to manufacture the housing of the restitution stations. According to T. B., kooky has also plans to operate their restitution stations as self-sufficient systems, powered by a decentralised supply of renewable energy in order to reduce electricity consumption. This could be provided by a wind power or photovoltaic system that feeds the surplus electricity into its own battery-supported storage system (personal communication, November 28, 2022).

A second relevant driver emerges within the cluster of environmental drivers, namely the reduction of a product's environmental footprint (D.E2). Kooky relies on e-mobility to perform all logistics tasks, which not only saves resources (fuel) but reduces the CO_2 footprint of their logistic operations.

Regarding the third environmental driver, no evidence was found that the kooky case addresses any resource-related risks (D.E3).

Looking at *Institutional* factors that drive the case, the European Union (EU) Directive on single-use packaging [13] and subsequent national laws on reusable packaging such as those in Germany [14] clearly are a driver. This law requires that all catering establishments offer reusable packaging. According to T. B., the EU has set the target of phasing out 20% (2030) up to 75% (2040) of disposable packaging (personal communication, November 28, 2022). The kooky concept enables its partners to comply with the new law (D.I1).

In Switzerland, there are no plans to ban single-use food packaging, but efforts are being made to introduce a punitive tax on packaging. Companies that sell packaging would be obligated to pay a penal tax of 0.01 CHF per package [12]. According to an estimation by M. H., some 25 mill. disposable cups are being used per month in Switzerland. This would sum up to penalties of approximately 2.5 mill. CHF per year, which can be avoided by implementing the kooky concept (personal communication, November 28, 2022). A similar concept of punitive taxes on packaging has already been realised in some German cities, e. g., the city of Tübingen [15]. The *Reuse* case allows companies to avoid these punitive measures (D.I2).

Kooky also has an answer to future legal obligations (D.I3). According to T. B., the EU could require the traceability of the cups in circulation in the future. Kooky considered this and can offer traceability due to the NFC chip in the kooky cup.

By introducing a reusable system, the partners receive a monetary incentive from the government (D.I4). For example, in Munich the incentive amounts to a one-time payment of 400 EUR, and in Basel there is also a financial contribution for partners who use the kooky concept (M. H., personal communication, November 28, 2022).

The analysis of drivers within the cluster *Stakeholder* showed that the usage of kooky products increases the level of reputation for companies (D.S1) and enables companies to adapt to changing customer behaviour and meet new customer requirements (D.S2), by providing a sustainable solution for disposable cups. T. B. explains, that in society the awareness for the necessity for sustainable processes has increased significantly and is also actively demanded by stakeholders (personal communication, November 28, 2022).

Since kooky is a first mover with its infrastructure, no evidence was found that following best practices of competition or the need to meet existing standards was a driver of kooky (D.S3).

Regarding the *Financial* drivers, some evidence was found. However, the kooky innovation is not driven by the need to avoid the volatility of the raw materials price (D.F1). Kooky rather provides a cost-saving opportunity to the society (D.F2). Disposal costs for the littering of takeaway coffee cups in Germany are estimated to be around 120 mill. EUR per year [16]. According to T. B., with an area-wide use of the kooky concept, these costs will be significantly reduced (personal communication, November 28, 2022).

The need to enhance competitiveness through innovation is not seen as a driver for kooky in the first place (D.F3), although kooky is aware that further innovation, such as the self-sufficient restitution infrastructure, will be necessary to stay ahead of the competition.

Similarly, the creation of value from waste and production streams does not affect the *Reuse* case of kooky (D.F4).

According to the CEO, T. B., financial success is the driving force of most start-up ideas and, of course, this is also true for kooky. However, overall financial drivers are rated less important. Kooky primarily wants to 'do good'. Due to this, the drivers of the cluster *Responsibility* (D.R1, D.R2) are considered key drivers. With its concept, kooky wants to assume social and ethical responsibility and has the desire to become a promoter of systemic change by increasing the awareness of waste reduction in society (T. B., personal communication, November 28, 2022).

7.2.3 Enablers of the Case

Enablers are factors that influence the successful implementation of a retention option. Their absence might be perceived as barriers. In the following text, the enablers that supported the implementation of the kooky case will be analysed. The results will be explained in detail and are summarised in Table 7.2.

Within the first cluster the analysis addresses enablers within the context of the *Firm*. Independent of the organisational structure of an entity – be it a startup or within a company – it becomes obvious that it is important to have a determined corporate culture (E.F1), but it is equally important to represent this culture to the outside world when introducing products and concepts that are 'new to the world'. According to T. B., kooky teamed up with well-known brands such as the WWF organisation to convey their philosophy and joined various podium discussions to gain a high social media presence (personal communication, November 28, 2022).

Another obvious main enabler for kooky was financial support or – in the case of a startup – the capital provided from outside (E.F1). T. B. states that investors such as UVC Partners and Verve Ventures as well as many individuals invested a total value of 6.2 mill. EUR in the startup (personal communication, November 28, 2022).

Table 7.2 Enablers of the *Reuse* case

Cluster	Code	Enabler	Context	Relevance
Firm	E.F1	Culture	A committed company environment and culture is important for every startup	++
	E.F2	Capital	To transition, investors invested a total value of 6.2 mill. EUR in the startup	+++
Value Chain	E.V1	Partnership	Since kooky is not in direct contact with the user, various partnerships with close relationships and low entry barriers are essential for the success	+++
	E.V2	Transparency	Enhanced information sharing and transparency are not represented in the first place	0
Technological	E.T1	Research	Some research was done, but does not count as an enabler in the first place	0
	E.T2	Mutuality	The design of the cups and lids is based on a joint decision-making with partners	+
	E.T3	R-technology	There was no special technical and technological know-how to realise the case	0
	E.T4	Outsourcing	Although research activities and production are outsourced, this is not a crucial enabler of the case	0
Customer	E.C1	Engagement	Increased customcombustioner engagement leads to a higher usage rate and is a key aspect of the success from the demand side as well as from an environmental point of view	+++
Material Supply	E.M1	Logistics	A reverse logistic network is implemented	+
	E.M2	Collection	A simple collection or take-back system is included in the core of the business model since the beginning	+++
	E.M3	Symbiosis	Engage in industrial symbiosis by using the synergies of large washing lines in canteens for efficient cleaning of components	++

0 representing insignificant importance, + indicating minor relevance, ++ reflecting notable relevance, and +++ signifying crucial importance

The kooky case is a Business-to-Business (B2B) case, since kooky does not directly work with the end-users of the coffee cups. Therefore, value creation takes place within the *Value Chain*, which is considered another important cluster for enablers: Partnerships (E.V1) are crucial to kooky's business model, namely the shops that offer the kooky concept or their headquarters. T. B. acknowledges that the cooperation with SBB and Valora was a relevant enabler (personal communication, November 28, 2022). In addition to serving customers, kooky seeks to support the goals of its customers to deepen the partnership. In this sense, the kooky concept supported the goal of the SBB to achieve waste reduction at Swiss train stations, which was and remains an essential point in the SBB agenda. The main factors that fostered the partnership with Valora were the availability and convenience offered by kooky. Since partnerships are a significant enabler for success, kooky focusses on close relationships and collaborations with low entry barriers for new partners (T. B., personal communication, November 28, 2022).

Within the cluster *Value Chain*, Transparency is considered a possible enabler. However, no evidence was found that enhanced information sharing is crucial for the success of kooky (E.V2).

The third cluster of enablers lies in *Technological Aspects*. The analysis showed that research had to be conducted to develop the infrastructure, especially the integration of the NFC-Chip in the cup. However, this was not considered an underlying enabler of this case (E.T1). Kooky does rely on a joint decision-making and includes the partner in the design process of cups and lids. This ability to mutuality is seen as a nice feature to have, which provides some additional competitive advantage (E.T2). The enabler 'R-technology' refers to the fact that a certain technology might be required to realise the retention option. No evidence was found for this (E.T3). Outsourcing (E.T4) on the other hand, can be observed: Since kooky does not have its own research department and there are no production cells, research activities and the production of cups and the restitution station are outsourced. However, this move is not considered a significant enabler for this *Reuse* case (T. B., personal communication, November 28, 2022).

Although kooky has no direct contact with end customers, their wishes must be considered when it comes to the enabling cluster of *customer*. Increased engagement and end customer satisfaction is the prerequisite for an increased usage rate of the cups. Only broad acceptance by end customers will lead to better environmental balance and enhance attractiveness on the market. With this, it is clear that customer engagement is a key enabler of the success of the case. To play this card, kooky attempts to exploit the current trend towards sustainable processes in society by pointing out the user's opportunity to actively participate in the trend. Furthermore, kooky plans to strengthen the relationship with end customers by introducing incentive systems such as loyalty programmes or gamification (M. H., personal communication, November 28, 2022).

Regarding the last enablers of *Material Supply*, another major enabler becomes evident. By providing a returnable cup system for takeaway cups, kooky has built RL into its core business. The collection of dirty cups and the return of clean cups to stores are provided by thoughtfully designed logistics processes, which are important but not crucial (E.M1). On the other hand, providing the collection infrastructure is one of the main enablers, as the restitution stations had to be designed and installed in the right places to make the case successful.

According to M. H., the cleaning of the cups in collaboration with canteens is an important enabler and was based on a strategic decision. This symbiosis (E.M3) secures mutual benefits, since it allows kooky to avoid costly investments in infrastructure. The canteens, on the other hand, increase the efficiency of their infrastructure and can realise new sources of income by collaborating with kooky (M. H., personal communication, November 28, 2022). Another possible symbiosis could be the use of local facility management at train stations or universities to empty and clean the restitution stations. This would allow kooky to concentrate their core business to logistics processes (T. B., personal communication, November 28, 2022).

7.3 Outcome and Impact of *Reuse*

7.3.1 Ecological Results

Due to the fact that kooky is still a growing startup, it is difficult to identify measurable impacts of its *reuse* case on a nationwide or even global level. However, kooky clearly has the selling proposition of reducing the ecological impact compared to disposable paper cups.

To evaluate the effective environmental impact of the kooky concept including its cups, kooky executed a product-based Life Cycle Assessment (LCA). According to Haessig et al., LCA is a methodology to calculate the environmental impact at the corporate level (corporate life cycle assessments) or at the product level (product life cycle assessments) [17]. kooky's product LCA was conducted in collaboration with the Institut für Wissen, Energie und Rohstoffe Zug (WERZ) at Eastern Switzerland University of Applied Sciences (OST). The underlying question of this LCA was to evaluate the number of cycles that a kooky cup must live to cause less environmental impact than a disposable paper cup [18]. Some of the main messages of the LCA report will be elaborated in the following.

In Switzerland, each resident is responsible for 700 kg of household waste per year [19]. This leads to more than 6 mill. tonnes of household waste per year. The share of takeaway coffee cups in these amounts is small, but is growing steadily. Based on German research data adapted for Switzerland, it is estimated that 260–290 mill. single-use cups are used and thrown away in Switzerland every year [20]. This equals an estimated 34 cups per inhabitant and generates 2,700 tonnes of paper cups waste. The potential to reduce the environmental impact by switching to a reusable cup system is great [21].

To calculate the impact of the kooky cup, two Key Performance Indicators (KPI) were used: The Global Warming Potential (GWP) and the Environmental Impact Points (EIP). GWP summarises the gas CO_2 and other gases with an impact on the climate and is measured in the mass of gases (kg CO_2 equivalents) over a certain period of time, usually 100 years [18]. EIP aims to evaluate environmental impacts in general. Therefore, this KPI takes into account many different environmental factors such as eutrophication, water, air and soil pollutants with ecotoxic or human-toxic effects. Also, resource consumption, energy, climate, land use, and many more aspects can be included. These areas of environmental impact are summarised and weighted according to the sustainability goals and commitments of the Swiss national policy [18].

Depending on the KPI used in the LCA, we get two slightly different answers as to the number of reuse cycles that bring an ecological benefit compared to disposable single-use paper cups. The results of the GWP are presented in Fig. 7.2, which states that the kooky cup with lid needs to be reused at least seven times to brake-even with the GWP of the disposable cup. The assessment also revealed that for cups without lid, the number increases to 17 cycles [18]. The LCA results highlight the ecological hot spots of the kooky cup, which lie in the production of the PP and NFC chips as well as in the electricity consumption during the cleaning process in the use phase.

The results of the EIP show an ecological break-even point for the cup with lid at 26 cycles of reuse. As with the GWP, avoiding the lids will substantially increase this number. In general, EIP highlights the same environmental hotspots as GWP. Production accounts for around 67% of the EIP. Within this, the NFC chips pose a higher impact in the EIP than in the GWP. Here, NFC chips account for approximately 40% of the impact. The use phase, including the washing process, accounts for around 44% of the EIP. In the EIP, many factors are taken into account in addition to the electricity consumption in the washing process, such as detergent, freshwater and Wastewater. That is why the washing process has a higher impact in the calculation of the EIP than in the GWP. In both cases, the switch to renewable energy for the washing process would reduce the impact by approximately 20–30%.

Currently, kooky has several tens of thousands of cups in circulation with a very high return rate. They could already reach the ecological break-even point for cups with lid, at either seven cycles (GWP) or even 26 cycles (EIP). In fact, the cup is designed to handle up to 500 cycles of reuse. Based on observations and practical experience so far, it is expected that the cups will achieve around 80–100 cycles through use. As shown in Fig. 7.2, the kooky cup, therefore, has the potential to reduce the environmental impact of disposable cups by 82% (GWP) or 41% (EIP). For single partners, the reduction rate was shown to be 100% of disposable cups when completely switching to the kooky concept. These figures show a significant positive measurable environmental impact of the case, not to mention the improvement it can bring to waste volumes and the avoidance of littering. The impact on the total waste volume is difficult to quantify. However, T. B. expects that the reusable cup system has the potential to save up to 30% waste volume in the public sector (personal communication, January 2, 2023).

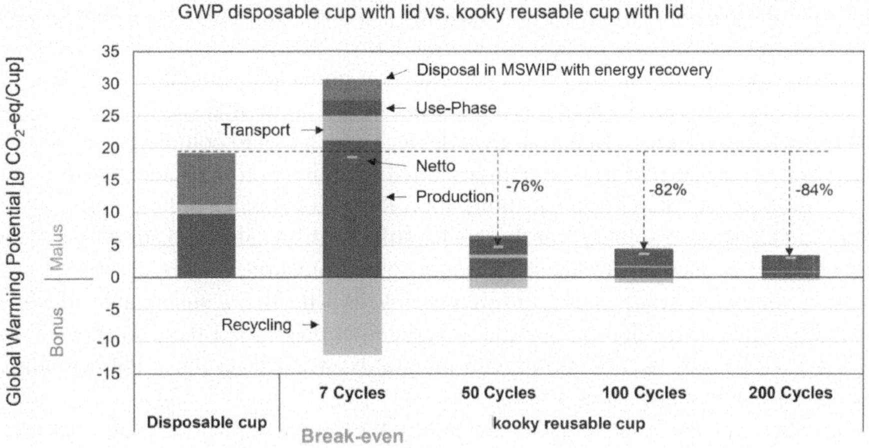

Fig. 7.2 GWP disposable cup with lid versus kooky reusable cup with lid (adapted from [18])

7.3.2 Social Results

As previously analysed, kooky fosters partnerships with its B2B customers by addressing their needs for waste reduction, securing legal compliance, and helping to live up to customer expectations. In order to highlight their contribution, kooky offers its customers a service to create reports on Environment, Social and Governance (ESG). This type of reporting is becoming increasingly important as investors, consumers, and other stakeholders are paying more attention to a company's impact on society and the environment. Reporting ESG can help companies communicate their responsibility and commitment to sustainability [22]. The focus on sustainability can strengthen a company's competitive advantage and acceptance in the society, and the support of the ESG reporting allows kooky to highlight and quantify the benefits of introducing a reusable system (T. B., personal communication, January 2, 2023).

Based on its partnerships with Valora, SBB and YORMA's, kooky has been able to introduce the reusable system in more than 250 locations so far and more are to come. They have increased awareness of disposable cups as an avoidable waste and have shown a viable solution for end-customers to participate in the sustainability movement. T. B. states that kooky has been able to attract more than 20,000 active and regular users so far, although users have to make an effort to download the app, register, deposit 1 CHF, scan the cup, find a restitution station, and drop off the cup. However, T. B. does not consider this a significant entry barrier, since digital tools and solutions have become increasingly accepted. T. B. is confident that the added value that a sustainable solution offers to users will exceed the effort (personal communication, January 2, 2023). This is especially true for certain groups of users that kooky tries to identify by analysing various perimeters such as demographic key figures or the population density per takeaway provider. Naturally, these insights and conclusions will lead to new application areas (so-called kooky spots) to further drive the penetration of the market (T. B., personal communication, January 2, 2023).

Regarding the social impact, critical voices may argue that the kooky concept undermines the idea of 'bring your own device' which would completely avoid any waste and should be preferred as a higher-rated retention option. According to T. B., kooky does not see itself as a substitute for a *Refuse* or *Reduce* option, because the idea is an alternative solution for those who still use disposable cups or do not want to carry their own cups with them (personal communication, January 2, 2023).

As a company, kooky seeks to hire employees who live a sustainable lifestyle themselves and carry the message of waste reduction beyond their workplace. In addition, kooky has been working with integrative workplaces since the beginning to create a positive impact on society.

In summary, it can be said that for kooky the social impact on society is very closely related to the ecological impact, but its contribution is also seen in its efforts to engage more and more people in the concept of sustainability and in the fact that kooky is a sustainable company itself.

7.3.3 Economic Results

Despite the positive ecological and social impacts and outcomes, no innovative idea will persist without economic success. In a startup, the payback might be even easier to track than for a project in larger companies, and kooky shared some valuable insights on this topic.

As mentioned above, there has been only one investment round for kooky so far. Investors such as UVC Partners, Verve Ventures and other individual investors have invested a total value of 6.2 mill. EUR in the business case. However, since the law on reusable packaging came into force in Germany in January 2023, kooky is planning another round of financing to raise the capital needed to penetrate the German market (T. B., personal communication, January 2, 2023).

At first, kooky used parts of their capital to hire employees and to research and develop infrastructure components consisting of cup, restitution station and app. Integration of the NFC chip into the cups was one of the main areas of technical development (T. B., personal communication, January 2, 2023).

In the course of time, kooky invested more capital in improving the components and implementing the infrastructure, because the key to success and therefore the focus was generally on improving usability and allowing more convenience for users through accessibility to the infrastructure. With regard to infrastructure implementation, the number of restitution stations in operation is around 300 by now.

Subsequently, a redesign of the restitution station led to a significant reduction in manufacturing costs. In addition, predictive maintenance was included to reduce the failure rates of the stations.

Other investments include the purchase of electronic cargo bikes that are used to collect cups. The collection concept itself was also revised and optimised to avoid additional costs.

Further investments were made in a marketing concept with the aim of achieving increased visibility and creating awareness. This has also contributed to positive brand development (T. B., personal communication, January 2, 2023).

Consequently, the investments made have led to an exponential increase in both the number of users and the number of cups returned. As a result, kooky was able to grow strongly in the Swiss market and gain market share. As kooky has gained importance, the negotiation power in purchasing cups and stations, as well as in cleaning, has also improved. Thus, kooky can secure price advantages and benefit from economies of scale.

Their attractiveness to partners has also increased, and kooky considers using this fact to realise a change in their pricing policy: In order to cover all the expenses for cleaning, collecting and overhead, kooky currently charges a service fee of 10% of the purchase price from the users. According to T. B., kooky would prefer that users use the reusable system as barrier-free as possible. For this reason, kooky considers the option to be financed from the partner side and no longer from the user side (personal communication, January 2, 2023).

Overall, it can be observed that the market success of kooky and the increasing number of users have had a noticeable effect on revenue. Kooky is on its planned path to break-even and become profitable.

7.4 *Reuse*: Summary and Learnings

The first finding of this chapter was the diverse definitions that exist with respect to the retention option *Reuse*. Depending on the framework, processes such as *Repair*, *Refurbish* or *Repurpose* are subordinated to the term *Reuse*. However, the focus within this chapter is placed according to the 10R framework and considers *Direct Reuse* only. *Direct Reuse* is defined as the second life cycle of a product under a change of ownership that barely needs any reprocessing to work as new for the same purpose. This means that minor repairs or cleaning are allowed, but no major repairs or refurbishments.

Drivers and enablers were worked out with respect to a specific *Reuse* business case. To provide an alternative solution for disposable cups, the startup kooky launched the first returnable cup system for takeaway cups with a proprietary restitution infrastructure. One of the main drivers for this use case was found within the environmental aspects, evolving from the overall idea of reducing the resources needed to produce disposable cups. Developing and implementing the *Reuse* concept was also driven by the desire of the founders to assume social and ethical responsibility and to become promoters of systemic change by increasing the awareness for finite natural resources. Another important driver in this case is the need for companies to comply with existing or future legal obligations regarding reusable packaging.

The presented *Reuse* case enables the sellers of takeaway drinks to comply with new legal requirements and avoid punitive measures.

Without strong enablers, a good idea will not sustain. Funding is an important enabler for most innovations, but becomes crucial for startups. The kooky idea was supported by investors, which enabled the development and the roll-out of the infrastructure. The success of the startup was also possible due to valuable partnerships with main customers, which enabled the rapid roll-out of selling and restitution points. Kooky fosters these partnerships by looking for mutual benefits and by offering a simple and sustainable all-inclusive solution to its customers. In addition to this, the case not only succeeded in direct customer relations, but a third major enabler was the fact that the presented *Reuse* case caught the awareness of end-users as well. Kooky definitely benefits from the trend towards sustainability in the society. In a *Reuse* case, a special focus must be placed on logistics topics. Material supply proves to be important but not critical, but the collection of used cups is a crucial enabler. Kooky defined a simple take-back system with a proprietary restitution infrastructure into the core of its business model. A fifth important enabler is the symbiosis that kooky engages in: To realise efficient and environmentally friendly cleaning, kooky takes advantage of the synergies of large washing lines by collaborating with canteens.

The results of the presented *Reuse* case are very positive and can even be quantified in many cases. The environmental impact of the reusable cup system is better compared to single-use cups as soon as 7 (GWP) or 26 (EIP) reuse cycles have been reached. With tens of thousands of cups in circulation and a high return rate, this can realistically be achieved. The social impact on society is very closely linked to the sustainable impact and the business success: By providing a reusable system, the awareness of waste avoidance will be increased. To drive this awareness even further, kooky invests effort and money in increased visibility, comfortable usability, and further growth.

Scaling also has a positive impact on the business side: It leads to increased revenue and reduced costs. Kooky is still in the early stage, but is on its way to profitability. The analyses show enormous economic potential.

Overall, it can be said that the kooky case shows encouraging results and can serve as a role model example for the retention option *Reuse*. The factors 'Environment', 'Institutional' and 'Responsibility' were clear drivers of the case and the realisation was enabled mainly by the factors 'Capital', 'Partnerships', 'Customer Engagement' and the logistic aspect 'Collection'.

References

1. Reike, D., Vermeulen, W. J. & Witjes, S. (2018). The circular economy: New or refurbished as CE 3.0? – exploring controversies in the conceptualization of the circular economy through a focus on history and resource value retention options. *Resources, Conservation and Recycling, 135*, 246–264. ISSN: 0921-3449.

2. De Brito, M. P. & Dekker, R. (2004). In Dekker, R., Fleischmann, M., Inderfurth, K. & Van Wassenhove, L. N. (Eds.) *Reverse logistics* (eds) (pp. 3–27). Berlin, Heidelberg: Springer. ISBN: 978-3-642-07380-9.
3. Bakker, C., Wang, F., Huisman, J. & den Hollander, M. (2014). Products that go round: Exploring product life extension through design. *Journal of Cleaner Production, 69,* 10–16. ISSN: 095-96526.
4. Silva, D. A. L., Santos Renó, G. W., Sevegnani, G., Sevegnani, T. B. & Serra Truzzi, O. M. (2013). Comparison of disposable and returnable packaging: a case study of reverse logistics in Brazil. *Journal of Cleaner Production. Cleaner Production: Initiatives and Challenges for a Sustainable World, 47*, 377–387. ISSN: 0959-6526.
5. Fleischmann, M. et al. (1997). Quantitative models for reverse logistics: A review. *European Journal of Operational Research, 103*, 1–17. ISSN: 0377-2217.
6. Agrawal, S., Singh, R. K. & Murtaza, Q (2015). A literature review and perspectives in reverse logistics. *Resources, Conservation and Recycling, 97,* 76–92. ISSN: 0921-3449.
7. Nasr, N. et al. (2018). Redefining value: the manufacturing revolution. Summary for policy makers ISBN: 978-92-807-3720-2. UNESCO.
8. Patagonia. (2023). *WORN WEAR - Patagonia*. Retrieved March 25, 2023 from https://www.wornwear.patagonia.com.
9. ebay. (2023). *eBay Inc. - Fast Facts*. Retrieved December 31, 2022 from https://www.investors.ebayinc.com/fast-facts/default.aspx.
10. kooky (2022) *Kooky - Das Mehrwegsystem für Take-away-Getränke*. Retrieved January 8, 2023 from https://www.kooky2go.com.
11. Valora Holding AG. (2023). *Valora brands Valora Holding AG*. Retrieved February 18, 2023 from https://www.valora.com:443/de/brands/.
12. Stadt Bern. (2023). *Sauberkeits-Charta, Stadt Bern*. Retrieved April 5, 2023 from . https://www.bern.ch/mediencenter/medienmitteilungen/aktuell_ptk/wirtschaft-und-stadt-wollen-abfall-gemeinsam-reduzieren.
13. European Union. (2023). *Directive (EU) 2019/904 of the European Parliament and of the Council of 5 June 2019 on the reduction of the impact of certain plastic products on the environment (Text with EEA relevance)*. Retrieved October 23, 2023 from https://www.eur-lex.europa.eu/eli/dir/2019/904/oj.
14. Bundesministerium der Justiz. (2023). *Verpackungsgesetz - VerpackG*. Retrieved March 26, 2023 from https://www.gesetze-im-internet.de/verpackg/__33.html.
15. Plasteurope.com. (2023). *PACKAGING TAX GERMANY: Judgment by Federal Administration Court: Tübingen waste levy is lawful / Municipalities should follow the example – Nabu*. Retrieved December 16, 2023 from https://www.plasteurope.com/news/PACKAGING_TAX_GERMANY_t252867/.
16. Verband Kommunaler Unternehmen e.V. (2023). *Einwegplastik und Zigarettenkippen in der Umwelt kosten Kommunen jährlich 700 Millionen Euro*. Retrieved April 8, 2023 from https://www.vku.de/presse/pressemitteilungen/archiv-2020-pressemitteilungen/einwegplastik-und-zigarettenkippen-in-der-umwelt-kosten-kommunen-jaehrlich-700-millionen-euro/.
17. Bundesamt für Umwelt BAFU. (2023). *Ö kobilanz (Life Cycle Assessment, LCA)* BAFU. Retrieved February 2, 2023 from https://www.bafu.admin.ch/bafu/de/home/themen/thema-abfall/abfallglossar/oekobilanz-lca.html.
18. Haessig, S., Moser, C. & Pohl, T. (2021). *Environmental impact evaluation of kooky cups* (University of Applied Sciences of Eastern Switzerland (OST)).
19. Bundesamt für Umwelt BAFU. (2021). *Abfallmengen und recycling 2020 im Überblick*. Bundesamt für Umwelt BAFU.
20. Kauertz, B., Schlecht, S., Markwardt, S. & Knappe, F. (2019). Untersuchung der ökologischen Bedeutung von Einweggetränkebechern im Außer-Haus-Verzehr und mögliche Maßnahmen zur Verringerung des Verbrauchs. *Umweltbundesamt.* ISSN: 1862-4804.

21. Wellbrock, W., & Ludin, D. (2021). *Nachhaltiger Konsum: Best Practices aus Wissenschaft, Unternehmenspraxis, Gesellschaft*. Springer Gabler Wiesbaden: Verwaltung und Politik. ISBN: 978-3-658-33353-9.
22. Courtnell, J. (2023). *ESG Reporting Preparation Guide: What is ESG Reporting?* Green Business Bureau. Retrieved April 8, 2023 from https://www.greenbusinessbureau.com/business-function/finance-accounting/esg-reporting-what-is-esg-reporting/.

Chapter 8
R3 Repair

Yannik Keller, Raphael Welter, and Roman Hänggi

Abstract *Repair* laws are being implemented, new *Repair* options are available, and sustainable *Repair* technologies are emerging. The need for a circular economy has led to new successful businesses. The use case of an office equipment provider shows that there is a trend toward greater sustainability. The company is taking steps to embrace this by implementing its own circular economy model, starting from a *Repair* approach. The main drivers and enablers of the business model are analysed. Furthermore, a positive impact on energy efficiency and carbon footprint is demonstrated. The company is now extending its model to *Refurbish* and *Recycle*, transforming the entire company.

8.1 Introduction to *Repair*

In 2021, the world generated 2.01 bill. tonnes of municipal solid waste. This is 0.74 kilogrammes per person per day [1]. Thus, electronic and electrical waste (e-waste) is estimated to contribute 57.4 mill. tonnes, which is greater than the weight of the largest man-made structure on the Earth, the Great Wall of China [2]. All this waste that ends up in landfills has raised questions about the linear economic model currently in place. How can this immense growth of waste and its impacts be reduced? One way to handle this is to repair defective products, so that the life of a product is extended. This prolongs the lifetime, i.e. use cycle, of a product, which prevents it from having to be thrown away, thus creating waste.

To increase the number of products that are being repaired, the campaign 'Right to Repair' was launched in 2019 [3]. The campaign started quickly and grew to

Y. Keller
Goldach, Switzerland

R. Welter
Chur, Switzerland

R. Hänggi (✉)
Brülisau, Switzerland
e-mail: roman.haenggi@ost.ch

A. Podleisek et al. (eds.), *Circular Economy: The Next Level of Company Success*,
https://doi.org/10.1007/978-3-031-66406-9_8

more than 100 organisations in more than 20 European countries. The main goal of the campaign is to improve the designs of the products so that the products can be repaired more easily. In addition, appropriate manuals should be available so that the products can be repaired. Another point of the campaign is to inform customers about the fact that the product can be repaired.

Due to the growing interest in sustainability, the European Union passed a regulation which laid down the requirements of Ecodesign [4]. Beginning in March 2021, manufacturers of household appliances such as dishwashers, washing machines, or refrigerators must provide spare parts for a period of seven to ten years after the purchase of the product.

Due to the immense amount of e-waste (e-waste) and the fact that it is the fastest-growing waste stream in the world [5], the governing coalition in Germany wants to do more than the regulation established in 2021. Their goal is to extend the regulation, so that laptops and mobile phones also have the right to be repaired. Specific actions would be that customers can replace defective parts, such as a battery, or that the warranty period is extended [6].

With the 'Right to Repair' and the need to have spare parts available for a long time after the last sale, manufacturers face various challenges. If spare parts must be reproduced after a longer period, the batch size will be lower, which normally results in higher production costs per part. It is also not an option to have the spare parts pre-produced in large amounts because of the inventory space that they would require.

There could be a solution with 3D printing technology that allows parts that are already past their active life in Product Lifecycle Management (PLM) to be created. This technology does not necessarily need to be used by manufacturers themselves. It could be used by a repair centre or even by the customer himself. The only thing a customer requires is a digital representation of the spare part, for example, a CAD model. This could be provided by the manufacturer or another database, as is already done in practice [7]. If customers expect to have a digital representation of spare parts to repair a product more quickly or repair it themselves, the entire market for selling spare parts may shift, resulting in fewer spare parts sales for the manufacturer. This threat of losing spare parts sales can be used as an opportunity to add customer value by providing them with a digital solution for spare parts. In this case, the exposure of the know-how and intellectual property rights must be considered.

In addition to the repair legislation, there are other forms of repair that are growing. One example is the event format 'repair cafes'. Repair cafes are an example of peer-to-peer, non-commercial repair workshops. At these volunteer meetings, broken things are repaired alone or in a group. It is ensured that the necessary tools and materials are available for the repairs [8].

There are already more than 3,200 repair cafes worldwide [9]. Self-repair could lead people to connect with their product, even when it is mass-produced [10]. This emotional attachment leads to further repairs in the future [11].

The listed trends in repair regulation and the existence of more than 3,200 repair cafes show that society wants to become more sustainable. But is repair the right way to deal with this immense amount of waste? Isn't it better when products don't need to be repaired because the durability of the product is longer? What can be done when a discarded product cannot be repaired anymore? Repair is certainly not the only factor that solves the waste problem, but it is part of a much larger concept, namely Circular Economy (CE).

Definition of Repair

A circular economy is based on three principles [12]:

1. Eliminate waste and pollution
2. Circulate products and materials at their highest value
3. Regenerate nature.

To comply with such principles, an optimal product design is necessary, the first step to a CE. When designing a new product, the three principles should be considered according to their ranking.

In the first step of designing a product, the elimination of waste and pollution is considered. The initial idea is to use materials that are safe and eco-friendly (Refuse (R0)) and still offer the same product function. If waste cannot be completely prevented, there is the option to reduce waste by designing smaller parts or making the production process more efficient (Reduce (R1)).

The second step is designing products such that they circulate at their highest value. Therefore, the design should be durable and designed for its useful life. The part should also allow for repair and maintenance (Repair (R3)) or product upgrade (Refurbish (R4)). In practice, this is done by designing the product modular. The design of a product in a modular way is also necessary for Remanufacture (R5) or Repurpose (R6).

When a product reached a state in which its life cycle can no longer be extended, the materials of the product should allow Recycle (R7) or Recover (R8).

In this chapter, the focus lies on Repair (R3), which belongs to the second principle 'Circulate products and materials at their highest value'. Drawing upon J. Kirchherr et al.'s definition, we define the retention option Repair (R3) as: Extending the use cycle of products or its parts through repair and maintenance of defective products so that they can be used with their original function. [13]. This definition differs from other R-frameworks like 3R or 5R discussed in the literature [14]. Differentiation from other retention options is shown in Table 8.1.

Although the definition of *Repair* is clear, it is not clear if a product is repaired by the user of a product. Certainly, the design of the product must allow for repairs. But even then, will the customer repair a product when he can do so? This question was addressed in a study in the UK [15]. The survey addressed the question of why people will not repair their devices. The main result was that people will not repair their devices when it is too expensive, too complicated or cheaper to buy a new product. Serviceability is an important requirement for manufacturers in the future; the design of the product must be developed so that a customer can repair the device easily and at low cost.

Table 8.1 Difference of *Repair* to other retention options

Retention option	Difference to R3 *Repair*
R2 *Reuse*	Can involve minimum amount of condition monitoring (cleaning or repackaging), where the main objective is not maintenance
R4 *Refurbish*	Restore functionality of a product or part, which is inferior or an upgrade to the original functionality of the product
R5 *Remanufacture*	Use parts of discarded product in a new product with the same original function
R6 *Repurpose*	Use discarded product or its parts in a new product with different function

Business Model – Product as a Service

The business model that allows the customer to not have to worry about cost and effort for repair is Product as a Service (PaaS). When a product needs to be maintained or repaired and the customer is not careless with the product, the salesperson is responsible for the maintenance or repair, making it cheaper and easier for the customer. On the other hand, the salesperson wants easy maintenance and a repair process such that its costs remain low. But even more, the long durability of the product is preferred, so that the product does not need to be maintained or repaired in the first place. And this is exactly what CE is about. The salesperson is incentivised to make the product more durable and to allow it to circulate at its highest value, which is the second principle of CE.

To better understand the PaaS approach, we take a look at an example with a bike. If you buy a bike, you look for the field of application and the price. You don't care too much if it's a belt-driven bike or a chain-driven bike. The technology used depends more on the field of application. On the other hand, with a PaaS approach, it matters which technology is used, because a chain-driven bike would need to be maintained by regularly oiling the chain. This could be done easily if the bike goes to the workshop after each use. But in modern approaches, where you can rent bikes with a smartphone in the city and leave the bike at a station or wherever you want on the street, the bike will not come back to the workshop. Moreover, it is being used by the next customer. In winter, when it is cold and the streets are salted to avoid ice, the chains will be even more affected by corrosion than a belt drive. Therefore, it just makes sense to choose a belt-driven bike in a PaaS approach. Here, the business model has led to a more durable product that complies with CE.

The business model PaaS is an approach of Reuse (R2). Where people can use the same resource over time, leading to more usage, more durable products, and fewer resources used in production. This turns out to be a keyword for the sharing economy, which could be a possible solution to greater sustainability. As the example with PaaS business model shows, Repair (R3) is linked to Reuse (R2) and can also be linked to R0 and R1 in the design phase. If a product cannot be repaired anymore, it is better if it can be recycled or at least recovered, meaning that CE is a holistic approach.

8.2 Analysis of a *Repair* Case

8.2.1 Description of the Case

This chapter looks at a business case that implements the CE principles in a holistic approach. For a detailed description and elaboration of the use case, interviews were conducted with R. S., the Sustainability Project Manager at the Lista Office LO.

Lista Office (LO) is a Swiss company that designs and implements office solutions for its customers. They sell chairs, tables, cabinet systems, lounges, and lights for the office area. They provide yearly office equipment for a room with approximately 300,000 m^2 and 330 employees. Their products, mostly made of steel, are developed and manufactured in-house in Switzerland or purchased from a partner. LO processes more than 2,500 tonnes of steel per year.

Founded in 1945 and expanded in the office market in 1980, LO has gone through many changes and trends in ergonomics and in desk sharing that emerged 15 years ago. When large banks with more than 1,000 employees saw that, for multiple reasons, such as working outside the office, holidays, or sick days, not every employee needed a desk. A trend has emerged that was supported by the need for home office during the recent pandemic.

In addition to the trend for sharing desks, there is a trend toward more sustainability. LO has a vision and a strategy that directly tackle sustainability. Therefore, an environmental management system was implemented according to ISO 14001, and suppliers and raw materials are chosen carefully. LO obtains materials such as fresh coniferous or deciduous timber, storm damaged wood, and sawmill timber waste from ISO 14001 certified suppliers and members of the WWF Wood Group, and applies the FSC label. The material used the most is steel, which does not emit chemicals or additives and is fully recyclable.

The production plants are powered by renewable energy using solar systems, which are installed on top of the production plants, or purchased green electricity. In the coating process, the contaminated air is purified. The Wastewater is cleaned by a wastewater treatment plant. On the customer side, the energy consumption of the electrically operated standing desk was reduced by 50%. All these points like material health, material reuse, renewable energy and carbon management, water stewardship, and social justice are considered in a 'cradle-to-cradle certificate,' which LO obtained in 2018 for its new product lines 'LO Next' and 'LO Extend.' The urge to obtain this certification came after participating in a tender where customers demanded more sustainability. However, it was not an easy decision to get a certificate because depending on the customer, different certificates are possible, there is no unified certificate on the market: cradle-to-cradle, Blue Angel, LEED, BIFMA, and other certificates. In which certificate would you invest your money? Standardisation is certainly desirable.

Having a long sustainable tradition and keeping up with customer expectations for more sustainability, LO decided to go one step further. LO developed its own CE concept (see Fig. 8.1), which transforms the whole company.

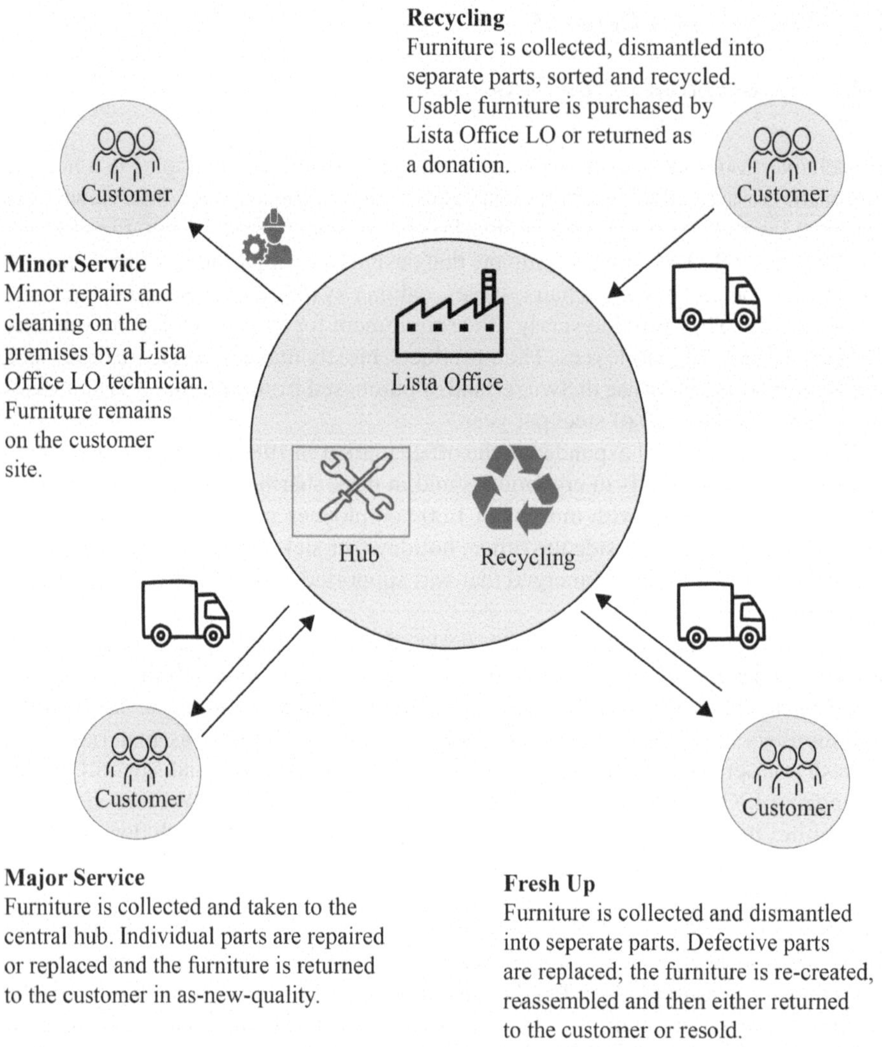

Fig. 8.1 The CE concept of Lista Office LO (own representation adapted from [16])

The concept of LO consists of a minor and a major service. This is the equivalent of R3. In these services, LO maintains and repairs furniture, such as broken drawer slides. If the circumstances allow for a minor service (small repair with little noise), the repair will be done on the customer's site. Otherwise, LO repairs the product in a central hub (major service). For the repair process, LO has a specific team and an existing logistic process. These repair processes have been in operation since the founding of the company because customers expect a service when the whole office is equipped with LO furniture. With this in mind, LO considers already in the development phase of a product how it could be easily disassembled later for *Repair*, *Refurbish*, or *Recycle*. For this reason, LO uses long-durability materials such as steel and these products are developed for lifelong usage and *Repair*, e.g. steel parts are not welded whenever possible. LO offers the customer a 5-year warranty.

Due to this long durability, the PaaS business model is an option for LO, where they call it 'Furniture as a Service'. This business model also fits the trend of sharing desks. A partner of LO already applies this business model. The company offers a furniture-equipped room as a rental. This offers the customer high flexibility and rapid adaptation to changing needs. In addition, it provides the customer with a low capital approach. On the other hand, the capital investment of such a service provider is high and Return on Investment (ROI) is expected to be very long-term. Currently, LO is not implementing such a business model. However, there could be future options to give partners a Business-to-Business (B2B) leasing option, which would fit the CE approach and retention option R2.

In the CE concept of LO, they added two retention options to the already existing *Repair* process. One is *Refurbish* (R2). Having already a Reverse Logistics (RL) process and a repair hub made it easy to implement a *Refurbish* process called 'Fresh Up'. This fresh-up process takes back furniture, which is then disassembled and the colour is removed. A new product is made by repainting and assembling the parts again. The product is then resold to the same customer or to a new customer.

A common problem is that customers do not want to pay the same price for a refurbished one compared to a new one. The general understanding is that a used product has lower quality, which is not necessarily true. Therefore, a lot of persuasion is required to get the customer to see that the products are more sustainable but cannot be cheaper due to the RL and rework processes.

This *Refurbish* process can be managed in the Enterprise Resource Planning (ERP) with some handmade processes. The pilot processes were successful, and LO takes back their furniture when it makes sense to organise this effort: a minimum quantity of 30 products must be given back to cover up transport expenses and to get a better ecological footprint. In some cases, LO even refurbished furniture that was not made by them, but that is not a standard process because spare parts are not available and administrative problems.

The second retention option that was implemented is *Recycle*. LO is taking back their products, disassembling and recycling them. In this way, the design must allow easy dismantling and good Recycling of materials. In Construction, attention is given to plastics so that easy sorting and recycling can be carried out. According to LO, 85% of the products can be recycled.

Although the change with this new concept of LO seems at first glance small, the changes which the company undergoes are tremendous. Taking back products from the client is not an easy thing. Questions like 'In which condition is the furniture?', 'Are pictures of the furniture good enough to take the products back?', 'How much does a customer get for his old furniture?' may arise and need to be addressed, and, especially, the financial compensation of the used products must be discussed with the customer. If a product is returned and can only be recycled, the customer pays for the Recycling process because the product will be dismantled, sorted, and recycled.

If a product is to be refurbished, LO must pay for the reverse logistic process, the dismantling, the repainting, and its general administrative cost. Furthermore, the customer with the used product wants to have compensation, and the new customer wants to pay less because the product was already used. This must be a calculation that is beneficial for LO. Therefore, convincing work must be done on the customer side by pointing out the environmental benefits like CO_2 reduction and fewer resources being used. Used products can't be taken back by the half price of the original price and refurbished products can't also cost the half price. Another issue that LO must consider is the saturated market in the product range 'storage. The cabinet system, which offers storage, can't be taken back without a second thought. Due to digitisation, less storage place is needed because demand is decreasing. You cannot pay for the return of a cabinet system if you cannot sell it later, this ends up as an inventory cost and later as a Recycling cost.

The change mostly affected the inside of the company, although there are only 2-3 employees in the core team involved. First, many new concepts had to be discussed with the executive board and then with other departments such as the purchase and production department, logistics with RL, and sales departments. This transformation process costs a lot of energy, for example, the discussion of whether CE is self-sustainable or if the approach leads to market cannibalisation. New products will be less produced due to the products that are repaired and refurbished. In general, the entire change process took some time, and even now, four years after reaching a certificate from cradle to cradle, the company is still in the early stages of CE. Big steps have been taken, the foundation is given, and now the process must be lived and continuously improved.

8.2.2 *Drivers of the Case*

In this section, we look at the relevant drivers, which leads the company to implement a CE concept. Those drivers are classified into five clusters: Environmental, Institutional, Stakeholder, Financial, and Responsibility. Those drivers are described in Chap. 3 and are assessed for the LO use case in Table 8.2.

During the implementation of the principles CE with the LO concept, there was a lot of scepticism around CE. The scepticism lowered with time as many customers demanded more sustainability. Customers asked what LO is doing with respect to sustainability, and they could answer with the cradle-to-cradle certificate, their concept CE and the ISO 14001 norm. Therefore, D.S2 (adapting to changing consumer behaviour, meeting customer expectations and increasing customer satisfaction) is the biggest driver for LO to become circular (+++).

Furthermore, the certificate, the implementation of ISO 14001, and the concept CE also decrease the resources in production (D.E1), decreases the Carbon footprint (D.E2), improves the corporate image of the company D.S1 and has a positive effect on social and ethical responsibility (D.R1). Those are factors that lead to

Table 8.2 Drivers of the *Repair* case

Cluster	Code	Driver	Context	Relevance
Environmental	D.E1	Resources	Decrease the resources used in production	+
	D.E2	Footprint	Decrease product carbon footprint	+
	D.E3	Risk	Resilience against resource-related risks (scarcity, supply challenges etc.)	0
Institutional	D.I1	Legal	Comply with legal requirements and norms	0
	D.I2	Punitive	Avoiding punitive measures	0
	D.I3	Juridification	Anticipate or meet future legal obligations	0
	D.I4	Tax	Taxation benefits, access to supportive funds, and subsidies	0
Stakeholder	D.S1	Image	Improve corporate image, legitimacy, or credibility	+
	D.S2	Customer	Adapt to changing consumer behaviour, meet customer expectations, increase customer satisfaction	+++
	D.S3	Practices	Meet the standards and follow the best practices of competitors	0
Financial	D.F1	Volatility	Price volatility of raw material	0
	D.F2	Savings	Cost saving opportunities	0
	D.F3	Competitiveness	Competitiveness through innovation and business development	++
	D.F4	Value	Create value from waste and production side streams	0
Responsibility	D.R1	CSR	Social and ethical responsibility	+
	D.R2	Promoter	Become a promoter of systemic change	0

0 representing insignificant importance, + indicating minor relevance, ++ reflecting notable relevance, and +++ signifying crucial importance

meeting customer expectations (D.S2) and also factors that the organisation wants to live by (sustainability and responsibility, vision, and strategy) and have then driver relevance (+).

Any project implemented by the company must be financially valuable, and this innovation of implementing the principles of CE also led to competitiveness. For companies further away, it is difficult to implement a sustainable process CE. The transport distance is too long to be CO_2 beneficial and they would have to build up locations where they could repair, refurbish or recycle. This offered LO a differentiation from other companies, and they become competitive through innovation (D.F3 ++).

D.F2 cost-saving opportunities were not made directly. The implementation of CE principles was an adaptation to changing consumer behaviour. D.F1 (price volatility of raw material) is not a financial driver either. Even when steel prices rose enormously in the middle of 2022, this was not a driver at the time of implementation. Also, creating value from waste (D.F4) is not a driver for LO. The recycling process is also a cost to the company and, consequently, the customer has to pay for it. In the refurbishment process, a lot of convincing work must be done to get the used products back for a fair price, which also does not make it a driver.

On the institutional side, there were no drivers for LO. Although they obtained a certificate, there is no norm or legal requirement that a company must have such a certificate.

To sum up drivers, the main driver is D.S2 (adapting to changing consumer behaviour) and D.F3 (competitiveness). Together with other drivers, it gives a good package to implement a CE transformation and positions the company better on the market side.

8.2.3 Enablers of the Case

The implementation of the R's depends not only on drivers but also on relevant enablers. Those enablers are classified into five clusters: Firm, Value Chain, Technological, Customer, and Material Supply. These enablers are described in Chap. 3 and are assessed for the LO use case in Table 8.3.

A factor that helped to enable a CE approach in LO was culture (E.F1). Before the concept of CE was planned, ISO 14001 was already implemented. LO has solar cells in both production plants and a water filtering system, making a CE approach just further helping to achieve the sustainability strategy (++). Another enabler that must be available is the capital necessary to support the transition (E.F2). By having 2-3 employees who work constantly in the CE core team and together with different departments, financial support is essential (++).

When a customer wants to give back a used product, a lot of convincing work must be done. Explaining why CE is important and at what price the product can be taken back helps the customer better understand the process. When the product will be recycled, the customer needs to pay, and when the product is refurbished, the customer gets a small compensation. Therefore, customer engagement (E.C1) is also relevant enabler (++).

An existing repair process, with a repair hub, helped implement additional retention options such as refurbishment or recycling. The available reverse logistic network (E.M1) supported the transition as well (+).

On the technological side, LO found a suitable solution to map the repair and refurbishing process in the ERP system (E.T3 ++). Partnerships on the side of the value chain are important. Support on the supplier side is necessary for certificates and an ISO 14100 process. Partnerships are also necessary for a future PaaS model

Table 8.3 Enablers of the *Repair* case

Cluster	Code	Enabler	Context	Relevance
Firm	E.F1	Culture	Committed company environment and culture	++
	E.F2	Capital	Available capital or financial support to transition	++
Value chain	E.V1	Partnership	Build closer relationships/invest in collaboration and partnerships	+
	E.V2	Transparency	Enhance information sharing and transparency, make use of digital technologies and standardise information	0
Technological	E.T1	Research	Experiment and research	0
	E.T2	Mutuality	Joint decision-making (customers and suppliers)	+
	E.T3	R-technology	Availability of technical and technological know-how for a suitable value retention in-house	++
	E.T4	Outsourcing	Possibility to outsource the activity	0
Customer	E.C1	Engagement	Increase customer engagement and awareness, provide incentives	++
Material supply	E.M1	Logistics	Available reverse logistic network	+
	E.M2	Collection	Create simple collection/take-back systems	0
	E.M3	Symbiosis	Engage in industrial symbiosis	0

0 representing insignificant importance, + indicating minor relevance, ++ reflecting notable relevance, and +++ signifying crucial importance

(furniture as a service). It is easier to make the change to CE possible when suppliers and customers work together. Therefore, the enabler E.V1 (Partnership) and E.T2 (mutuality) have a relevance of +.

8.3 Outcome and Impact of *Repair*

8.3.1 Ecological Results

By building 333 m^2 of solar panels on top of the production plants, it was possible to generate around 62,000 kWh of annual electricity. Furthermore, LO has installed a Wastewater treatment plant, which cleans oils, fats, and chemicals. Last Lo has implemented an air filter to purify the contaminated air from the process powder coating. Their products have a five-year warranty, a label that the products are economically friendly painted, and 85% of the materials are recyclable. Since 2018, the new product lines ‘LO Next’ and ‘LO Extend’ have additionally a certificate of cradle to cradle (bronze). To represent a pioneer in sustainability development, LO is a member of Swiss cleantech and participates in EnAW's (Swiss Private Sector Energy Agency) climate protection programme.

With the implementation of the ISO 14001 standard years ago, LO is on its way to fulfil its sustainability mission and strategy. Their environmental officer observes environmental facts, which is required by the ISO 14001 standard. In the 2021 report, published annually by LO, it is stated that energy efficiency increased by 129.7% compared to 2013. Furthermore, CO_2 emissions were reduced by 47.0% compared to 2013.

In 2022, they plan to replace some LED light bulbs and do further process optimisation in manufacturing. The aim is to reduce energy consumption by 1% and reduce CO_2 emissions by 2% by creating a CE concept with *Reuse*, *Repair*, *Refurbish* and *Recycle*. Although this concept was implemented some years ago, CO_2 reduction and energy consumption cannot be quantified on a product level. On the other hand, there is the question of how accurately must a CO_2 emission reduction and an energy consumption be considered at the product level or even at a retention option? It would be beneficial if the quantified impact on the product level could be communicated to the customer in the future, given that the calculation must be practical for LO.

LO is just in the beginning stages to become circular. Processes are established, successful pilot projects happened, and an exact measurement of impacts is a project for the future years and the ongoing continuous improvement process.

8.3.2 *Social Results*

For society in general, which is concerned about responsible consumption and production (i.e., United Nations Sustainable Development Goal (SDG) number 12), LO finds an active contribution by implementing ISO 14001. Energy efficiency increased by 129.7% and CO_2 reduction by 47%, which are also numbers that society benefits from.

Another social factor that helps customers and users is the products of LO themselves. Products such as seats, height-adjustable tables, lights, and comfortable lounges are good for ergonomics. But not only do users (employees) benefit, employers can also benefit. A study by the Fraunhofer Institute [17], which summarised the results of 250 case studies, showed that employers benefit in many ways from good ergonomics, wherein 95% of the studies (mostly studies in the USA) a productivity improvement of 20–30% was achieved.

8.3.3 *Economic Results*

Sustainable market positioning in LO is going on for a long time. The strategy of being sustainable, by implementing ISO 14001, obtaining certificates such as cradle to cradle and implementing the principles CE is certainly a good path, as more and more customers demand a certificate or want to know what LO is doing for sustainability. The sustainability approach also led to a good image of the corporate identity, which is used in marketing according to the motto: Do good and talk about it.

Implementing the CE principles occurred only a few years ago, some pilot projects were successfully executed, which meant that they were not only sustainable, but also financially valuable. A challenge for LO is to change the expectations of customers so that they see the benefits of CE and are willing to pay for repaired or refurbished products. But if the concept is already financially valuable and a trend for CE can be observed, positioning the market with the concept CE makes sense.

8.4 *Repair*: Summary and Learnings

The LO approach is a good way to implement the principles CE. They develop, manufacture and sell office furniture and then consider the retention options R1 (Reduce), R2 (Reuse), R3 (Repair), R4 (Refurbish), and R7 (Recycle). They start with the design phase, where they try to reduce the steel material so that the furniture is still stable. They design the furniture modular, meaning as less welding connections as possible, and use materials and connections that can easily be dismantled, sorted, and recycled again. Steel design has long durability and would be a perfect precondition for a future rental model (furniture as a service). Furthermore, the pre-existing repair process with its RL helped implement additional retention options, such as refurbishing and recycling. The use case of LO clearly shows that CE is a holistic concept, which starts in the design phase, where all retention options need to be considered. Furthermore, the concept of CE can be complemented by other sustainability approaches, such as certificates or the ISO 14001 standard.

It must be noted that the entire concept of CE is a major transformation for a company. Many processes must be defined, different departments involved, and people trained. Implementing a concept CE is not a small project, but a change that affects the entire company.

Another good approach to adding to R3 (Repair) is the PaaS model with R2 (Reuse). By offering a product for rent, the ownership stays by the company; then the company wants to have a long durability product or an easy repair because he must pay for the repair service. The customer will repair the device because it is cheap (he does not have to pay for the repair) and it is easy (the company needs to arrange the repair service). That is why the PaaS model leads to a more sustainable approach in general. The PaaS model could also be financially valuable when renting income can be used to cover (partially) the fixed cost for a company, leading to more Resilience.

Another insight gained is that society at large is increasingly demanding greater sustainability. Therefore, there will be repair laws for having spare parts available for some years, as is already the case for household devices such as refrigerators. The long availability of spare parts could be addressed with technologies like 3D printing.

References

1. The World Bank (2022). *Trends in solid waste management*. https://www.datatopics.worldbank.org/what-a-waste/trends_in_solid_waste_management.html.
2. WEEE-Forum. (2022). *International E-waste day: 57.4M Tonnes expected in 2021 — WEEE forum*. https://www.weee-forum.org/ws_news/international-e-waste-day-2021/.
3. *Who we are - Right to Repair Europe*. https://www.repair.eu/about/ (2022).
4. European Comission. (2019). *Commission Regulation (EU) 2019/2022 of 1 October 2019 laying down ecodesign requirements for household dishwashers pursuant to Directive 2009/125/EC of the European Parliament and of the Council amending Commission Regulation (EC) No 1275/2008 and repealing Commission Regulation (EU) No 1016/2010 Text with EEA relevance*. http://www.data.europa.eu/eli/reg/2019/2022/oj
5. World Economic Forum. (2022). *A new circular vision for electronics*. https://www.weforum.org/reports/a-new-circular-vision-for-electronics-time-for-a-global-reboot/.
6. ARD-aktuell. (2022). *Mehr Tempo beim Recht auf Reparatur*. tagesschau.de. https://www.tagesschau.de/wirtschaft/verbraucher/recht-auf-reparatur-verbraucherschutz-nachhaltigkeit-elektroschrott-101.html.
7. Traceparts (2022). *Free 3D models, CAD files and 2D drawings - TraceParts*. Design library - traceparts. https://www.traceparts.com/en.
8. Jonas, M., Nessel, S. & Tröger, N. (Eds.) (2021). *Reparieren, Selbermachen und Kreislaufwirtschaften: Alternative Praktiken für nachhaltigen Konsum*. Wiesbaden: Springer Fachmedien Wiesbaden. ISBN: 978-3-658-31568-9.
9. Repair Café (2022). *Home — Repaircafe*. Retrieved September 27, 2024 from https://www.repaircafe.org/en/.
10. Rogers, H. A., Deutz, P. & Ramos, T. B. (2021). Repairing the circular economy: Public perception and participant profile of the repair economy in Hull, UK. *Resources, Conservation and Recycling, 168*. ISSN: 0921-3449.
11. Meissner, M. (2021). Repair is care? - Dimensions of care within collaborative practices in repair cafes. *Journal of Cleaner Production, 299*. ISSN: 0959-6526.
12. Ellen MacArthur Foundation. (2022). *What is a circular economy?* What is a circular economy. https://ellenmacarthurfoundation.org/topics/circular-economy-introduction/overview.
13. Kirchherr, J., Reike, D. & Hekkert, M. (2017). Conceptualizing the circular economy: An analysis of 114 definitions. *Resources, Conservation and Recycling,* 127, 221–232. ISSN: 0921-3449.
14. Reike, D., Vermeulen, W. J. & Witjes, S. (2018). The circular economy: New or Refurbished as CE 3.0? – Exploring controversies in the conceptualization of the circular economy through a focus on history and resource value retention options. *Resources, Conservation and Recycling,* 135, 246–264. ISSN: 0921-3449.
15. Jaeger-Erben, M., Frick, V. & Hipp, T. (2021). Why do users (not) repair their devices? A study of the predictors of repair practices. *Journal of Cleaner Production,* 286, 125382. ISSN: 0959-6526.

16. Lista Office LO. (n.d.). *Lista Office LO - Multizone concept* Become stronger - zone by zone. Retrieved February 2, 2023 from https://www.lista-office.com/en/business-solutions/competences/multizone-concept/.
17. Daub, U., Ackermann, A. & Kopp, V. (2019). *Ergonomie-benefits*, 28. Stuttgart: Fraunhoferr-Institut für Produktionstechnik und Automatisierung IPA. http://dx.doi.org/10.24406/ipa-n-559153.

Chapter 9
R4 Refurbish

Denis Hammer, Patrick Looser, and Roman Hänggi

Abstract The establishment of a circular economy and the integration of all stakeholders within it require a shared understanding and alignment of objectives between customers and manufacturers. In addition, it involves the design of products that facilitate component replacement. In the chosen case study, an electric tricycle manufacturer for mail delivery opted to implement a comprehensive *refurbishment* process. The company successfully uses reused components to overcome supply bottlenecks. Positioning itself as a pioneer and advocate for the circular economy, the company aims to set new standards through innovation. It emphasises that this economic model is viable and should be pursued, even if it results in a zero-sum game. The primary drivers identified for this initiative are the environment and individual responsibility. The company culture and a return programme initiated by a major customer also played crucial roles. Despite the fact that the second life programme currently does not cover its costs, it has created numerous opportunities. However, the company is steering this business concept beyond its financial objectives. Their mission is to transform the company towards a circular economy, significantly reducing waste. Consequently, the company has achieved very good results, saving 50% of grey energy and reusing 21 tonnes of materials annually.

9.1 Introduction to *Refurbish*

The Theory of Reasoned Action (TRA) was developed by Fishbein and Ajzen [1] to help researchers look at behavioural antecedents. This theory posits that human behaviour is mainly driven by two factors. The first, known as attitude, encompasses

D. Hammer
Winterthur, Switzerland

P. Looser
St. Gallen, Switzerland

R. Hänggi (✉)
Brülisau, Switzerland
e-mail: roman.haenggi@ost.ch

A. Podleisek et al. (eds.), *Circular Economy: The Next Level of Company Success*,
https://doi.org/10.1007/978-3-031-66406-9_9

an individual's inclination to react positively or negatively towards various entities, such as objects, people, institutions, or events. The second factor, termed subjective norm, involves the perceived social influence on the individual, which includes the perceived expectations of others and the accompanying pressure to behave in certain ways in response to those perceptions.

Using this theory as a foundational principle, an examination of the impact of economic history on the development of humanity can be undertaken. During the twentieth century, capitalism transformed the average individual into a consumer with an insatiable appetite for more goods [2].

The seminal book 'Propaganda' by Edward Bernays advised businesses to 'keep in constant communication through advertising and propaganda' because the challenge was now to create customers rather than just solve problems [3]. With regard to Ajzen's TRA, one could conclude that people were exposed to considerable propaganda in the 20th century, which in turn massively influenced consumer behaviour now.

Therefore, both consumers and manufacturers must play their part in the Circular Economy (CE). Especially in today's consumer society, users need to set a new standard for what they are willing to accept from manufacturers in terms of sustainability and vice versa. For example, even if France promotes and allows the sale of products without wrapping or with reusable packaging, it is of no use if the consumer does not follow this path [4].

One good way to meet the needs of today's consumer is to *Refurbish* used products. This increases the quality of the product and resells it at an affordable price. The current ecological crisis that the world is experiencing has changed the global landscape and, with it, the policy-making priorities of various heads of state. Unfortunately, empirical evidence shows that absolute decoupling of the Gross Domestic Product (GDP) from environmental impact is unlikely to occur [5, 6]. Resource use and GDP are the main driver of environmental impacts. The growth of GDP is driving us to consume more and more resources [5]. And because every capital system needs growth to survive, we are likely to increase environmental pressures [6].

To address this issue, we can optimise our resource use, essentially minimising waste. Although this may not lead to complete decoupling, we can still achieve a substantial relative decoupling and transform how we utilise these resources. An effective approach is the reuse of products. Often, items are replaced with newer models not because they have reached the end of their useful life but simply because of frequent upgrades. A study in Spain [7] showed that there is a tendency to store unused small information and communication technology devices at home, even if these Electrical and Electronic Equipments (EEEs) are in good shape. 73.9% of the respondents kept the devices at home and did not use them. Most of them because of emotional attachment. It was also recognised that only about 13% buy a new device because the old one is broken or has reached the end of its life.

The authors of a white paper [8] from Utrecht University discuss the market, culture, technology, and regulation as the four main barrier that prevent or slow the development of the CE. We see that there are several factors why part of our society will not or cannot extend the useful life of the product.

The concept of CE has become increasingly popular in recent years as a more sustainable way of doing business [9]. An important component of CE is the principle of reusing products and components. This not only reduces waste and pollution, but also conserves resources and reduces emissions. Used products should not be viewed as waste, but as an opportunity to *Reuse* in different ways. By repairing and refurbishing products, we can keep them in circulation longer and delay their replacement with new products. To make the transition to CE, we need to change our mindset and start seeing used products as an opportunity.

Refurbished products are often seen in electronic devices these days. At Revendo, for example, you can buy used cell phones that have been refurbished with new components. Additionally, personal computers are excellent devices for this retention strategy. They consist of components that even amateurs can replace. In most cases, the PC is brought up to a better standard by buying a new main processor or graphics card. For larger devices, such as production machines, it is usually weighed whether it is worth buying a new machine or whether a manufacturer's retrofit is sufficient [10]. Nowadays, you can also buy refurbished products from major manufacturers such as Apple, Sonos, Nintendo or Microsoft [11]. But not every device can be purchased without hesitation, and printers and televisions are not recommended [12]. In the case of the printer, for example, previous life may have worn out the inner chambers of the device, so that it no longer works as previously thought.

Furthermore, used and refurbished products can sometimes be cheaper than new versions from brands that are more expensive to start with [13]. One of the challenges of this approach is the fact that there are potentially many different 'kinds' of refurbished products available. Some may be refurbished by a third party with the original manufacturer's authorisation. In other cases, the products may have been stolen, refurbished, and sold as such [14]. Therefore, it is important to know exactly what you are getting and to check whether the manufacturer is legally allowed to resell the product as refurbished.

Definition of *Refurbish*

In the context of the 10R framework, refurbishment is the process of restoring or upgrading a product with the same or newer advanced components to restore functionality that may be less than the original specification [15]. During this process, ownership may change. According to Chap. 3 similar buzzwords to this are: Reconditioning, Retrofit, Refresh and Redesign.

Generally, reconditioning extends the life of a product beyond its original intended life while providing the predicted performance. Furthermore, transparency in the communication of the process should be maintained, as Wallner et al. [16] have shown that for customers, replacing contamination-sensitive parts is most important for refurbished products. An example of this is the ear plugs used in headphones. These were used in the ear, which can discourage consumers from reusing them. Therefore, the plastic that touched the ear should be replaced. The differences from neighbouring Retention Options (Rs) in the medium loop R of the R Framework are explained in more detail in Table 9.1.

Table 9.1 Difference of *Refurbish* to other retention options

Retention option	Difference to R4 *Refurbish*
R2 *Reuse*	May include a minimum of condition monitoring (cleaning or repackaging), but product cannot be improved in function or condition
R3 *Repair*	Maintenance of broken parts to retain or restore the products original functionality
R5 *Remanufacture*	Use parts of discarded product in a new product with the same original function
R6 *Repurpose*	Use discarded product or its parts in a new product with different function

The practice of refurbishing has increased in recent years as the issue of sustainability has gained global attention. Consumers are interested in eliminating waste and extending the life of products as much as possible. Refurbished goods often cost less than new products, which is another incentive for consumers.

Businesses also benefit from refurbishing goods, as it is often less expensive if the product design was chosen correctly. By modifying products with retention options, they stay in use longer, reducing the environmental impacts associated with manufacturing new products or disposing of old ones.

Overall, refurbishing is a promising way to reduce environmental impact while meeting consumer demand for quality products at a lower cost.

9.2 Analysis of a *Refurbish* Case

9.2.1 Description of the Case

Kyburz Switzerland was founded in 1991 and is based in Freienstein-Taufen, Switzerland. With around 170 employees, they mainly develop and produce electric tricycles. Common uses of these vehicles are for post offices to deliver letters and packages or for seniors to get to another place quickly. All vehicles are powered by battery modules that will be recyclable up to 91% in the next few years. It produces around 2,500 electric vehicles every year. Its clients include several post offices around the world and private customers, mostly retired. The first senior vehicle, the ‘Kyburz Classic’, helped maintain personal mobility, especially in more rural areas. The breakthrough then came in the Business-to-Business (B2B) sector with the ‘DXP’ mail delivery vehicles. In 2015, these were used by more than 5,000 postal operators in Switzerland, Europe, New Zealand, and Australia. In addition, there were various trolleys for autonomous delivery systems. For example, the ‘auto-etrolley III’ is currently being used in a pilot project in Norway to deliver letters and parcels. It drives itself to the customer’s door, avoids obstacles, and contacts the customer via smartphone to let them know that something is ready to be picked up. They also develop different ‘eRods’, which are street squads for different environments. The standard variant is a street-legal sports car, but there is also the off-road variant and other prototypes.

The company is the leading supplier in the European market for electric vehicles for logistics companies and aiming to create sustainable solutions for the urban transport of goods. The product line will soon include driver-less autonomous electric cars like the aforementioned 'auto-etrolley III'. This new technology might be attractive for post offices and retirees alike.

They continue to set the bar for sustainability. As a result of their reliability and high quality products, they last for many years. They lasted for so long that they did not anticipate that the first Swiss Post vehicles, which returned around 2019, would still be fully functional. Although these were usually in use for about seven years in all weathers, be it snow and salt, sand or bumpy terrain. Their products were not designed from the outset to give a second life. Looking for ways to reduce waste, as the company culture is very sustainable, they saw an opportunity to implement different retention strategies and extend the lifetime of their vehicles and components. This is how the 'MultiLife' concept was born (Fig. 9.1). The MultiLife philosophy permeates the whole business, allowing for a major improvement in the eco-balance of all services and products. The case study shows how a sense of duty to sustainability opens new doors and how the implementation of refurbishment has led to this.

After the first lifecycle, DXP mail delivery vehicles return after seven years of service in most cases. These are first carefully inspected as a whole in the company's '2ndLife hall' in Switzerland. Many components are removed and inspected. The chassis also has to be checked for rust, which is the main factor in deciding whether the entire vehicle is disassembled and the components themselves are stored or whether a refurbished 2ndLife vehicle is created. The B components are refurbished as well as possible and reassembled. Batteries below a capacity of 80% are replaced with other used batteries. Finally, the yellow post covers are replaced with white ones for private customers. In the coming chapters, the case study will revolve around this concept of 2ndLife. The following processes in the MultiLife concept are not part of the *Refurbish* strategy, but are briefly illuminated because of the innovation in battery

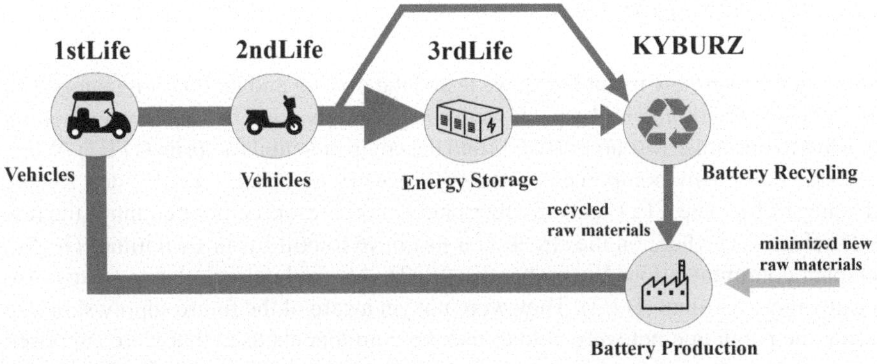

Fig. 9.1 KYBURZ Switzerland's MultiLife concept for battery sustainability (own representation adapted from [17])

recycling that arose from it. It is also intended to show that different Rs can be used after one is implemented and that even the resulting waste can still be processed in some ways.

Depending on the battery capacity, the battery is transferred to the third life or recycled. In 3rdLife, batteries are mostly used as portable energy storage for their own trailers or sold as energy storage for home use. As part of an internal study at Kyburz Switzerland in 2019, they found that lithium-ion batteries installed in their vehicles can be recycled up to 91% without the use of chemicals. To do this, the battery must be discharged to two volts. After being cut open, the individual layers are removed and dissolved in water. Only the electrolyte cannot be recycled at this time. The MultiLife concept reduces the grey energy of vehicles and key components by 65% and conserves resources.

Throughout the case study, the CEO, K. M., and the head of the 2ndLife business segment, L. B., were interviewed for the case study. The case study was created with a top-down approach. This means that we already had a hypothesis from previous research, which was expressed in the form of the enabler and driver frameworks in Tables 9.2 and 9.3, in order to understand the approach of the company towards circularity. We used mainly the why and how questions to find out why the decision was made and what resulted from it. A constant exchange of information was conducted to shed light on the entire process and to provide insight into the results. Based on the information collected, a comprehensive picture of the retention strategy chosen and its performance was constructed. To gain a deeper understanding of the project's background, along with the interviews, we also consulted the 255-page history of the company. A key objective of the case study was to identify those enablers and drivers that were relevant to the company and how they affected the company's environmental, social, and economic results in a way that was relevant to the company.

9.2.2 *Drivers of the Case*

Based on the overview in Table 9.2, we tested the driving and facilitation factors. For the company, *Environmental* aspects were a major motivator. They are always looking for ways to minimise resources (D.E1) and the environmental footprint (D.E2). When they saw how many components were still in working order, they decided to take advantage of it. The CEO said: 'Switzerland is not a resource-poor country, the raw material is everywhere on the street, you just have to build your own mine' (K. M., personal communication, November 30, 2022). As a side effect, the company also became more resilient (D.E3). They were not yet aware of the future supply shortage during the pandemic but were able to use the components used that were supposed to be used for the main production refurbishment programme as well. The benefit seen from this reinforced the belief in the programme.

Table 9.2 Drivers of the *Refurbish* case

Cluster	Code	Driver	Context	Relevance
Environmental	D.E1	Resources	Decrease the waste of resources and create your own mine	+++
	D.E2	Footprint	Decrease product carbon footprint and grey energy	++
	D.E3	Risk	Resilience against resource-related risks	+
Institutional	D.I1	Legal	Comply with legal requirements and norms	+
	D.I2	Punitive	Avoiding punitive measures	0
	D.I3	Juridification	Anticipate future legal obligations	0
	D.I4	Tax	Taxation benefits, access to supportive funds, and subsidies	0
Stakeholder	D.S1	Image	Improve corporate image, legitimacy, or credibility	0
	D.S2	Customer	Adapt to changing consumer behaviour	+
	D.S3	Practices	Set the best practices for competitors	++
Financial	D.F1	Volatility	Price volatility of raw material	+
	D.F2	Savings	Cost saving opportunities	0
	D.F3	Competitiveness	Competitiveness was increased by being a pioneer	++
	D.F4	Value	Less waste through reusing depreciated products	++
Responsibility	D.R1	CSR	Social and ethical responsibility	++
	D.R2	Promoter	Become a promoter of systemic change	+++

0 representing insignificant importance, + indicating minor relevance, ++ reflecting notable relevance, and +++ signifying crucial importance

We could not find significant evidence that *Institutional* reasons led the company to use refurbishment. However, the company follows ISO 14001 (D.I1). The overall objective of such certification is to promote environmental protection, reduce environmental impacts, and thus properly implement environmental goals – in line with economic, social, and political requirements. As a result, a certain part of the profit must be invested in sustainable R&D projects, which indirectly also brings tax benefits (D.I4).

Since the image (D.S1) has already been built for sustainability, it does not need to be polished. There were also no requirements from external *Stakeholders* like customers (D.S2) pushing for it. The customer segment has not really changed. Although sustainability has become the norm, there are very few customers who actually act out of intrinsic motivation. It was noted that used parts can be used for large orders if the customer does not object. However, a small discount is given here.

One of the main drivers was to be a pioneer and, therefore, to show what could be the new standard in the industry (D.S3). This puts pressure on competitors to also enter CE. Depending on the industry and product, refurbishments are offered in various forms, but not with such a complex product on this scale.

Financial aspects were not really an issue in the integration of the 2ndLife process. Price volatility was not felt at the beginning, but even more so during the pandemic. Chip and board shortages led us to pay more attention to the process (D.F1). They are now happy to have taken this step years ago. Savings were also not an issue (D.F2). On the contrary, the first economic audit showed that the project would not be profitable. However, since corporate culture incorporates sustainability as a core element, the step was taken nonetheless. Competitiveness (D.F3) in itself was not an impetus for the integration of the R. A lot of money had to be invested to set up the process on this scale. The resulting publicity made it possible to gain competitiveness. It was also possible to deliver a large order faster than the competitor, because the 2ndLife system made it possible to avoid delivery bottlenecks. As an example, Hungary wanted to buy 800 vehicles, which had to be built over several years due to material shortages. 2ndLife components were able to place the order as the company offered to equip a new chassis and battery with the used components. We see that this played indirectly on the driver D.F3. For D.F4, they wanted to recycle material and its residual value as much as possible. The company saw that the components still had a long life and saw the value in them, which made the development of the 2ndLife process possible. Since used products usually have less value in the eyes of customers, companies only want new vehicles or a big discount on vehicles with used components in them. To counteract this, customers were promised a longer warranty on these components. This was accepted for value compensation.

Ethical motivations, such as reducing the harmful effects of environmental pollution, were a key driver (D.R1). The crux of electromobility would be, among other things, the battery, and thus one wanted to get as much out of the products as possible. But the *Responsibility* does not stop there. The company sees itself as a pioneer in the area of sustainability (D.R2). Steps are being taken to become more sustainable and to promote sustainability. Refurbished 2ndLife vehicles will be available as part of the CE-Symposium to build innovative vehicles or equipment for participating teams. Therefore, the challenge theme revolves around CE.

9.2.3 Enablers of the Case

The most important enabler is the company itself (see Table 9.3). This is shown by the entire culture (E.F1), which is built on sustainability. Suppliers are also ensured to have a similar company culture. For example, care was taken not to use batteries with cobalt. There is also sufficient capital to support this (E.F2), because without it, it would not be possible to take a step like this. There was even a large barrier in enabler E.F1. The CEO really wanted to integrate the materials used in CE. However, the person responsible for the internal feasibility study showed in several reports that

Table 9.3 Enablers of the *Refurbish* case

Cluster	Code	Enabler	Context	Relevance
Firm	E.F1	Culture	Committed company environment and culture	+++
	E.F2	Capital	Available capital or financial support to transition	++
Value chain	E.V1	Partnership	Build closer relationships/invest in collaboration and partnerships	+
	E.V2	Transparency	Enhance information sharing and transparency, make use of digital technologies and standardise information	0
Technological	E.T1	Research	Experiment and research	+
	E.T2	Mutuality	Joint decision-making (customers and suppliers)	0
	E.T3	R-technology	Availability of technical and technological know-how for a suitable value retention in-house	+
	E.T4	Outsourcing	Possibility to outsource the activity	0
Customer	E.C1	Engagement	Increase customer engagement and awareness, provide incentives	++
Material supply	E.M1	Logistics	Available reverse logistic network	++
	E.M2	Collection	Create simple collection/take-back systems	+
	E.M3	Symbiosis	Engage in industrial symbiosis	++

0 representing insignificant importance, + indicating minor relevance, ++ reflecting notable relevance, and +++ signifying crucial importance

it made more financial sense to shred the vehicles. Apparently, the project manager at the time drew the system boundaries in such a way that this result was achieved. In the end, the person concerned resigned and the project was built anyway, which was only possible with sufficient liquid funds.

There were no barriers in the *Value Chain* (E.V1 and E.V2). The company itself was already sufficiently closely connected with its partners and fully transparent from the start. Now that we have learnt from the process, we can improve the product design and show the manufacturers the remaining gaps in the process.

Technologically, no barrier was identified. Through service, people already knew the process and how to use refurbishment (E.T3). A vehicle and battery test bench and the necessary protocols needed to be developed to implement Rs at scale (E.T1). Nothing was found on Mutuality (E.T2). There were plans to outsource services to other companies (E.T4). However, it was decided against this to better control *Environmental* driver (D.E). For example, components most commonly used will be integrated into CE because the components could also be used in new constructions. Customers of the company are fond of the project (E.C1). Particularly financially weak private individuals can now afford a sustainable product. The aforementioned CE-Symposium can also create an incentive to become interested in the topic. The '2ndLife hall' was also converted into a showroom, so that everything can be viewed, all for free, so that as many people as possible can be reached.

Building up the value retention strategy to this extent works only if an encompassing return process with the customer is established (E.M1). As early as 2010, discussions were held with Swiss Post to take-back the vehicles in a return programme. In this way, they knew exactly how many vehicles could be refurbished each year. This knowledge also meant that there were no problems in the take-back system (E.M2), which is an important component. Since the company does not want to produce waste, it is looking for possible ways to return each component back to the economy. Symbiosis also plays an important role in this (E.M3). For example, although the framework conditions were weak, the company tended to create them itself. Batteries with partly depleted capacity were resold for home storage and an innovative battery recycling plant was created for defective batteries. Used motors are sent back to the manufacturer instead of being disposed of, as this allows the manufacturer to get the raw materials back with less energy. More ideas are being sought and are to come in the future.

9.3 Outcome and Impact of *Refurbish*

9.3.1 Ecological Results

The ecologically positioned company is making great strides in the right direction through its refurbishment programme, and is contributing to reducing the impact on the environment and encouraging the use of CE practices through its multi-life concept.

This programme can *Refurbish* more than 400 electric vehicles per year and helps to extend the life of these vehicles so that they can be used again instead of going to landfill. More precisely, the first lifecycle of seven to nine years can be extended by the 2ndLife *Refurbish* programme by approximately another seven years.

Consequently, this means that the vehicle and its components can remain on the road for an average of 14 years, instead of only seven years, as originally believed. Table 9.4 shows some of the components that are disassembled and their further use. 'Usability' here means reusability for other vehicles without major repairs. The next column 'B-goods' indicates components that must first be repaired before they can be re-installed. However, depending on the component, the intended use may also change. The main example here are batteries which become battery storage units. The last column shows how many percent do not have another life due to wear and tear and are therefore recycled or disposed of.

The grey energy of a former postal vehicle can thus be reduced by about half. As the refurbishing process uses less energy than the production of new components, it means that the amount of electricity needed to dismantle an old vehicle would be a fraction of what is needed to manufacture a new model. Currently, it is unknown exactly what energy and emission savings will result from the programme, but this

Table 9.4 Some disassembled components and their usability (based on L. B., personal communication, November 8, 2022)

Component	Usability (%)	B-goods or other use (%)	Professional recycling or disposal (%)
Motor	65	30	5
Control	95	0	5
Batteries 100–240 Ah	30	65[a]	5
Chassis	30	15	55
Tire	40[b]	0	60
Ignition lock	0	0	100

[a]Energy storage
[b]At least 80% tread

will be assessed through internal studies in the next few years to create transparency in the process.

In 2021, a total of 21,474 kg of components were reused from old vehicles that would otherwise have been thrown away. This resulted in 403 new 2ndLife DXP vehicles from used components. In comparison, they produced 2,169 new vehicles in the same year.

Indirectly, the idea of refurbishing has also contributed to the recycling of batteries, as it was some sort of enabling factor. The aim was to return every component and every small part of the vehicles to the economy. Therefore, many batteries that could no longer be installed in vehicles were seen, but they were not to be thrown away.

Batteries with more than 80% capacity are used as components in 2ndLife vehicles. If the capacity is lower, they are used as energy storage in other trailers made by the company or sold as home storage. Batteries with defective cells or batteries at the end of their useful life should be recycled as much as possible. With the idea of recycling their own batteries, an internal study was carried out. In an amazing twist of fate, the company discovered that the lithium-ion batteries they are using (lithium iron phosphate) can be recycled by using only water. The voltage of the battery needs only to be raised to two volts to enable this. 91% of the raw materials can be recovered in this way. If you include the reuse of batteries for a 3rdLife, the company achieves 65% grey energy savings.

All these individual use cases and migrations to other Rs could only be started through the broad-based *Refurbish* process. Based on the vast amounts of components and materials, this laid the foundation for further retention options.

Despite how far the company has come, the management of the company is not yet satisfied. Further steps towards circularity are planned. It is evident that ecological results have been achieved and that most of the components used can now be reintroduced into the economy again, but this entire issue has also revealed that there is always the possibility of going one step further.

In the coming years, each scrapped component will be analysed separately to determine whether more can be extracted from it. For example, they are currently in contact with the engine manufacturer. The engine itself consists of a simple aluminium and steel part. Instead of placing the engine block as a whole into a furnace and thus using a lot of energy to extract raw materials, the engine manufacturer has been offered the opportunity to return the individual components in a divided form. In this way, the supplier knows which metals and alloys he is getting back and can thus return the material to the economy in a more energy-efficient way. The exact figures are not known here, but it is hoped that this will generate added economic value.

More data could not be obtained here because the ecological study was unfortunately discontinued.

9.3.2 Social Results

Currently, not much is known about the social results that can be achieved through rehabilitation. However, there are a few things that can be inferred from the available evidence. The customer often has anxiety about buying refurbished goods, and there are a number of reasons for this. One reason for customer anxiety is the sense that refurbished goods are inferior and should be avoided if possible. As an example, one worries about the remaining quality of the components. However, what has endured seven tough years as a post-service vehicle will last for years under milder conditions in 2ndLife. Therefore, vehicles are tested on an in-house test stand, and every component is inspected. This is to meet customer needs. In addition, a two-year factory warranty is provided. This helps mitigate these concerns. The 2ndLife vehicle also meets the customer's needs in terms of range; they can still achieve about 80% of the range of new vehicles. You can choose between 100 and 160 Ah batteries. Together with a driving performance of 45 km/h and a payload of 30 kg in the front and 90 kg in the rear, all customer needs are met. This allows you to achieve a lot on short distances. The only exception, which also applies to new vehicles, is that a vehicle licence category A1 is required to drive electric tricycles. In general, the electric vehicle meets the needs of many customers and presents an ideal opportunity to continue to reduce our environmental footprint.

As more and more people become aware of the need to protect the environment and reduce the impact we have on the planet, many are looking for alternatives to reduce their Carbon footprint. To encourage people to consider refurbished electric vehicles as a viable alternative to more traditional modes of transport, they opened the showroom to everyone so that they could test their electric vehicles for themselves. People were able to meet with the sales team and discuss the various models and what they were capable of. Having the opportunity to test the vehicles themselves gave people confidence in choosing an electric vehicle over a more traditional gas-powered car. But transparency does not stop there, the showroom stretches over the entire production of used vehicles. Completely free of charge, you can book guided tours to see the whole process down to the smallest detail. This is intended on the one hand to show how sustainably the company thinks and acts and, on the other hand, to encourage people at every stage of their lives to think about CE. From gymnastics clubs to schools and companies, many have already booked a tour. The aim is to have a long-term impact on society. Over the last few years, excluding the years of the epidemic, an average of 60 tours have been conducted per year.

To further spread the programme, 2ndLife vehicles are available for the CE symposium. This event was introduced by the company in 2020 to share ideas in CE through high-level presentations and workshops by experts. Apart from the 100 or so interested parties from the business world, an open challenge has been introduced since 2021, in which the vehicle is to be converted according to a specified sustainable theme. Each team is sponsored with a tricycle and with 1,000 CHF. In addition to the fact that the tricycle can be kept afterwards, there are cash prizes worth 6,000 CHF. So far, an average of 16 teams have participated in the challenge each year. The company also has an impact on local communities. For example, it outsources simple parts to the Mörtplatz Foundation, a training centre for young people with mental and social difficulties. Around 45 people are employed in 11 workshops and studios. Vocational training is carried out in a protected environment, followed by integration into the primary labour market.

9.3.3 Economic Results

As production had to increase, a new hall had to be built to do so. The new hall came just in time for the realisation of the retention option, as it allowed the entire process, including the vehicle showroom, to be integrated into the old facility. The initial investment was not really one single invest, but rather ongoing investment costs that are expected to amount to 100,000–200,000 Swiss Franc (CHF) per year by 2022, which have been accumulating since 2019.

The battery recycling process, a result of the retention strategy, was initially designed to cost around 1 mill. CHF, with the company and Empa, the Swiss Federal Laboratories for Materials Testing and Research, each contributing similar amounts.

In the CEO's opinion, CE is more important than profit. This can be seen here, since the time and personnel invested barely cover the cost of 2ndLife's operations. Due to the low prices, which amount to approximately 1/3 to half of the cost of a new car, vehicles always find buyers. The price of the DXS for private individuals starts at 15,290 CHF. As a used vehicle, it can cost between 6,050 and 7,590 CHF in Switzerland, depending on the model. A total of 2,572 vehicles were built in 2021, of which 403 were refurbished DXP.

There has been a noticeable increase in the number of vehicles sold to Eastern Europe. The favourable prices also attracted many retirees to the company. In Switzerland, for example, about half of 2ndLife's customers are pensioners. Abroad it looks a bit different, because there are mainly companies that take the vehicles. It is estimated that about 80% are companies. Table 9.5 shows that 900 reconditioned vehicles have been sold to date, plus another 304 vehicles sold ex-plant. In total, out of 3,504 repossessions since 2019, 1,204 have been sold, 201 DXP are still active

Table 9.5 Used vehicles sold globally (based on L. B., personal communication, November 8, 2022)

Country	Number sold until 14.12.2022
Switzerland	576
Hungary	130
Germany	103
Austria	50
Slovenia	33
Thailand	3
Nauru	3
Poland	2
Total	900

in the shed, and 2,101 vehicles have been withdrawn from service. Parts that are dismantled also flow into service partner networks, own repairs, retail sale of components, and into the large order in Hungary, where 800 vehicles are on order. In this large order, only the chassis and battery are newly manufactured; the remaining components are taken out of the 2ndLife process.

However, this R contributed primarily indirectly to the company's intangible assets. For example, the CEO applied for a patent on the battery recycling system, although he wanted to make the knowledge freely available to everyone. He also began cooperating with other companies on this.

As a positive side effect, the company now maintains contacts with major automotive manufacturers and has also had a strong media presence, which has improved the company's image. Interestingly enough, many companies that were not particularly interested in CE are now using electric vehicles to show themselves as green to the outside world as if they were interested. The mainstream has arrived and they just want to strengthen their image. But it really should be the other way around; with an intrinsic desire to act sustainably, they should be on the lookout for green alternatives, as this case study demonstrated.

9.4 *Refurbish*: Summary and Learnings

As we can see, multi-component products on a large scale can be accomplished using the refurbishment process. The concept does not have to be limited to electronic equipment; it can also be used for other products. The important thing is that both the customer and the manufacturer have to be on the same page. The manufacturer has to be able to recognise that the refurbished product is not the same high quality as a brand new one, but will be good enough for customers at a reasonable price. If used components are installed in new products, customers also saw a reduction in value here. To counteract this, the warranty for these components was extended. This meant that the vehicles retained their value and could be sold at full price without suffering any further losses.

There is a barrier to the future second life of a device if the product is designed in a way that makes it very difficult for the components to be replaced. Therefore, already in the design stage, attention should be paid to what may result after the first life of the product. However, customers also need to give back products and use products from others. The company recognised this from the outset and stipulated in the contract for the first major order that the vehicles must be returned to the company. This reverse supply chain was the first important step in the development of the 2ndlife process. In this way, the company knew how many vehicles would be returned each year and was able to set up the process in this dimension. Only through symbiosis between both parties a truly CE can be achieved. Kyburz Switzerland Switzerland has shown that it can score in this regard.

Although an economic feasibility study was conducted and found to be negative, the company still decided to implement the 2ndLife process on a large scale. Above all else, the sustainable *Environmental* idea (D.E) and the *Responsibility* (D.R) to conserve resources and minimise the footprint of the product were driver for the go-ahead for the retention option.

As a result of further enabler, the process itself was strengthened and integration was even further accelerated. Several factors contributed to increasing reliance on the process, including volatile raw material prices or parts that were no longer available for new vehicles due to shortages. Thus, in the event that supply bottlenecks occurred, it was convenient to rely on the reused components that resulted from the refurbishment to bridge them. This gave rise to a wider range of opportunities and possibilities that were not available previously. During the time of the pandemic, when everyone was experiencing supply problems, they were able to bridge the gap between supply and demand for a customer. Given these benefits, there was little reason for anyone to refuse to take this approach.

However, *Institutionally* (D.I), things seem rather poor. Although political pressure is being applied to make the environment more sustainable, there has been no real legislation to hold big companies accountable. The company indicates that taxes (D.I4) can be reduced by investing profits, but they are not really noticeable when it comes to the sustainability they demonstrate. In any case, the reality is that in CE as a whole, only slow progress is made until there is some kind of legislation in place.

In the *Stakeholder* area (D.S), the company clearly wants to establish new standards for the industry. As a pioneer and promoter of circular economy (D.R2), the company wants to show that this economic model works and should be pursued, even if it is ultimately only a zero-sum game.

As to enabler for effectively implementing a retention option, a lot depends on corporate culture (E.F1). A company's own corporate culture tells us whether it is motivated by an external desire to appear green or whether it is motivated by an intrinsic desire to be sustainable. The entire retention process probably only went so well because the company was intrinsically motivated to be green.

Capital is a catalyst for this process, as large-scale CE projects are only possible when you have liquid funds to implement them. Experiments and research were conducted (E.T1), to ensure that the products were of the highest quality. Furthermore, it is important to note that the contract signed with the postal service for the return programme (E.M1 and M2) was significant enabler to the success of obtaining this large number of electric vehicles.

On the ground, their efforts have not yet translated into financial success for 2ndLife vehicles, but integration has opened up many more opportunities. As far as the company was concerned, this was not their primary goal anyway. A much more important factor was closing the loop in CE. However, the pursuit of the *Refurbish* strategy increased competitiveness and resulted in a large order for 800 vehicles for Hungary. This order has not yet been financially taken into account.

There have been three years since the setup of the 2ndLife process, the revenue barely covers the costs at this early stage. This is mainly due to the drive to find and implement more retention option. So, development costs are high at this stage. However, in this case, the idea of how to deal with the waste of the refurbishment strategy led to the discovery of a method in which batteries could be recycled using only water to recover 91% of their raw materials. This method was recently tested and will eventually become part of future batteries in other electric vehicles. In this way, it was also possible to attract a lot of attention and establish contacts with car manufacturers.

In terms of ecological impact, some positive results can be seen. The company has been able to save 50% of grey energy. It is also able to reuse 21 tonnes of materials annually rather than throwing them away. It is intended that this will protect the environment in the long run because less waste will end up in landfills and less mining will take place for virgin materials. Other companies also advertise that these vehicles reduce their ecological footprint.

References

1. Ajzen, I. & Fishbein, M. (1980). *Understanding attitudes and predicting social behavior* (296 pp). Prentice-Hall. ISBN: 978-0-13-936443-3.
2. Higgs, K. (2021). *A brief history of consumer culture*. https://thereader.mitpress.mit.edu/a-brief-history-of-consumer-culture/.
3. Bernays, E. L. (2007). *Propaganda: Die Kunst der Public Relations* (158 pp). Orange Press. ISBN: 978-3-936086-35-5.
4. Bizjak, D. & Piotr, B. (2021). Explained: Europe's new waste prevention and Reuse laws.
5. Parrique, T., Barth, J., Briens, F., Kerschner, C., & Kraus-Polk, A. (2019). *Decoupling-debunked: Evidence and arguments against green growth as a sole strategy for sustainability*.
6. Woerden, W., Groot, T., van de Pas, R. & Bloemen, S. (2021). *Living well on a finite planet. Building a caring world beyond growth*. Amsterdam. https://www.commonsnetwork.org/product/living-well-on-a-finite-planet-building-a-caring-world-beyond-growth/.

7. Bovea, M. D., Ibáñez-Forés, V., Pérez-Belis, V. & Juan, P. (2018). A survey on consumers' attitude towards storing and end of life strategies of small information and communication technology devices in Spain. *Waste Management, 71,* 589–602. ISSN: 0956-053X.
8. Kirchherr, J., Hekkert, M., Bour, R., Kostense-Smit, E. & Muller, J. (2017). *Breaking the barriers to the circular economy.* Deloitte and Utrecht University. https://circulareconomy.europa.eu/platform/en/knowledge/breaking-barriers-circular-economy.
9. Bocken, N. M. P., Niessen, L. & Short, S. W. (2022). The sufficiency-based circular economy—An analysis of 150 companies. *Frontiers in Sustainability, 3.* ISSN: 2673-4524.
10. Nevatio. (n. d.) *What you need to know about machine Retrofits—Nevatio* Retrieved December 5, 2022 from https://www.nevatio.com/resources/machine-retrofits.
11. Faulkner, C. (2020). *Where to buy refurbished products.* The Verge. Retrieved December 5, 2022 from https://www.theverge.com/21530290/refurbished-tech-products-gadgets-where-to-buy-amazon-ebay.
12. Garriss, K. (2022). *Refurbished tech market growing as a way to save cash, and help the environment.* FOX23 News. Retrieved December 5, 2022 from https://www.fox23.com/news/refurbished-tech-market-growing-way-save-cash-help-environment/E6UP2RJN7FDYTKPN7HENEDDVX4/.
13. *Why Refurbished* Apple. (2022). https://www.apple.com/shop/refurbished/about.
14. Bonnington, C. (2014). How to avoid accidentally buying a stolen phone. *Wired.* ISSN: 1059-1028. https://www.wired.com/2014/10/how-to-avoid-a-stolen-phone/.
15. Reike, D., Vermeulen, W. J. & Witjes, S. (2018). The circular economy: New or refurbished as CE 3.0? – Exploring controversies in the conceptualization of the circular economy through a focus on history and resource value retention options. *Resources, Conservation and Recycling, 135,* 246–264. ISSN: 0921-3449.
16. Wallner, T. S., Magnier, L. & Mugge, R. (2022). Do consumers mind contamination by previous users? A choice-based conjoint analysis to explore strategies that improve consumers' choice for refurbished products. *Resources, Conservation and Recycling, 177,* 105998. ISSN: 0921-3449.
17. *KYBURZ MultiLife optimiert die Ökobilanz von E-Mobilien nachhaltig.* (2022). KYBURZ Elektrofahrzeuge und Transportlösungen. https://kyburz-switzerland.ch/.

Chapter 10
R5 Remanufacture

Raphael Meyer, Odin Peter, and Katharina Luban

Abstract *Remanufacture* is a retention strategy that prolongs the lifespan of a product by adding lifecycles to the original product lifecycle. *Remanufacture* needs less resources, is cheaper and helps to improve a company's reputation, because product components are reused and reprocessed. In the following chapter, drivers and enablers for the successful implementation of a specific use case about this retention strategy are identified, and the impact and outcome are analysed in detail.

10.1 Introduction to *Remanufacture*

According to Chap. 3, Remanufacture (R5) represents the fifth retention option and belongs to the medium loop Rs with the goal of extending the useful time span of a product concept. *Remanufacture* does not prolong the actual time a specific tangible product is used, but rather ensures that an identical product can be remanufactured using the same components. Once the useful life of a specific tangible product is over, it is disassembled, its components are inspected and cleaned, reprocessed or replaced, and then used to remanufacture a new version of the same product [1]. As a result of these process steps, the geometric form of the product and the components is maintained and the added value of the components is preserved [2, 3]. The option of *Remanufacturing* enables another or even several new lifecycles of a product [1]. Figure 10.1 shows how this extends the overall useful life in a Product Lifecycle (PLC). The remanufactured product is at least 'as good as new' [4] and might even be better than the first product. Since there is always a leap in time between the time a

R. Meyer
Freienbach, Switzerland

O. Peter
Mönchaltorf, Switzerland

K. Luban (✉)
Sevelen, Switzerland
e-mail: katharina.luban@ost.ch

A. Podleisek et al. (eds.), *Circular Economy: The Next Level of Company Success*,
https://doi.org/10.1007/978-3-031-66406-9_10

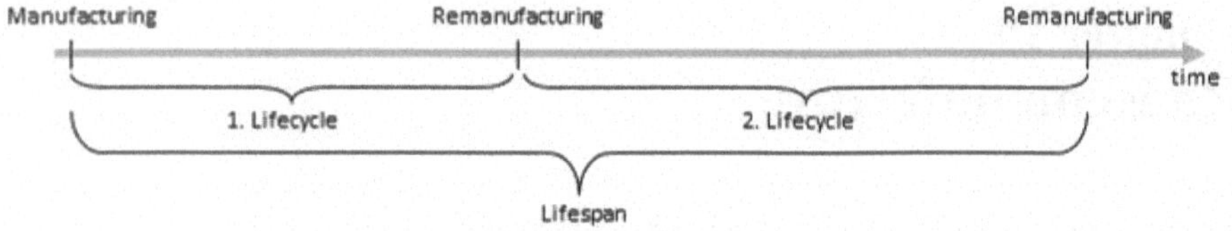

Fig. 10.1 Lifecycle of a remanufactured product

product was manufactured and the remanufacturing, more advanced manufacturing technologies can be used for the later version [3]. In addition, known weak points of a product concept can be improved by using updated or qualitatively enhanced components. As a result, the quality, performance or expected duration of the next lifecycle of remanufactured products can even exceed that of the original product [5].

Remanufacture Process

According to Andrew et al., the *Remanufacture* process consists of five key stages: grading, disassembly, inspection, components reprocessing and reassembly/testing [6, 7]. The process is circular because the goal is to keep the product in the loop. This aligns with the framework of Reike et al. [8] where remanufacturing is defined as a loop between 'use – end of use – production'. Figure 10.2 shows this loop and the five steps it consists of. The illustration was created on the basis of [6, 9].

Grading: The remanufacturing loop always starts with the inspection of the state of appearance of the products used. Heavily damaged products are sorted out because they cannot be remanufactured in an economical way. Depending on the degree of damage to the product, they can be put in another loop of the Circular Economy (CE) model, for example *Repair* or *Recycle* [10]. The quality of the products used that are classified as remanufacturable can vary and may be classified into different quality groups. The best quality group gets the highest priority for remanufacturing [6], but also lightly damaged products can be used for remanufacturing.

Disassembly: In a next step, the product needs to be disassembled into its modules and individual components. Depending on the product, different tools are involved in the dismantling. Usually, general-purpose tools, such as power drills, are sufficient. For complex products or hazardous parts, robot arms might be required [6]. Disassembling is not always a mere reverse assembly of the product. Often, more steps need to be taken, because there is no easy reverse operation for assembly operations such as gluing, riveting, pressing, and welding [10]. If it is not possible to fully disassemble a product, another retention strategy can be considered for the specific module in question.

Inspection: Before the inspection, the individual components must be cleaned, including degreasing, deoiling, derusting, and removing old paint. Cleaning parts

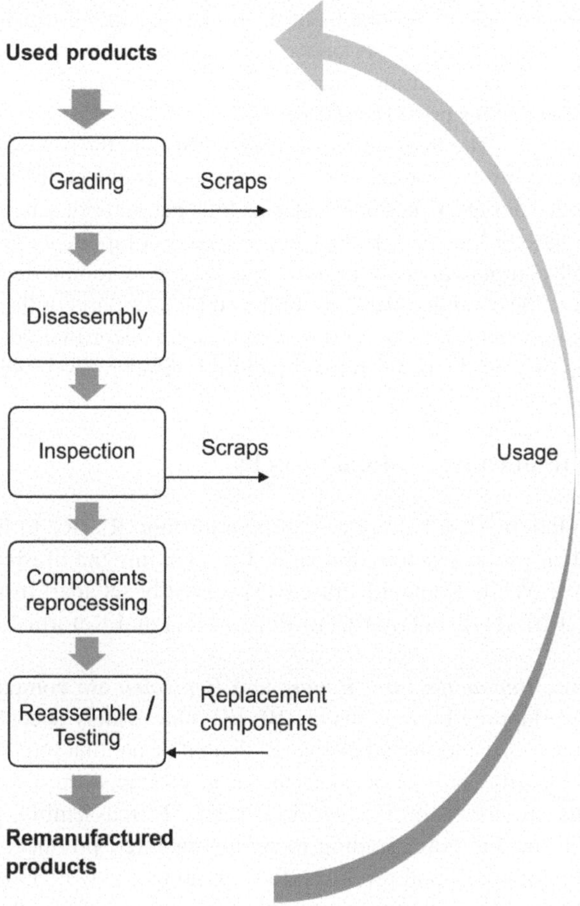

Fig. 10.2 Remanufacture process (own representation adapted from [6, 9])

is essential and can be a deciding factor when evaluating the process in economical and ecological dimensions. The disassembled and cleaned components must then be inspected and evaluated for reusability [10]. During this inspection, components are either classified as remanufacturables or as scrap. For scrap, another retention strategy, such as *Recycling* is still possible. The quality of components that are classified as remanufacturables can differ and therefore are classified into different quality groups. These groups will be treated differently in the following step.

Components Reprocessing: Components that were classified as remanufacturables during inspection will be considered in this process step. The components of the highest quality receive the highest priority as they retain the highest value [6]. The goal of the reprocessing is to ensure a 'like-new' condition of the components [10]. Components of the highest quality group may only need cleaning and surface finishing. Poorer quality components may also need to be repaired, involving metal cutting processes such as turning, milling, and grinding, which will cause longer processing time [6].

Reassembly/Testing: The final step is the reassembly and the product testing. Similarly to the disassembly, general-purpose tools such as power drills are very often sufficient, but for complex products or hazardous parts, robot arms might be necessary. For reassembling the product, reprocessed components are used. Since usually not all components of a product are graded as remanufacturable in the inspection step, the scrapped parts are replaced by new parts or by using components from other used products. In this way it might occur that out of three used products only two products are remanufactured. After reassembly, a functional test run of each product is performed [6].

Differentiation from Other Medium Loop Rs

Five Retention Options (Rs) belong to the medium loop Rs (R2 to R6), as shown in Sect. 3.1.3. Their purpose is to retain value by extending the lifespan of products or components [9]. While Remanufacture (R5) is clearly set apart from Reuse (R2) and Repurpose (R6), its distinction from Repair (R3) and Refurbish (R4) is more nuanced.

In the literature *Remanufacture*, *Repair* and *Refurbish* are compared in different ways with overlapping boundaries [1, 10–13]. One widely accepted difference between *Remanufacture* and the other two Rs is that remanufactured products lose their individual identity. This is because the used products are disassembled and their components are sorted into categorised pools. In reassembly, parts are randomly selected from this pool, leading to remanufactured products that combine elements from multiple original products, thus creating a new identity. In contrast, *Repair* maintains a product's identity by simply addressing faulty parts, while *Refurbish* updates components, but keeps the product's inherent composition and identity intact.

The loss of identity of a product becomes an issue in terms of traceability, but when evaluating the Rs in terms of environmental and economic impact, value retention is even greater with *Remanufacture* clearly surpassing the options *Refurbish* and *Repair*. But not only is the retention of the value greater with *Remanufacture*. This option also allows for the highest expectation about the length of the new lifecycle. This is due to the inspection of all parts during the remanufacturing process to ensure their 'like-new' condition. There are no signs of wear or tear anymore; otherwise, the parts are replaced. In addition, the known weak points of the product are eliminated during the process, which benefits the quality and performance of remanufactured products. Therefore, the expected lifetime for remanufactured products is highest, followed by refurbished and repaired products.

To reach this longer lifecycle, a lot of work is required, described as work content. As a consequence, remanufacturing with its five key steps (see Fig. 10.2) requires the highest amount of work to complete the process. Compared to this, *Repair* usually

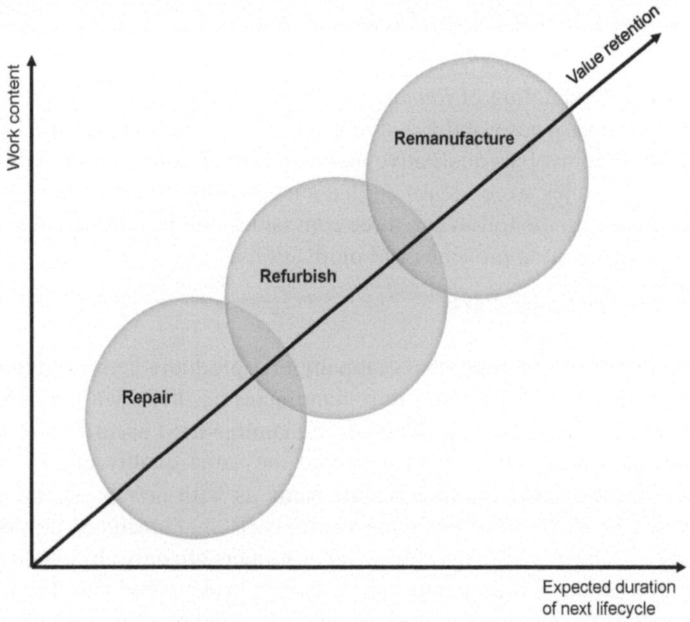

Fig. 10.3 Distinguish between Remanufacture, Refurbish and Repair (own representation adapted from [10, 13])

accounts for the lowest effort [13, 14]. Figure 10.3 illustrates these hierarchies in the dimensions 'value retention', 'expected duration of new lifecycle' and 'work content'. The model is based on [10, 13].

The above analysis can be used to derive implications for business models. Manufacturers can give longer warranties for remanufactured products than for refurbished or repaired products. Remanufacturing ensures a performance, quality, and life-time at least as good as the original product. Typically, for refurbished products, the warranty applies to all major wear parts. For repaired products, the warranty only applies to the repaired parts [13].

Differentiation to Manufacture

If it can still be done in a profitable way, remanufacturing is usually considered to be more sustainable than manufacturing: The impact on the environment is less harmful compared to conventional manufacturing [10]. Products that are remanufactured instead of manufactured require 70–85% less material and 50–80% less energy. In addition to these environmental benefits, remanufacturing is also 20–80% cheaper than conventional manufacturing [10, 15]. The main reason for these benefits can be found in the remanufacturing process. The economic and environmental value of the product can be preserved because the geometric form of the product is kept [3]. However, it also needs to be mentioned that there is a direct dependence between remanufacturing and manufacturing. Without manufacturing the original product, remanufacturing is not possible.

10.2 Analysis of a *Remanufacture* Case

To deepen the understanding of *Remanufacture* as a value retention strategy within CE, the methodology of case studies was chosen. A case study is a detailed study of a specific subject that involves qualitative methods [16]. To identify a worthwhile use case, several companies were identified that successfully implemented the concept of remanufacturing. In the following, three companies will be introduced in different industries with different approaches and motivations.

Example 1: Engine Spare Parts

The first example is an international company that produces heavy equipment and engines. They have been successful in remanufacturing their own machinery parts for more than 50 years. Used products are remanufactured according to the same specifications as new products, and therefore the same quality can be achieved. Performance and expected lifetime are the same as with new parts and therefore remanufactured products have the same warranty. Remanufactured products from this company are cheaper than new ones, but customers are only eligible to purchase remanufactured products if they return their used or broken products. The main goal of this company is to offer high-quality, inexpensive, and readily available products.

Example 2: Industrial Bearings

This international company is a manufacturer of industrial bearings. They have been remanufacturing industrial bearings for more than 25 years. After a certain period of use, industrial bearings show signs of wear and tear, such as rust, indentation, and cracks. Therefore, a thorough analysis of the current state of every single bearing is needed. They are disassembled and good quality components are used to build new bearings. Despite the high work content involved in this process, remanufactured bearings are still cheaper than newly manufactured bearings and have shorter lead times. In addition, remanufacturing helps to prolong the bearing life. This example company also emphasises the high potential of analysing used bearings in detail: In the remanufacturing process, known weak points are eliminated with the goal of extending the usable time of remanufactured bearings.

Example 3: Medical Products

Another example of remanufacturing is a European company that specialised in remanufacturing medical products. The company does not own products, but its business model is based on remanufacturing in a sector characterised by single-use products. This company was founded 25 years ago with the goal of reducing the carbon footprint of the healthcare sector by implementing CE. All medical products must be certified and legal standards must be met. The company has certified processes and shows that remanufacturing is not only an option to reduce costs, but it is also ecological and more sustainable.

General Motivation for Using *Remanufacturing*

All companies presented are examples for the retention strategy *Remanufacture*. However, the approaches of these companies are quite different. The reason for using remanufacturing can be intrinsic (company driven) or extrinsic (government, suppliers, customer and partner driven). Nevertheless, whatever drives the case, it is compulsory that it must make sense regarding profitability, and accordingly, none of the analysed companies has implemented remanufacturing without any economic benefit. The profitability of a remanufacturing project depends on many factors, such as the number of pieces that are remanufactured or logistic aspects. Remanufacturing can directly influence profitability by higher margins or a cheaper price, and therefore, win new customers. Market shares can also be increased by putting emphasis on environmental, social and marketing aspects. In essence, there are three different motivations for using remanufacturing:

- **Environmental:** lower material usage, lower energy consumption, lower CO_2 emissions
- **Economical:** cheaper than newly manufactured products, winning new customers, higher margins
- **Social:** better reputation, taking responsibility.

This first overview gives a general understanding of why implementing a remanufacture process can make sense for companies. However, the success factors and enablers are still blurry and not yet well understood. Therefore, a deep dive was conducted into one other use case, working out drivers and enablers.

10.2.1 Description of the Case

For a deeper understanding of how a company can successfully apply remanufacturing, the single case study method was applied [16]. A fourth company was chosen, which was not yet described in the previous chapter.

The case company is a Swiss company that successfully applies remanufacturing and is one of the pioneers in the field. The 130 year old company is a manufacturer of furniture and currently employs around 350 employees. The company focusses on office furniture, serving mainly the Business-to-Business (B2B) sector. The best running products are office chairs. Since the company was founded, innovation has always been a major topic. Only a few years after the beginning, the first patent was filed. With this drive for innovation, international success came. The use of the latest industrial production methods in combination with high-tech parts and unusual design proved to be extremely successful in domestic and international markets. The company clearly commits itself to the principles of sustainable company management, rather than 'growth at any price'. As a consequence, the company is constantly moving between tradition and transformation, trying to strike the right balance between financial success, social focus, and ecological responsibility. These three aspects are the same aspects that were identified as the three main motivations for remanufacturing, namely, environmental, economical, and social aspects. Therefore, it follows naturally that 16 years ago the case company expanded with a new business unit 'Remanufacturing'. In this business unit, they focus on ecological but also on economic aspects. The aim is to prevent furniture from being thrown away at the end of its lifecycle. With *Remanufacturing*, the case company can prolong the lifespan of used furniture and, furthermore, it can improve the original product with improved or new features.

For a better understanding of how the case company applies remanufacturing successfully, drivers and enablers were analysed by conducting interviews with company employees. The interviews were conducted with D. S., the company's public relations manager on November 24 and 29, 2022 and with A. D., the company's head of remanufacture on November 29, 2022.

10.2.2 Drivers of the Case

Following the framework developed in Chap. 3, the case was analysed with respect to the drivers and enablers. As a first step of the use case analysis, factors that drive remanufacturing for the case company were identified and summarised in the following paragraphs. Table 10.1 shows an overview of all the drivers discussed and their relevance to this case of *Remanufacturing*.

Environmental

One of the biggest strengths of remanufacturing compared to conventional manufacturing is the fact that less resources are required to produce the same product. Many parts can be used again after the product is disassembled and do not need to be manufactured again. Naturally, decreasing the resources used in production (D.E1) is a quite strong driver. For the same reason, there is also less CO_2 emitted to remanufacture a product compared to manufacturing and therefore the carbon footprint of the products is reduced (D.E2), which is important for the case company. Not only because customers demand it, but also because it fits their commitment to the principles of sustainable company management, which is a big part of their company philosophy and strategy. The case company has been producing in a climate-neutral way for seven years. They produce energy efficiently and handle resources with care. However, CO_2 emissions cannot be completely eliminated. Therefore, the generated emissions are compensated through cooperation, incurring additional costs.

As analysed previously, remanufacturing helps reduce material use. When it comes to furniture, the main resources are wood, metals, thermoplastics, and for upholstery parts, it is wool, synthetic, and natural fibres. The case company relies on close supplier relationships and practices insourcing to achieve the highest possible control over the supply chain. For example, eight years ago, they expanded their business and built their own business unit to trade solid wood. They cover their own needs and sell to other wood processing companies. Therefore, according to D. S., the case company never really had a problem with the availability of resources (D.E3),

Table 10.1 Drivers of the *Remanufacture* case

Cluster	Code	Driver	Context	Relevance
Environmental	D.E1	Resources	Decrease the resources used in production	++
	D.E2	Footprint	Decrease the carbon footprint of the product	++
	D.E3	Risk	Resilience against resource-related risks (scarcity, supply challenges, etc.)	+
Institutional	D.I1	Legal	Comply with legal requirements and norms	0
	D.I2	Punitive	Avoiding punitive measures	0
	D.I3	Juridification	Anticipate or meet future legal obligations	0
	D.I4	Tax	Taxation benefits, access to supportive funds and subsidies	0
Stakeholder	D.S1	Image	Improve corporate image, legitimacy, or credibility	+
	D.S2	Customer	Adapt to changing consumer behaviour, meet customer expectations, increase customer satisfaction	+++
	D.S3	Practices	Meet the standards and follow the best practices of competitors	0
Financial	D.F1	Volatility	Price volatility of raw material	0
	D.F2	Savings	Cost saving opportunities	++
	D.F3	Competitiveness	Competitiveness through innovation and business development	++
	D.F4	Value	Create value from waste and production side streams	++
Responsibility	D.R1	CSR	Social and ethical responsibility	+++
	D.R2	Promoter	Become a promoter of systemic change	++

0 representing insignificant importance, + indicating minor relevance, ++ reflecting notable relevance, and +++ signifying crucial importance

which therefore is not considered a driver in this case (personal communication, November 29, 2022).

Institutional

When it comes to the manufacture of furniture, especially office furniture, there are many legal standards and requirements that must be considered during the design and production phase. Load capacity, anti-tilt, fire protection, and ergonomics are topics of these standards and requirements. Over time, these requirements change, and old furniture may not meet all requirements. During the remanufacturing process, the furniture can be updated and the characteristics can be improved. Therefore, institutional drivers such as complying with legal requirements (D.I1), avoiding punitive measures (D.I2) or meeting future legal obligations (D.I3) were identified as potential drivers. However, this is more a theoretical application, which, according to D. S., rarely happens in reality (personal communication, November 24, 2022). Therefore, institutional drivers play only a minor role in remanufacturing.

Stakeholder

The case company is clearly a pioneer in the remanufacturing of furniture. Following the best practice of competitors (D.S3) is therefore not a relevant driver. Being a pioneer, it was quite difficult at first to find customers for their remanufacturing programme. It was something new and unusual. However, that period did not last very long. Quite rapidly, the case company was able to build up a reputation for remanufactured furniture, which also improved its image (D.S1). However, according to A. D., this is a side effect and was not originally a driver to implement remanufacturing (personal communication, November 29, 2022). However, nowadays the company sees the value of having a good reputation as a remanufacturer of furniture. Consequently, they actively communicate about their remanufacturing process, their abilities, and their successes. This attracts many new customers. Lately, the case company experienced that these customers are not only interested in remanufacturing. It is not uncommon for companies to have large inventories of old furniture that is no longer used. Recently, there have been many requests if the company can make something from this old furniture. Most of the time, a solution can be found with remanufacturing. As a result, the complexity of the case company's processes increases, but on the other hand, they also increase the know-how about remanufacturing different types of furniture.

In some cases, remanufacturing is not an option. For example, if the furniture no longer serves a purpose in its current form because its design or function is outdated. For these products, the case company created a new business branch. Old and unused furniture forms the basis for new pieces of furniture with a new form and purpose. In the 10R framework of Reike et al. [8], this new branch is best described as a mix between *Repurpose* and *Reuse*, depending on the project. Although we draw a clear line between these retention options, it is worth mentioning that without the knowledge that was built up in *Remanufacturing*, this innovation would never have happened, due to the similarity of know-how and technologies in remanufacturing. Therefore, it is safe to say that meeting customer expectations and adapting to changing consumer behaviour (D.S2) was and still is one of the biggest drivers of remanufacturing for the case company.

Financial

As described in the examples in Sect. 10.1, remanufacturing can be significantly cheaper than manufacturing (D.F2). This can also be observed in the case study. According to the case company, the market price difference can be up to 50% compared to conventionally manufactured products. The resulting economic benefits for customers are a prerequisite for acceptance and successful implementation of remanufacturing. Serving these customers provides a competitive advantage for the company compared to their competitors (D.F3) and is the basis for further innovation. Value creation from waste products (D.F4) is intrinsically implemented in the idea of remanufacturing.

Responsibility

As stated before, social and ethical responsibility (D.R1) is a main concern of the case company. Their whole culture is built around that thought. In recent years, people became aware of the environmental problems of this world. More and more people consider environmental topics such as CE to be important. The case company clearly sees this external culture as a driver for their environmentally friendly remanufacturing programme. This is true for the owners of the company, but every employee is committed to that philosophy as well. The commitment to the principles of sustainable company management has been recognised with many awards in the field of sustainability. Extensive reporting on social and ethical topics are further indicators that the case company successfully implemented a culture of responsible business management. Remanufacturing fits perfectly to this company's internal culture and therefore it can be stated that *Culture* is a main driver (D.R2). Their success as a pioneer in that field was a big image push, and they play a role in promoting a systemic change towards a CE.

As a summary of the drivers for remanufacturing, it can be said that the most important aspects are the willingness to assume social and ethical responsibilities, as well as the willingness and ability to adapt to changing consumer behaviour.

10.2.3 Enablers of the Case

In a next step of the use case analysis, enabler for the remanufacturing case company were identified. They are summarised in the following paragraphs. Table 10.2 shows an overview of all possible enablers and their relevance with regard to the retention option *Remanufacture*.

Firm

When talking to employees and when looking at the website of the case company, it becomes clear that its culture not only is a very important driver, but also counts as a main enabler (E.F1). As mentioned above, the general drift in society toward more sustainability is a driver for remanufacturing. If this external culture is implemented in the internal company culture, it can also serve as an enabler. In the case company, everyone is committed to the company's economical, ecological, and social goals. This culture is tangible throughout the entire company, but starts with the company's owners, who set strict guidelines and specifications where the company should go. Therefore, when remanufacturing was implemented as a business unit, management support secured initial funding (EF2). However, the case company does not subsidise any department for a long time. Nowadays, the business unit 'Remanufacturing' is profitable.

Table 10.2 Enablers of the *Remanufacture* case

Cluster	Code	Enabler	Context	Relevance
Firm	E.F1	Culture	Committed company environment and culture	+++
	E.F2	Capital	Available capital or financial support to transition	++
Value chain	E.V1	Partnership	Build closer relationships/invest in collaboration and partnerships	0
	E.V2	Transparency	Enhance information sharing and transparency, make use of digital technologies and standardise information	0
Technological	E.T1	Research	Experiment and research	0
	E.T2	Mutuality	Joint decision-making (customers and suppliers)	++
	E.T3	R-technology	Availability of technical and technological know-how for a suitable value retention in-house	+++
	E.T4	Outsourcing	Possibility to outsource the activity	0
Customer	E.C1	Engagement	Increase customer engagement and awareness, provide incentives	++
Material supply	E.M1	Logistics	Available reverse logistic network	0
	E.M2	Collection	Create simple collection/take-back systems	0
	E.M3	Symbiosis	Engage in industrial symbiosis	++

0 representing insignificant importance, + indicating minor relevance, ++ reflecting notable relevance, and +++ signifying crucial importance

Value Chain

No evidence of enablers regarding the value chain could be found (E.V1, E.V2). This is the case because, for remanufacturing, less resources are needed and therefore fewer materials need to be transported. However, the products used must be transported from the customer to the remanufacturing plant, but transportation, compared to traditional manufacturing, is less critical due to the smaller amounts of resources and parts required and, compared to the total costs, transportation makes up only a small part. Therefore, the aspects of the value chain cannot be ignored, but in this specific use case, they are not considered an important enabler.

Technological

A. D. (personal communication, November 29, 2022) mentions another essential factor that allowed the case company to successfully implement remanufacturing of furniture: All technical and technological know-how (E.T3) was and is available within the company due to the fact that manufacturing of all essential components and the final assembly is in-house. In-house competencies help to design new furniture considering remanufacturing processes from the start. Beyond a considerate product design, remanufacturing does not need special research or experiments (E.T1).

Due to the fact that all essential processes such as disassembly, grading, and cleaning are done in-house, there was no need to outsource or collaborate with suppliers to realise the remanufacture business case (E.T4). Outsourcing would also be difficult to reconcile with the goal of manufacturing locally, which is one of the key characteristics of the case company's sustainability programme.

Finally, it can be stated that neither suppliers nor customers played a significant enabling role to develop the case, so 'mutual decision making with suppliers or customers' is not considered a relevant enabler (E.T2).

Customer

Today, the involvement and participation of customers (E.C1) is important for many companies. With new ideas and more difficult orders, customers drive the case company to realise more complex projects. As a result, internal knowledge and competences are enhanced. High customer engagement and awareness can be an enabler in staying in the business of remanufacturing, but in the case of the case company it was not an enabler in the first place.

Material Supply

Regarding logistics and collection (E.M1, E.M2) no evidence was found to classify this as important enablers. The case company remanufactures furniture for major customers. Systems such as a take-back system or a reverse logistic system are therefore not necessary, since the required amount of products can be achieved with only one or a few customers. However, if a company wants to implement remanufacturing working with more than one major customer, Reverse Logistics (RL) systems could become critical enablers. For the case company, remanufacturing only becomes worthwhile in an economic sense, when reaching a considerate number of remanufactured units. Without the large customers that provide this number of units, take-back systems could become inevitable. This system of remanufacturing old, used products for large customers is an industrial symbiosis (E.M3). The customer is happy to replace its old furniture with remanufactured, like-new furniture. The case company only needs to acquire a small number of these customers for remanufacturing large amounts of furniture. Most of the resources required for remanufacturing are parts of the old furniture. And since the ownership of the furniture never changes to the case company, most of the needed resources are available without the need to buy them. Therefore, the monetary investment for remanufacturing is small, which benefits the case company.

To conclude the analysis of the enablers, it can be stated that culture (E.F1) and capital (E.F2) were necessary prerequisites for success. The value chain aspects were not identified as crucial. From the technological aspects, the availability of internal know-how facilitated the realisation, but other aspects such as research, decision-making with partners, and outsourcing were not relevant. When working with only one customer in a symbiosis (E.M3), the need for the otherwise important logistics aspects and aspects of product recording is eliminated.

10.3 Outcome and Impact of *Remanufacture*

To evaluate the remanufacturing case presented in this chapter, the outcome and impact of the case shall be evaluated as objectively as possible. To do so, the interview partners were asked for direct and indirect results and long-term effects of implementing this case.

10.3.1 Ecological Results

One of the reasons for implementing CE is the improvement of ecological impact, which can be evaluated with respect to resource consumption and waste reduction.

Less Resources

During the remanufacturing process, components are disassembled and reprocessed before being used again in assembly. Compared to conventional manufacturing, this is a resource- and emission-saving way of fabricating 'as-new' products, because these components do not need to be newly manufactured. Depending on the number of components repeatedly used in the remanufactured product, their geometry and condition, emissions, and resources can be saved. Of course, this will vary from product to product.

Furniture is made from wood, metals, thermoplastics, wool, as well as synthetic and natural fibres. Tastes change over time, and so does the design of furniture. Old pieces of furniture are usually more massive and therefore also heavier than new pieces of furniture. Resources were used very generously in the past. For the remanufacturing of furniture, the former design is an advantage: The wooden parts of the furniture have high stability even after the steps of a reprocessing process such as sanding. Replacement of wooden parts in remanufactured furniture is therefore necessary only in exceptional circumstances. According to the case company, compared to virgin manufacturing, remanufactured furniture therefore uses *100% less wood.*

For furniture upholstery parts, materials such as wool, synthetic, and natural fibres are used. According to D.S., signs of wear and tear in the upholstery parts are almost always present in any used furniture (personal communication, November 29, 2022). Therefore, these materials cannot be reduced in the remanufacturing process, and the same number of resources are used for upholstery in manufacturing as in remanufacturing. No general statement can be made about other materials used in furniture, like metals or thermoplastics, because this depends on the type of furniture that is remanufactured as well as its condition. According to the company's business report, considering all applied materials to manufacture their products, remanufacturing furniture in comparison uses *85% less resources.*

Less Waste

Another objective of CE is the reduction or even avoidance of waste. Considering remanufacturing as an alternative to discarding complete products, all components that are used again can be considered as waste reduction. In this sense, 85% less waste is caused. However, the company that implements remanufacturing will most certainly increase the amount of scrap material. The reason for this lies in the parts of the disassembled products that cannot be used again. A look at the waste statistics of the case company shows this development: With the growth of the business unit 'Remanufacturing', the amount of non-recyclable waste increased. Therefore, according to the company's business report, large remanufacturing projects can cause up to *20% more non-recyclable waste.*

10.3.2 Social Results

The evaluation of the outcome and impact of CE projects also includes social aspects, which will be analysed in this chapter.

Industrial Symbiosis Through Customer Exchange

One of the most important drivers of the presented case was the desire to meet the expectations of customers and adapt to changing consumer behaviour. By completing many successful remanufacturing projects, also from public customers, the case company was able to build up a reputation as a very capable and competent remanufacturer. It is no surprise that word spread and the case company's remanufacturing programme was mentioned in many newspaper articles. Recently, the case company received an important award in sustainability, and their remanufacturing programme was one of the main reasons for that. As a result of the increased reputation as a remanufacturer, the image of the whole company improved. More and more potential customers are becoming aware of the existence of remanufacturing as a value retention option to boost CE. They consider it an option for their old furniture. Furthermore, the case company actively communicates about its projects and its abilities. This attracts many new customers who want to make use of these abilities. Since there exists a variety of old furniture, the customer demands are increasing. With increasing customer requirements, remanufacturing projects are becoming more and more complex and difficult to master. The case company needs to create new areas of expertise to successfully master all these new projects. Thus, innovation increases. This innovation can help the company stay successful in completing remanufacturing projects and further increase their reputation as a remanufacturer.

All these described aspects and connections are shown in Fig. 10.4 as a causal loop diagram.

Social Change Away From Throwaway Society

Today, most people have a basic understanding of what CE is, although a social change away from the currently practised throwaway society or Linear Economy (LE) cannot be observed in many countries. According to the Swiss Federal Office for Environment, the per capita quantity of municipal solid waste increased from 659 kg in 2000 to 700 kg in 2020. This makes Switzerland one of the highest waste producers in Europe [17]. The case company is primarily active in the B2B sector, so

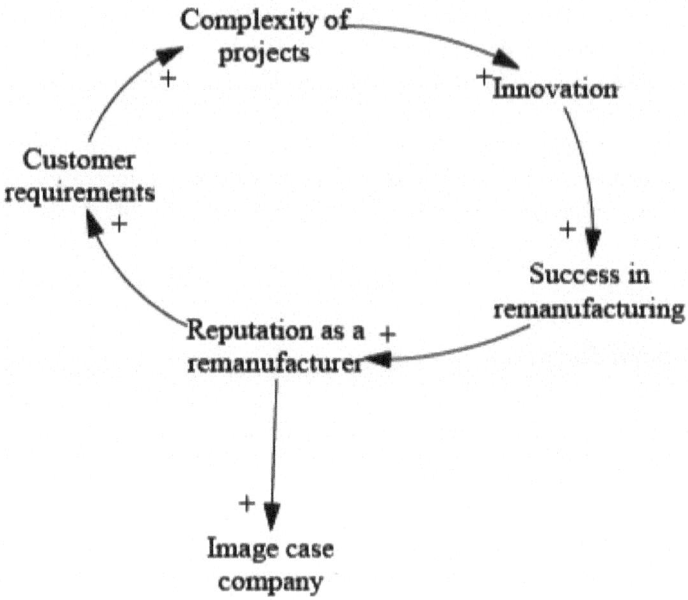

Fig. 10.4 Industrial symbiosis through customer exchange

no impact on municipal solid waste will be noticeable due to their remanufacturing programme. However, remanufactured furniture from the case company is mostly used in public spaces like theatres or spaces with a lot of people passing through, such as offices or hotels. In these public spaces, the furniture will be seen by many people. Remanufactured furniture might become fashionable. Consequently, the case company does contribute to the social change away from the throwaway society with their remanufacturing programme.

10.3.3 Economic Results

Finally, the evaluation of a CE case study also must consider economic aspects, without which a case cannot be evaluated from a business perspective.

Less Cost

It is easy to state that a substantial portion of the material cost can be saved compared to conventional manufacturing, given the fact that remanufacturing is about reusing old components. Remanufacturing companies do not provide detailed figures on the economic benefits. However, the fact that remanufactured products are often offered at a lower price can lead to the conclusion that remanufactured products have lower overall cost. According to the case company, the market price of remanufactured furniture, compared to conventionally manufactured furniture, can be up to *50%*

cheaper. Needless to say that these savings can only be reached under certain circumstances: First, the cost of collection and remanufacturing must not exceed the savings in material. Also, the amount of remanufactured products is a decisive factor: According to A. D., remanufacturing a single piece of furniture is almost never economically beneficial (personal communication, November 29, 2022). In the case of high value items, this might be different, but to reach the full economic potential of remanufacturing business cases, large quantities of used furniture are needed.

New Business Models

Even in the case that remanufacturing is not a viable option, it triggers new possibilities to earn money. Once the product no longer serves its purpose in its current form and has no market potential, other retention options can be applied. As mentioned above, the case company created a new business branch by building new pieces of furniture from old, unused furniture by changing the design or dedicating the furniture to another purpose. Although in this case less parts can be reused compared to remanufacturing, this is another option to apply CE and to fulfil customer requirements that cannot be fulfilled with remanufacturing. D. S. states, that this new business idea helps the company to expand the scope of competence and to even further increase their reputation (personal communication, November 29, 2022). However, without remanufacturing, they would never have created this new business model, which is currently quite successful. Therefore, one possible outcome of remanufacturing is not only to expand business, but also to *trigger new business models* as an additional benefit.

Since this new business model is also triggered by increasing customer requirements, this aspect also belongs to the industrial symbiosis through customer exchange. Therefore, Fig. 10.4 can be expanded with a new loop. Figure 10.5 shows the new loop with other business models.

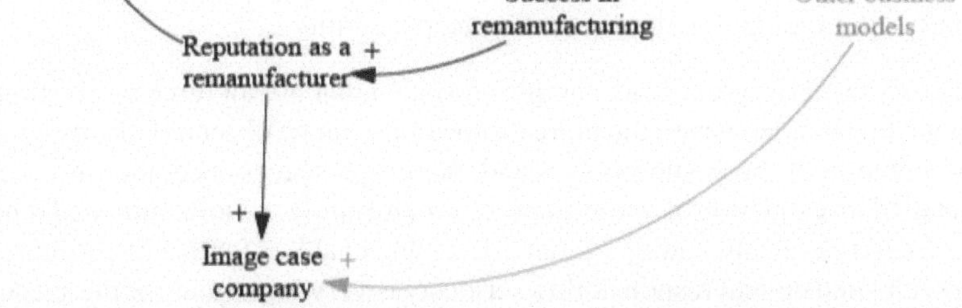

Fig. 10.5 Industrial symbiosis through customer exchange with new business models

10.4 *Remanufacture*: Summary and Learnings

Remanufacture (R5) embodies the fifth retention option of CE and belongs to the medium loop Rs with the goal of prolonging the lifespan of products or parts of the product. The main goal of remanufacturing is to transfer the inherent value of the used products as much as possible to the remanufactured products. *Remanufacturing* is a process of disassembly and re-build. Parts are graded, disassembled, inspected, reprocessed, or replaced, and then reassembled. The finished product is at least 'as good as new', but remanufacturing can even go one step further: Known weak points in a product can be eliminated and more advanced technologies can be used compared to the original manufacturing. As a result, the quality of remanufactured products can be improved compared to the original quality of the product. This results in improved characteristics or longer lifespan.

Remanufacturing is undoubtedly more sustainable than manufacturing. Less material and less energy are needed, although it is difficult to quantify the amount. As a result, remanufacturing under the right circumstances is substantially cheaper than conventional manufacturing. The reason for these benefits is that the economic and environmental value of the product can be preserved because the geometric form of the product is maintained. However, these economic benefits can only be achieved when a certain minimum number of the same products are remanufactured.

A special aspect of Remanufacture, which draws the boundaries to Repair and Refurbish, is that for remanufacturing complete disassembly is needed. Each part of the product is inspected and, if necessary, repaired or reconditioned. This results in a high work content and therefore high costs. If reconditioning or repair is not possible, parts are replaced with parts from other disassembled products or with newly manufactured parts. After the reassembling and testing of the remanufactured product, the process is finished. Due to the high content of work, remanufacturing is generally more expensive than repairing or refurbishing, but the results obtained with respect to performance, quality, and expected useful life are better.

A thorough literature review showed that there are mainly three motivations for using remanufacturing:

- **Environmental:** lower material usage, lower energy consumption, lower CO_2 emissions
- **Economical:** cheaper than newly manufactured products, winning new customers, higher margins
- **Social:** better reputation, demonstrating responsibility.

The analysis of different real examples revealed that not all three motivations are needed for success. However, the more motivations can be observed, the more stable is the foundation of the business cases and, therefore, success becomes more likely.

A detailed investigation of a single case from a furniture manufacturer and remanufacturer led to interesting findings about drivers and enablers for this retention option. Taking social and ethical responsibility seriously is very important for the successful implementation of remanufacturing. A commitment from management to sustainable company management is very beneficial, as remanufacturing is a perfect match for that goal.

Another main driver is a certain flexibility and willingness to adapt to changing consumer behaviour. Consumers play the role of a motor in pushing the company to more complex projects and, therefore, to more innovation.

For a successful implementation of remanufacturing, having all the needed technologies and know-how in house is very beneficial and minimises risks. If this is given, remanufacturing can serve as an expansion of the already existing business model with low risks because big investments in technology and know-how are not needed. Another important enabler is a living company culture around the principles of sustainable company management.

The outcomes and impacts were identified in three fields. The most important *ecological result* is that less resources are needed compared to conventional manufacturing. Up to 85% of the resources can be saved by remanufacturing furniture. However, this rate is highly influenced by the product design and the materials used. Considering that the alternative to remanufacturing is to throw away the products, these savings in resources also mean reductions in waste volume. When it comes to *social results*, it is obvious that remanufacturing triggers innovation. The intensive exchange with customers, the participation in awards, and the fact that the products are used in public spaces foster the perception and acceptance of CE concepts. When it comes to *economical results*, the business case holds as long as the cost of disassembly does not exceed the impressive savings realised by reducing the need for new materials. This cost of disassembly can be influenced by the appropriate design of the product. A very interesting economic aspect is the fact that remanufacturing can serve as an enabler for other Rs and business models. The case company expanded their business by creating new pieces of furniture from old, unused furniture that is not suitable for remanufacturing. This is an additional model that works only in combination with remanufacturing because the same know-how is required. Regarding the *economic results* of remanufacturing, remanufactured products are very beneficial to customers, since they are up to 50% cheaper compared to manufactured products while offering the same level of quality and warranty.

In summary, remanufacturing is already a widely used retention strategy. Most of the main motivations of companies are economic reasons, but more and more companies are becoming aware of the ecological and social potential that remanufacturing brings. Therefore, in a future with scarce resources, where social pressure towards a more sustainable economy constantly grows, remanufacturing will certainly be used more frequently than today. The potential of remanufacturing is massive, the necessary enablers are within the sphere of influence of the companies and they can benefit socially, economically and ecologically.

References

1. Ismail, N. H., Mandil, G. & Zwolinski, P. (2014). A remanufacturing process library for environmental impact simulations. *Journal of Remanufacturing 4,* 2. ISSN: 2210-4690.
2. Stark, R., Seliger, G. & Bonvoisin, J. (Eds.) (2017). *Sustainable manufacturing: Challenges, solutions and implementation perspectives.* Cham: Springer International Publishing. ISBN: 978-3-319-48513-3.
3. Hammond, R., Amezquita, T., Bras, B. & Woodruf, G. (1998). Issues in the automotive parts remanufacturing industry – A discussion of results from surveys performed among remanufactureres. *International Journal of Engineering Design and Automation, 4.* https://api.semanticscholar.org/CorpusID:11191043.
4. Albrecht, S., Fischer, M., Leistner, P. & Schebek, L. (Eds.) (2021). *Progress in life cycle assessment 2019* (239 pp.). Cham: Springer. ISBN: 978-3-030-50519-6.
5. Yang, J. & Zhang, A. (2022). *From zero waste to material closed loop: The way towards circular economy* (1st edn.) Singapore: Springer Singapore. ISBN: 978-981-16-7683-3.
6. Andrew, M. & Ibrahim, R. N. (2013). Remanufacturing process and its challenges. *Journal of Mechanical Engineering and Sciences 4,* 488–495. ISSN: 2231-8380.
7. Liu, S.-C., Shi, P.-J., Xu, B.-S., Xing, Z. & Xie, J.-J. (2005). Benefit analysis and contribution prediction of engine remanufacturing to cycle economy. *Journal of Central South University of Technology 12,* 25–29. ISSN: 1993-0666.
8. Reike, D., Vermeulen, W. J. & Witjes, S. (2018). The circular economy: New or Refurbished as CE 3.0? – Exploring controversies in the conceptualization of the circular economy through a focus on history and resource value retention options. *Resources, Conservation and Recycling, 135,* 246–264. ISSN: 0921-3449.
9. Vermunt, D. A., Negro, S. O., Verweij, P. A., Kuppens, D. V. & Hekkert, M. P. (2019). Exploring barriers to implementing different circular business models. *Journal of Cleaner Production 222,* 891–902. ISSN: 0959-6526.
10. Kauffman, J. (2013). *Handbook of sustainable engineering.* New York: Springer. ISBN: 978-1-4020-8938-1.
11. Nee, A. Y. C. (2014). *Handbook of manufacturing engineering and technology.* New York: Springer. ISBN: 978-1-4471-4669-8.
12. Karvonen, I., Jansson, K., Tonteri, H., Vatanen, S. & Uoti, M. (2015). Enhancing remanufacturing – studying networks and sustainability to support Finnish industry. *Journal of Remanufacturing, 5,* 5. ISSN: 2210-4690.
13. Ijomah, W. L. & Danis, M. (2022). In Goodship, V. & Stevels, A. (Eds.) (2012). *Waste Electrical and Electronic Equipment (WEEE) handbook* (pp. 145–162). Woodhead Publishing. ISBN: 978-0-85709-089-8.
14. Ghadimi, P., Gilchrist, M. D. & Xu, M. (Eds.) (2022). *Role of circular economy in resource sustainability.* Cham: Springer International Publishing. ISBN: 978-3-030-90216-2.
15. Xu, B.-S., Liu, S.-C. & Wang, H.-D. (2005). Developing remanufacturing engineering, constructing cycle economy and building saving-oriented society. *Journal of Central South University of Technology, 12,* 1–6. ISSN: 1005-9784, 1993-0666.
16. Baxter, P. & Jack, S. (2015). Qualitative case study methodology: Study design and implementation for novice researchers. *The Qualitative Report.* ISSN: 2160-3715, 1052-0147.
17. Bundesamt für Umwelt BAFU (2022). *Rohstoffe, Abfall und Kreislaufwirtschaft: Das Wichtigste in Kürze.* Retrieved December 27, 2022 from https://www.bafu.admin.ch/bafu/de/home/themen/thema-abfall/abfall-/-das-wichtigste-in-kuerze.html.

Chapter 11
R6 Repurpose

Nicolas Hofer, Roger Rinderer, Morena Di Silvestre, and André Podleisek

Abstract Currently, *Repurpose* is a retention option with limited successful industrial examples. It necessitates the development of a new ecosystem, a process that is inherently complex and time-consuming. Its broad application in the economic sphere is still in its infancy. Despite this, a selection of examples are presented, providing clarity on the concept's practical aspects. Two cases are then discussed in more depth, namely reWood, which has made *Repurpose* its core business model, and FREITAG, which stands out as a paragon, successfully translating the *Repurpose* ethos into economic viability.

11.1 Introduction to *Repurpose*

Within the hierarchy of the 10R Framework presented in Sect. 3.1.3, Repurpose (R6) between Remanufacture (R5) and Recycle (R7), represents the last Retention Option (R) of the medium loop Rs. According to Reike et al. [1] and Potting et al. [2], *Repurpose* is defined as the use of a discarded product or its parts in a new product with a different function. Given this definition, it is crucial to examine certain key aspects to accurately identify what distinguishes a genuine *Repurpose* case within Circular Economy (CE). As Schild [3], pointed out, two main concepts define *Repurpose*: A new function of the product, addressed by the *purpose* in repurpose, and the fact that the product comes from something already used before, addressed by the *re* in repurpose. On the one hand, this definition requires that a product or its components

N. Hofer
Zürich, Switzerland

R. Rinderer
Flums, Switzerland

M. Di Silvestre
Tamins, Switzerland

A. Podleisek (✉)
Wettswil am Albis, Switzerland
e-mail: andre.podleisek@ost.ch

A. Podleisek et al. (eds.), *Circular Economy: The Next Level of Company Success*,
https://doi.org/10.1007/978-3-031-66406-9_11

be discarded, indicating that they have completed their initial lifecycle. In other words, the product no longer meets the need of the consumer. On the other hand, the definition also hinges on the product losing its original intended function. The product is adapted to a new role in its subsequent lifecycle and is kept in the economy longer [1, 4, 5].

However, not all *Repurpose* cases fit this description. For example, drug repurposing in the medical sector. Sildenafil, originally intended to treat angina and hypertension, was later recognised for its effectiveness in treating erectile dysfunction [6]. Similarly, Amantadine was first used to combat influenza, but is now also prescribed for Parkinson's disease [7]. These instances illustrate that, although new uses/functions were identified, the drugs did not lose their original function. Additionally, once a specific drug is administered to a patient, it cannot be used by another individual. A second use cycle is not possible. Thus, such types of repurposing diverge from the intended CE notion of *Repurpose*.

11.1.1 Differentiating Repurpose from Other Rs

It can be quite difficult to differentiate between *Repurpose* and *Reuse*. For example, an empty water bottle is used as a flower vase. Is the bottle being *reused* or *repurposed*? The answer depends on the definition of the intended function. Let us assume that the intended function of this bottle is simply to contain 'something', using it as a vase aligns with *Reuse*. However, if the function is defined as storing drinking water and the bottle is then used as a vase, it falls under *Repurpose*. The original function has been altered, signifying a transition in product use.

Another example is the circular lifecycle of flat glass units with the intended function of thermal insulation. If the disassembled glass units continue to function as components of the exterior facade, it is *Reuse* (see Fig. 11.1). In contrast, if the glass units no longer meet insulation requirements and have been adapted for interior use, now serving as partitioning windows or walls [8], it is *Repurpose*. Therefore, a deeper examination of what the product's intended function truly is sometimes necessary to correctly categorise the action.

Distinguishing *Repurpose* from other strategies is more straightforward. Within the medium loop Rs, *Repurpose* is distinct in its approach. While *Remanufacture*, *Refurbish*, and *Repair* aim to conserve the original form and function of the product, *Repurpose* focusses on keeping the form of the product or its components in use, with the function deviating from the original. This is the key difference and demonstrates the unique role *Repurpose* plays in CE.

In contrast to long loop Rs, such as *Recycle* and *Remine*, which aims to reclaim value in terms of materials to be used again, or *Recover*, which aims to reclaim value in terms of energy, *Repurpose* and other medium loop Rs focus on preserving the value of the product or its components. This difference is depicted in Fig. 11.2, where two approaches are contrasted for handling post-consumer plastic bottles. On the left, the *Remine* approach of Tide Ocean SA [9] is highlighted. It involves the collection

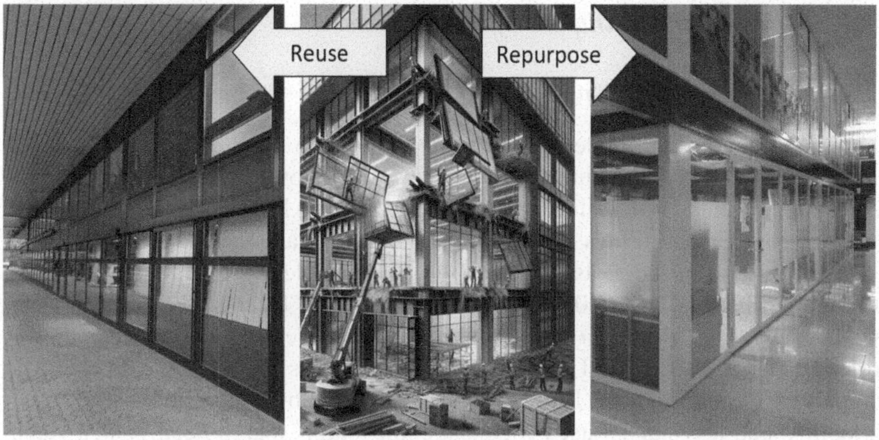

Fig. 11.1 Reuse versus Repurpose of flat glass: On the left, Reuse if the flat glass is used in the facade; on the right, Repurpose if the flat glass is used as interior partitioning walls

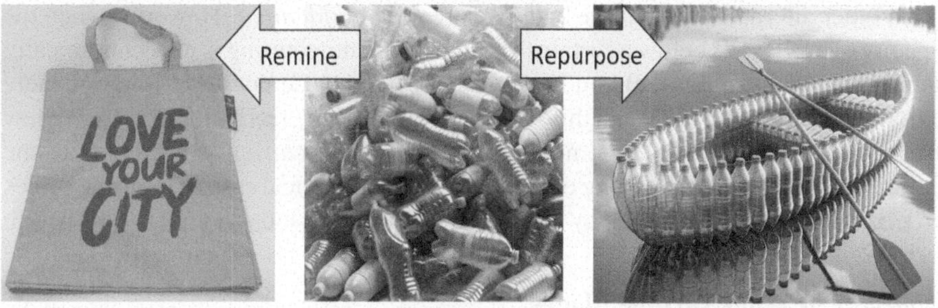

Fig. 11.2 Remine versus Repurpose of PET bottles: On the left, a bag produced by Tide Ocean SA from remined ocean-bound bottles [9]; on the right, bottles repurposed into a canoe

of ocean-bound plastics from islands, beaches, and landfills and the processing into new products, such as the showcased bag. On the right, the *Repurpose* approach of Ambiba and Nature [10] is shown. The organisation creatively reassigns the bottles a new function by using them to build canoes. This approach not only conserves the embodied energy and material of the bottles but also leverages their existing shape, demonstrating an innovative method of extending the use cycle of a product beyond its intended use.

11.1.2 Practical Examples of Repurpose

Repurpose of Lithium-Ion Batteries

The first compelling example of *Repurpose* in practice is observed with Lithium-Ion Batteries [11]. As the automotive industry shifts towards electric vehicles (e-vehicles), the demand for Lithium-Ion Batteriess (LIBs) has increased significantly. This surge in demand brings both opportunities and challenges. Estimates suggest a substantial rise in the global stock of electric vehicles, potentially reaching 50 mill. up to 130 mill. by 2030 [12, 13]. A crucial point for these batteries is when their capacity falls below 80%, a commonly accepted threshold for replacement. It is projected that by 2030, up to 11 mill. tonnes of spent LIBs will be discarded [12]. The complex nature of recycling LIBs [14], due to their high power capacity, technological variability, and less standardised chemical composition, makes *Repurpose* a viable alternative to retain some of the value of these discarded batteries. Repurpose is also the compelling alternative when examined through the lens of the Circular Value Hill, a topic explored in Sect. 3.1.2.

An other alternative is that the batteries are *Remanufactured*. However, when the capacity of a LIB falls below a certain threshold, remanufacturing is not economically viable [15]. In this case, LIB can be repurposed for a different application, namely stationary energy storage; and this with less effort, i.e., technological know-how or even capital investment. The main cost of *Remanufacturing* is attributed to the testing, grading, and repackaging of the LIBs [15]. A practical case of multiple cycles for LIB from electric vehicles by *Refurbish*, *Repurpose* and *Recycle* is outlined in Sect. 9.2.

Repurpose of Wind Turbine Blades

Wind turbines, with a typical lifespan of 20 to 25 years, are projected to contribute around 43 mill. tonnes of composite waste by 2050 [16]. *Recycling* this composite material presents multiple challenges [17, 18], as does *Refurbish* or *Remanufacture*, due to logistical issues related to size, accessibility, and reassembly [19]. *Repurpose* emerged as a viable and creative alternative. The lightweight and high strength of Wind Turbine Blades (WTBs) render them ideal for various engineering and infrastructure applications. Discarded WTBs could be repurposed into house roofs [20], main grinders for bridges [21, 22], urban bike shelters [23], or even lounge chairs (see Fig. 11.3). These examples also highlight another aspect of *Repurpose*, demonstrating that it can involve using parts or segments of a product, not exclusively the entire item.

Repurpose in Fashion and Household

In addition to the cases presented, there are more examples, such as the Quake Catcher Network, where acceleration sensors from discarded phones or laptops are used to detect earth quakes [24, 25], or the inventor Topher White, who created rainforest sound monitoring using discarded mobile phones [26]. Other examples exist in the

Fig. 11.3 Creative applications of repurposed wind turbine blades by ANMET: On the left, a repurposed WTB as a lounger chair; on the right, a repurposed WTB as a bridge grinder in Poland (photo courtesy of ANMET [22])

Fig. 11.4 Repurpose spawns creativity: On the left side, chair made out of jeans. On the right side, backpack made out of T-shirts made by Nicolas Heeb [30]

manufacturing sector, where steel cutting tools are repurposed to wood cutting tools to bypass the costly sharpening process.

However, within the fashion and household domain, *Repurpose* has found a multitude of applications, driven mainly by consumer creativity (see Fig. 11.4). Everyday items that are often overlooked can be transformed into something entirely different. This includes using disposable razors for arts and crafts, converting tissue boxes into decorative storage, reimagining cardboard tubes for various Do-It-Yourself projects [27] or converting wood pallets into a piece of furniture. Wine bottles are repurposed into elegant candle holders or quirky planters, while old books find a new purpose as unique shelving units or craft materials [28]. Even broken plates can be creatively upcycled into mosaic art, and washing machines can be reborn as eclectic pieces of furniture or garden planters [29].

11.2 Analysis of Two *Repurpose* Cases

11.2.1 Description of the Repurpose Case reWood

ReWood, a Swiss company founded in 2018 in Biel, operates with a team of seven employees [31]. The core of reWood's strategy revolves around breathing new life into discarded wood, transforming it into unique and durable furniture pieces (see Fig. 11.5). The company aims to reduce the number of trees logged while offering its clients stylish, unique, and above all sustainable furniture. In particular, the company exclusively works with waste wood, completely avoiding the use of new wood. Additionally, the company strives to transform consumer perceptions of consumption by focussing on the sustainable and circular philosophy embodied in its products. ReWood actively encourages individuals to make mindful purchasing choices, opting for sustainable goods while recognising and advocating for the advantages these choices offer.

Drivers of the ReWood Case

According to C. G. (personal communication, December 2021), the concept of reWood stemmed from a deep-seated frustration with what they perceived as wasteful practices within the industry. The founders observed an overwhelming amount of wood waste, from discarded pallets to scraps, that was routinely consigned to disposal, irrespective of the business's scale or sector. Moved by the scale of this waste, the founders recognised an opportunity to innovate and cultivate a more sustainable practice (D.F3). ReWood's vision was crystallised around the principle of turning

Fig. 11.5 Unique furniture from repurposed wooden crates (photo courtesy of reWood [31])

waste into value, transforming neglected wood into distinctive furniture that combines elegance with a storied past. Building on the definition in Sect. 11.1, we can contextualise reWood within the R-Framework: The concept can be distinguished from wood *Recycling* because the original form, for example, a plank, is retained and fundamental to the newly created product. It is also different from *Repair* or *Refurbish*, because the plank is used as a component in a new type of product. Furthermore, it is not *Remanufacturing*, because the plank can be modified or turned into a completely different part of.

Their drive was not only to create a sustainable business, but to influence the entire furniture industry, that is, to set a new standard for competitors. The goal is to catalyse a systemic change (D.R2) by showcasing the viability and beauty of repurposed materials, thus fostering a shift in consumer behaviour towards more eco-conscious purchasing decisions (C. G, personal communication, December 2021).

The operational strategy of reWood diverges from conventional production norms, making certain drivers less pertinent to its business model. For example, driver D.E1, which is centred on reducing resource consumption during production, has minimal immediate significance. The foundational raw material of reWood, waste wood, serves a secondary purpose beyond its initial use. This driver is more applicable to primary manufacturers, where reduction of wood used would be a direct concern. The company's process, by virtue of repurposing existing materials, intrinsically conserves resources and sidesteps the typical resource intake of traditional manufacturing.

Furthermore, D.E3, the resilience against resource-related risks, has a diminished applicability. ReWood's approach, which uses waste materials, inherently protects the company from fluctuations and uncertainties in the primary raw material markets.

Institutional drivers (D.I1–D.I4) also appear to be extraneous. The company was not established as a reaction to institutional pressures or directives; it was born out of an innovative response to the problem of waste. Consequently, drivers related to compliance, legal obligations, or government incentives were no catalysts for reWood's foundation.

Stakeholder-related drivers (D.S1–D.S3), such as those concerned with corporate image or consumer behaviour adaptation, also did not influence the initial phase of the company. ReWood did not emerge from the need to reform an existing corporate image or align with shifting consumer trends; it was created anew with a distinct mission and vision.

However, the financial and responsibility drivers are indeed instrumental in the establishment of reWood. Creating value from waste, becoming a competitive business, committing to social responsibility, and promoting systemic change were key motivators. These drivers underscore reWood's foundational ethos: creating economic value while advancing environmental stewardship. The business was not driven to change; it was created as a change agent, embodying financial prudence and ecological consciousness from the outset.

The insights provided by C. G., will be used in Sect. 11.3.2 to populate Table 11.1 to reflect the factors that influencing the *Repurpose* cases. This approach diverges from our standard analyses conducted in other R-case studies due to the specifics of reWood's business model and the available data. Unlike other instances where direct confirmation of driver relevance was obtained, our assessment here is based on informed argumentation.

Enablers of the ReWood Case

At the core of reWood lies a deep-rooted commitment to sustainability that is reflected in all aspects of their operations. From the sourcing of materials to the finished products, reWood takes a comprehensive approach to environmental responsibility, demonstrating its steadfast commitment to sustainable practices and committed culture (E.F1). ReWood's operational success is underpinned by two fundamental factors: strategic sourcing of waste wood and a deep commitment to customer engagement. The company sources its wood from the local industry. It is generously supplied, often at no cost or at a reduced price (C. G., personal communication, December 2021), a sign of a strong partnership and the establishment of a beneficial symbiosis (E.V1, E.M3).

Customer engagement is another critical pillar of reWood's business model, especially considering their reliance on online sales. They have an open-door policy at their workshop, inviting customers to be part of the creation process, thus fostering a sense of community and transparency (E.V2 and E.C1). This strong customer relationship is further evidenced by the organic promotion of reWood's brand on social media, demonstrating customer loyalty and trust in the brand. Transparency in their operations builds trust and loyalty, which is also echoed through social media promotion by their customers.

Design innovation also plays a significant role in reWood's strategy. The company adeptly adapts its products to the varying characteristics of waste wood, embracing its natural properties when creating furniture. This approach conserves energy by avoiding extensive material processing, and also turns potential design challenges into creative opportunities. For doing so, reWood engages with customers to identify the best combination of available wood and design choices (E.T2). The availability of raw materials is a crucial factor and in reWood's case intrinsically tied to the wastefulness of others. To address this, reWood employs a flexible approach in their furniture designs, making the most of diverse wood types, thus ensuring a steady flow of materials for their unique products. Despite these strengths, reWood faces a significant challenge in scalability. This issue is not unique to reWood and has been echoed in broader academic discourse. For example, Morseletto points out the complexities of *Repurpose* processes [5], and Kyungeun reviewed the benefits and drawbacks of upcycling [32].

Key Learnings

Through the case study of reWood, it becomes evident that the company's success is intricately tied to its unique value proposition and its steadfast commitment to sustainability. These attributes are not mere facets of their operation; they are the foundation of their value to customers, offering a compelling alternative in a market saturated with conventional offerings. Although reWood's local production model inherently faces challenges related to scalability because of the available raw materials, its agility and flexibility in production are key to navigate this constraint. By tailoring their products to the types of wood available, reWood has turned potential limitations into a distinctive competitive advantage, showcasing adaptability as a cornerstone of modern sustainable business practices.

11.2.2 Description of the Repurpose Case FREITAG

Company Description and Background

FREITAG, established in 1993 in Zurich, has evolved from a practical solution for cyclists into a global symbol of sustainable fashion. The brand's journey began when brothers Daniel and Markus Freitag crafted the first prototype of what would become their iconic messenger bag (see Fig. 11.6). Inspired by the constant flow of trucks through Zurich's transit routes, they found a new application for old truck tarpaulins, bicycle tubes, and car seat belts. Their initial design not only addressed the need for a durable and water resistant bag, but also laid the foundation for the innovative approach of FREITAG in repurposing industrial materials.

Over the years, FREITAG has expanded its product line beyond messenger bags. Today, the brand offers 33 bags and 34 accessories, such as notebook and phone cases, agendas, wallets, and key holders. Each product, crafted by a workforce of around 250 employees, stands as a unique piece, testament to the potential of repurposed materials. Their commitment to sustainability is further emphasised in their FREITAG manifesto, which proclaims 'We think and act in cycles' [33], highlighting their dedication to fostering a CE. The phone case, for example, is made from old ski boots. The Institute for Materials Science and Plastics Processing (IWK) at the

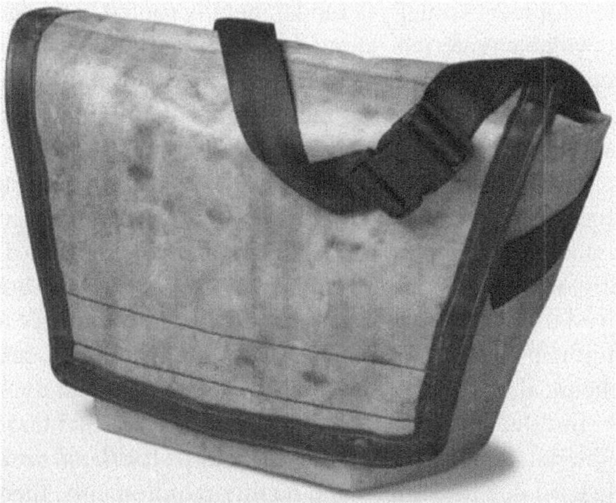

Fig. 11.6 Prototype of the FREITAG messenger bag from 1993 made out of used truck tarpaulins, bicycle inner tubes and car seat belts (photo credits: FREITAG [33])

Eastern Switzerland University of Applied Sciences (OST), together with FREITAG, have developed a process to recycle old ski boots into a sleek and modern cover for smartphones [34, 35]. The product is created not only from recycled materials but is also fully recyclable. In a commitment to this cyclical approach, FREITAG has introduced a return programme, ensuring that once these stylish covers have served their purpose, they can be returned and transformed into new protectors.

FREITAG's remarkable growth is quantifiable, with yearly sales of around 500,000 bags and accessories. This success is further underscored by their direct operation of 29 stores, in addition to a network of 260 retail partners in 23 countries.

FREITAG's Production Process

FREITAG's production process begins with truckspotting, where dedicated individuals known as truck spotters search Europe for used truck tarpaulins, the key resource. Approximately 340 tonnes of these tarpaulins are sourced annually. Once in the factory, the tarpaulins are dismantled and any unusable parts like eyelets and straps are removed. The tarpaulins are then washed in washing machines using rainwater. After being dried and sorted by colour, the tarpaulins are handed over to the bag designers. Then these designers select and cut the most visually appealing parts of the tarpaulins. The final phase involves sewing the pieces into unique products. The completed items are then subjected to a quality check before being dispatched for sale in stores and online.

11.3 Outcome and Impact of *Repurpose*

The outcome and impact has been analysed for FREITAG from which we derived the drivers and Enablers. FREITAG's journey over the past 30 years has been marked by a steadfast commitment to repurposing materials, focussing on quality and durability, and celebrating the unique character of each product. The company's product development strategy is fundamentally rooted in the use of repurposed, recycled, and recyclable materials, along with circular design principles that ensure longevity and repairability, taking into account the end of the Product Lifecycle (PLC) from the outset. FREITAG claims that with this in mind, the ecological footprint of a product is reduced by up to 80% [36].

The commitment of FREITAG to circularity is further exemplified in its production processes and material choices. Repurposing used truck tarps into durable bags reduces CO_2 emissions by 22% compared to the use of new materials [36]. Additionally, the incorporation of recycled PET from used bottles into their bags, which is dyed using a more environmentally friendly spin-dye technology, reflects their commitment to purposeful material selection. This process significantly reduces water, chemical, and energy use compared to traditional dyeing methods.

In 2022, FREITAG's material input was 365,005 kg, with the majority of this material, about 75.8%, being truck tarps for *Repurpose* [36]. As not all tarpaulin can be used due to cuttings and quality requirements, there is still production waste. At its Zurich production site, this waste was reduced by 28.9% to 161 tonnes in 2022 compared to the previous year. Part of this reduction was achieved in the *Repurpose* process of truck tarps, which represent 75% of their operational waste, by improving the cutting and using cutouts for smaller accessories [36].

The transformation of worn-out ski boots into smartphone cases using recycled Thermoplastic polyurethane (rTPU) shows the effort of FREITAG to create products that are not only recycled but also recyclable up to seven times [36].

FREITAG's approach to sustainability is not just about repurposing waste or using recycled materials; it is about rethinking the entire lifecycle of a product. From design to production and beyond, every step is carefully considered to minimise environmental impact and maximise product lifespan. This holistic approach to sustainability, combined with innovative design and material use, has positioned FREITAG as a leader in sustainable product development and a model for others in the industry.

11.3.1 Drivers and Enablers of the FREITAG Case

In contrast to the other cases discussed in this book, the case of FREITAG was a desktop research without interaction with company representatives. However, considering the detailed information available on the company's website [37], its Impact Report [36], and the joint projects of OST and FREITAG, we believe that our analysis is substantiated. Still, we would like to point out that this is only our view of the specific drivers and enablers influencing FREITAG's business model and strategy.

It is clear that the origin of FREITAG and its subsequent evolution suggest intrinsic drivers that align with sustainable innovation rather than a reaction to regulatory changes, making drivers such as D.I1 to D.I4 less immediately relevant. The company's roots in repurposing – beginning with the creation of a durable, water-resistant bag from truck tarpaulins – demonstrate an early alignment with driver D.F4, creating value from waste streams. This foundational step indicated a keen foresight into consumer behaviour changes, making D.S2 a possible driver, as the brand anticipated the market's readiness for sustainable, robust products.

In its expansion to a wide range of products, including new material combinations such as recycled ski boot smartphone cases, FREITAG continues to innovate, aligning with driver D.F3, as it seeks to maximise the utility of waste materials and increase their competitive edge. Their reported decrease in production waste and their focus on reducing ecological footprint suggest that D.E2 also serves as a driver, propelling their strategy to introduce new products that adhere to the principles of CE.

The enablers for FREITAG appear to be deeply embedded in their culture (E.F1) of sustainability and innovation, as evidenced by their collaboration with the IWK. The company's dedication to a 'cycles' mindset is an enabler in itself, fostering engagement (E.C1) within its workforce and with its consumers.

11.3.2 Drivers and Enablers of Repurpose

Based on the research and interviews with reWood and the desk research performed on FREITAG, as discussed above, we have created summary tables consisting of both repurpose cases, one for drivers (Table 11.1) and one for enablers (Table 11.2).

Table 11.1 Drivers of the *Repurpose* cases

Cluster	Code	Driver	Context	Relevance
Environmental	D.E1	Resources	Decrease the resources used in production	0
	D.E2	Footprint	Decrease product carbon footprint	+
	D.E3	Risk	Resilience against resource-related risks (scarcity, supply challenges ect.)	0
0 due to the unique approach of using discarded materials right from the start; footprint consideration were true for both companies				
Institutional	D.I1	Legal	Comply with legal requirements and norms	0
	D.I2	Punitive	Avoiding punitive measures	0
	D.I3	Juridification	Anticipate or meet future legal obligations	0
	D.I4	Tax	Taxation benefits, access to supportive funds, and subsidies	0
Both company's foundation was driven by innovation in response to waste-related challenges rather than external institutional factors				
Stakeholder	D.S1	Image	Improve corporate image, legitimacy, or credibility	0
	D.S2	Customer	Adapt to changing consumer behaviour, meet customer expectations, increase customer satisfaction	++
	D.S3	Practices	Meet the standards and follow the best practices of competitors	0
The companies were not established with the intention of reforming an existing corporate image or conforming to changing consumer trends; however, they were founded with a mission and vision to wow a new type of consumers from the outset				

(continued)

Table 11.1 (continued)

Cluster	Code	Driver	Context	Relevance
Financial	D.F1	Volatility	Price volatility of raw material	0
	D.F2	Savings	Cost saving opportunities	+
	D.F3	Competitiveness	Competitiveness through innovation and business development	+++
	D.F4	Value	Create value from waste and production side streams	+++
Both companies differentiate through a unique offering and value creation from waste. Cost saving for the input materials compared to virgin material is a favourable side effect and has been rated with '+'. Price volatility was not deemed an initial driver for starting *Repurpose*				
Responsibility	D.R1	CSR	Social and ethical responsibility	+++
	D.R2	Promoter	Become a promoter of systemic change	+++
Responsibility aspects were a key driver for founding reWood, they are also embedded in FREITAG's manifesto [33]				

Table 11.2 Enablers of the *Repurpose* cases

Cluster	Code	Enabler	Context	Relevance
Firm	E.F1	Culture	Committed company environment and culture	+++
	E.F2	Capital	Available capital or financial support to transition	0
The entire business model of reWood and FREITAG is a demonstration of their corporate values that emphasise ecological preservation and a genuine dedication to creating a sustainable company atmosphere and culture, based on the reuse of discarded materials				
Value chain	E.V1	Partnership	Build closer relationships/invest in collaboration and partnerships	+++
	E.V2	Transparency	Enhance information sharing and transparency, make use of digital technologies and standardize information	+++
ReWood is built on strong partnerships with local industry players for sourcing of waste wood at no cost or at a reduced price, plus an open-door policy and customer involvement in the creation process strongly reflected by E.V2 and also E.T2. FREITAG emphasises the circular community and partnerships for material improvement and is building its new product lines on collaboration				

(continued)

Table 11.2 (continued)

Cluster	Code	Enabler	Context	Relevance
Technological	E.T1	Research	Experiment and research	+
	E.T2	Mutuality	Joint decision-making (customers and suppliers)	+++
	E.T3	R-technology	Availability of technical and technological know-how for a suitable value retention in-house	++
	E.T4	Outsourcing	Possibility to outsource the activity	0
ReWood's design innovation showcases E.T1 (experimenting), E.T2 and E.C1, as the company collaborates with customers to adapt product designs to the varying properties of waste wood, resulting in very personal furniture pieces. FREITAG is working with suppliers on better truck tarps and has built very specific know-how in tarp quality management, design, and, recently, take back				
Customer	E.C1	Engagement	Increase customer engagement and awareness, provide incentives	+
Customer engagement is a key enabler for both companies. Due to the uniqueness of each manufactured product, customers establish a personal relationship to 'their' product				
Material supply	E.M1	Logistics	Available reverse logistic network	0
	E.M2	Collection	Create simple collection/take-back systems	+
	E.M3	Symbiosis	Engage in industrial symbiosis	++
Both companies have benefitted from a relatively low hurdle to material access (E.M2). They are actively creating a symbiosis for material input. FREITAG is now establishing a new take-back system, but this was not key for the existing product lines. Enaging in symbiosis is a key aspect for both, with limited scalability for reWood				

In the tables, drivers or enablers that do not align with reWood's or FREITAG's operational framework or the companies' founding principles are marked with '0', indicating their peripheral relevance to their corporate strategy. Drivers or enablers that are deemed critical to the companies' strategic goals and operational decisions are denoted with '+++' for high priority.

11.4 *Repurpose*: Summary and Learnings

ReWood and FREITAG are prime examples of companies that have successfully made the concept of *Repurpose* their business models. Both companies have managed to find new applications for discarded products, giving them new life and function, and demonstrate that sustainability can go hand in hand with economic growth.

Both companies provide value to customers through unique *Repurposed* products and commit themselves openly to contributing to a CE. They face common challenges, notably the availability of raw materials, which is crucial for their production processes. FREITAG has established a network of truck spotters to overcome this challenge, while reWood has found local partners.

Key differences emerge in their approach to scalability and market presence. FREITAG has achieved international reach with stores worldwide, while reWood operates within the local Swiss market via an online shop.

Drawing from the use cases of FREITAG and reWood, key aspects that *Repurpose* businesses should consider in practice include delivering customer value through unique and sustainable products, ensuring long-term use or the possibility of additional cycles, and focussing on the availability and longevity of raw materials. Scalability, although challenging, has been shown to be attainable, as evidenced by the global expansion of FREITAG. Flexibility and creativity in product design are also essential, allowing businesses to adapt and evolve within a CE.

The success of these two companies is underpinned by a deep-seated commitment to repurposing materials, a drive for innovation and competitive edge, and a commitment to reducing their environmental impact, all of which are facilitated by a strong, sustainability-focused company culture and close relationships with customers and partners. The key to a successful *Repurpose* business lies in creating a valuable offer that maintains its worth in the economy, ensuring the uniqueness and sustainability of products, supporting future recyclability, thus generating a true CE in the long term. However, both companies also demonstrate the limitations of today's *Repurpose* business models of a very specific, narrow market. The new product lines of FREITAG, mobile phone cases from rTPU, touch a new segment, but still within fashion/accessory. Its new take-back concept could demonstrate the successful implementation of multiple cycles.

References

1. Reike, D., Vermeulen, W. J., & Witjes, S. (2018). The circular economy: new or refurbished as CE 3.0?–Exploring controversies in the conceptualization of the circular economy through a focus on history and resource value retention options. *Resources, Conservation and Recycling, 135*, 246–264. ISSN: 0921-3449.
2. Potting, J., Hekkert, M., Worrell, E., & Hanemaaijer, A. (2017). *Circular Economy: Measuring innovation in the product chain PBL publication number: 2544*. The Hague: PBL Netherlands Environmental Assessment Agency.
3. Schild, T. (2020). *Design for Repurpose towards a circular economy*. Retrieved December 28, 2021 from http://resolver.tudelft.nl/uuid:96a8271f-5ee3-4bbe-8fec-1400c2719ef1
4. United Nations Environment Programme. (2021). *UNEP Circularity Platform. Repurpose*. https://buildingcircularity.org/repurpose/
5. Morseletto, P. (2020). Targets for a circular economy. *Resources, Conservation and Recycling, 153*, 104553. ISSN: 0921-3449.
6. Ghofrani, H. A., Osterloh, I. H., & Grimminger, F. (2006). Sildenafil: from angina to erectile dysfunction to pulmonary hypertension and beyond. *Nature Reviews Drug Discovery, 5*, 689–702. ISSN: 1474-1776.

7. Lee, H. -M., & Kim, Y. (2016). Drug repurposing is a new opportunity for developing drugs against neuropsychiatric disorders. *Schizophrenia Research and Treatment, 2016*, 1–12. ISSN: 2090-2085.
8. Geboes, E. (2020). Architectural Flat Glass: understanding the current barriers of the closed-loop alternatives to support the transition towards a circular life-cycle of flat glass. http://rgdoi.net/10.13140/RG.2.2.21229.00488
9. Tide Ocean SA. (2024). *100% recycled ocean-bound plastic*. https://www.tide.earth/en/tide-ocean-material/
10. Madiba & Nature. (2024). *Réalisations* madibanature. https://madibanature.wixsite.com/madiba
11. Ahuja, J., Dawson, L., & Lee, R. (2020). A circular economy for electric vehicle batteries: driving the change. *Journal of Property, Planning and Environmental Law, 12*(3), 235–50. https://doi.org/10.1108/JPPEL-02-2020-0011
12. *Meeting the EV challenge: Responsible sourcing in the electric vehicle battery supply chain* Trafigura. https://www.trafigura.com/brochure/meeting-the-ev-challenge-responsible-sourcing-in-the-electric-vehicle-battery-supply-chain/
13. *International Energy Angency* IEA. (2022). https://www.iea.org/data-and-statistics/data-tools/global-ev-data-explorer. https://doi.org/10.1016/j.seppur.2016.08.031
14. Swain, B. (2017). Recovery and recycling of lithium: A review. *Separation and Purification Technology, 172*, 388–403. https://doi.org/10.1016/j.seppur.2016.08.031
15. Chen, M., et al. (2019). Recycling end-of-life electric vehicle lithium-ion batteries. *Joule, 3*, 2622–2646. ISSN: 25424351.
16. Liu, P., & Barlow, C. Y. (2017). Wind turbine blade waste in 2050. *Waste Management, 62*, 229–240. ISSN: 0956-053X.
17. Leahy, P. G. (2020). End-of-life options for composite material wind turbine blades: recover, repurpose or reuse? http://rgdoi.net/10.13140/RG.2.2.16039.37287
18. Ragaert, K., Delva, L., & Van Geem, K. (2017). Mechanical and chemical recycling of solid plastic waste. *Waste Management, 69*, 24–58. ISSN: 0956053X. https://linkinghub.elsevier.com/retrieve/pii/S0956053X17305354
19. Ramesh, M., Palanikumar, K., & Reddy, K. H. (2017). Plant fibre based bio-composites: Sustainable and renewable green materials. *Renewable and Sustainable Energy Reviews, 79*, 558–584. ISSN: 1364-0321. https://www.sciencedirect.com/science/article/pii/S136403211730730X
20. Bank, L., et al. (2018). Concepts for reusing composite materials from decommissioned wind turbine blades in affordable housing. *Recycling, 3*, 3. ISSN: 2313-4321.
21. Suhail, R., et al. (2019). In *Analysis and Design of a Pedestrian Bridge with Decommissioned FRP Windblades and Concrete*.
22. ANMET. (2024). *Wind Turbine Blade Recycling* Anmet. https://www.anmet.com.pl/.
23. Eilers, V. H. (2022). *Vindmøllevinge får nyt liv på Aalborg Havn* Energy Supply DK. https://www.energy-supply.dk/article/view/699757/vindmollevinge_far_nyt_liv_pa_aalborg_havn.
24. Sumy, D. (2016). *Quake catcher network*. Retrieved October 15, 2022 from https://serc.carleton.edu/earthconnections/descriptions/174872.html.
25. Cochran, E. S. (2018). To catch a quake. *Nature Communications, 9*, 2508. ISSN: 2041-1723.
26. TEDxCERN. (2023). *Topher White: What can save the rainforest? Your used cell phone*. https://www.youtube.com/watch?v=xPK2Ch90xWo.
27. Nystul, J. (2022). *Repurposing Disposable Items: 63 Ideas To Reduce Waste* Section: Bright Ideas. https://www.onegoodthingbyjillee.com/repurposing-disposable-household-items/.
28. Gerhard, N. (2022). *78 Hacks for Repurposing Random Old Household Items* https://www.familyhandyman.com/list/79-old-items-totally-fit-for-repurposing/.
29. Beaty, V. (2022). *100 Ways to Repurpose and Reuse Broken Household Items* DIY & Crafts. https://www.diyncrafts.com/6081/repurpose/100-ways-repurpose-reuse-broken-household-items.
30. Heeb, N. (2024). *Online art, clothes*. https://100percentshit.com.
31. reWood. (2021). *Über uns*. https://rewood.ch/about-us/.

32. Kyungeun, S. (2022). *A review on upcycling: Current body of literature, knowledge gaps and a way forward.* In *ResearchGate 17th International Conference on Environment, Cultural, Economic and Social Sustainability (Venice Italy, 2015)* (pp. 28–40). https://www.researchgate.net/publication/299559229_A_Review_on_Upcycling_Current_Body_of_Literature_Knowledge_Gaps_and_a_Way_Forward.
33. FREITAG. (2022). *Über FREITAG. Manifest* https://www.freitag.ch/de/manifesto.
34. OST. (2023). *iPhone Hüllen* OST. https://www.ost.ch/de/projekt/iphone-huellen-1780.
35. FREITAG. (2023). *The circular apres-ski boot cellphone cases*. https://www.freitag.ch/en/circular-phone-case.
36. FREITAG LAB. AG. (2023). *Impact Report 2022*. https://media.freitag.ch/sites/default/files/230927_freitag_impact_report_2022_de.pdf.
37. FREITAG. (2022). *Über FREITAG. Geschichte*. https://www.freitag.ch/de/history.

Chapter 12
R7 Recycle

Raffael Römer and André Podleisek

Abstract *Recycling* is the oldest and most widely used retention option in the context of Circular Economy. *Recycling* of raw materials was already a decisive economic factor in ancient times. Using examples from the literature, this chapter provides a profound introduction to what *recycling* means and how it can be applied. Using asphalt *recycling* as use case, it describes which enablers and drivers have an influence on the success of this business model. Overall, institutional drivers such as the legal framework and punitive measures provide strong incentives for *recycling* asphalt pavements. In addition, a dedicated culture in a firm towards reducing the environmental impact enables the change towards a green economy. In the specific case, however, this led to the paradox situation that sustainability goals for CO_2 reduction are adverse to the expansion of the Circular Economy through *recycling*.

12.1 Introduction to *Recycle*

Recycling has been practiced by mankind since time immemorial. Archaeological finds prove that even in the Stone Age, worn out flints were reprocessed into smaller tools [1]. In the roman era, recycling became an important part of the economy and critically affected production and consumption [2]. The demand for available and affordable raw materials has preserved this economic sector until today. Circular Economy (CE) ideas and the increasing focus on energy and resource efficiency led to an additional boom in the recycling market.

Today, recycling is performed for various materials. The decisive factor for whether and to what extent a material is recycled is the quantity and degree of purity with which it is used [3]. This has a direct impact on the energy required for recycling, including return logistics. If the energy input for recycling is lower than

R. Römer
Küsnacht, Switzerland

A. Podleisek (✉)
Wettswil am Albis, Switzerland
e-mail: andre.podleisek@ost.ch

A. Podleisek et al. (eds.), *Circular Economy: The Next Level of Company Success*,
https://doi.org/10.1007/978-3-031-66406-9_12

for the production of a new product, there is an attractive environment for recycling, and the material could be kept in the loop. In fact, one could say that natural stocks (i.e. geologic reserves) are transferred to anthropogenic stocks (i.e. capital goods in our society) through recycling [4].

Typical products that can undergo a recycling process are metals, polymers, organic and inorganic materials [5] – basically all materials which are extracted from nature with the exception of those consumed in any way, e.g. energy carriers like oil or natural gas, as their chemical structure is destroyed, e.g. into CO_2 and water.

Definition of *Recycle*

Recycling is the process of recovery and reprocessing of waste material with the intend to use it in a new product. Recycling consists of the collection of waste material and the forming of new raw material that can be used to manufacture new products. The goal of every recycling process is to generate a material that is equivalent to a raw material mined from natural resources, and that can be used to substitute raw materials obtained from the nature such as wood, coal, mineral ores, petroleum and natural gas. Recycling plays a major role in reducing the amount of waste deposited in landfills, and therefore reduces the pollution of land, water and air. Recycling represents a classic end-of-life strategy to make the bound raw materials of the obsolete product available again. The material is retained in the CE.

Recycling typically involves technological equipment to shred, sort, melt and capture materials. The 'structure of the products is not maintained' [6].

The literature distinguishes between *pre-consumer* and *post-consumer* recycling. According to standard DIN EN 45557:2020 [7] pre-consumer recycling material is defined as 'material diverted from the waste generated during a manufacturing process', while post-consumer recycling material is defined as 'material generated by households or by commercial, industrial, and institutional facilities in their role as end-users of the product which can no longer be used for its intended purpose.' Pre-consumer recycling should therefore not be considered as part of CE, it is an optimisation of a production process. Under the principles of a CE, pre-consumer waste ceases to exist as a separate category. Instead, it is an integral aspect of the optimisation of production processes, where manufacturing practices are fine-tuned to minimise by-products and surplus. Any inescapable excess is viewed not as waste, but as a crucial resource, recycled and channeled back into new manufacturing processes [8].

***Recycle* in Context**

Recycling is one of the major factors in CE, as the possible applications of recycled raw materials are comparable to primary raw materials. In contrast to medium loop Rs (Reuse (R2), Repair (R3), Refurbish (R4), Remanufacture (R5) and Repurpose (R6)), Recycle (R7) can contribute to technological progress as new and more efficient products can be produced from recycled raw materials. Recycled materials can be used for completely different products with any specification, independent of the products the material originates from.

The production of raw materials from natural resources is often energy intense and has a big environmental impact (pollution, waste, land consumption by open pit mining) [4]. Recycling contributes significantly to reduce these impacts and could further reduce the economic and political dependence on other world regions.

Repurpose (R6) is often performed on a Micro or Meso scale. Finding a new use for obsolete products is unlikely in an industrial scale. To at least partially keep the original geometry of a product requires a lot of creativity. Recycling on the other hand can be performed in an industrial scale based on proven procedures.

Recover (R8) is the next lower rated retention strategy after Recycling. If prerequisites for *Recycling* are not given, the bound energy of a material can still be used. If materials are present in a very heterogeneous form (e.g. household waste), or the forming of materials requires a change in the chemical structure of the material (e.g. thermosetting plastics), *Recover* can be a more suitable retention option compared to *Recycling*.

A number of techniques is used including incineration of organic material, and anaerobic digestion of biologic materials. Energy recovery through chemical or biochemical processes always requires a change in the chemical structure of the starting materials.

Economics of *Recycling*

Recycling is an important process in many industries, and it is often driven by financial incentives. When a material can be recycled in a way that is financially viable and aligned with the political framework, it will be recycled. This is particularly true for metals, which can be recycled at high rates for certain metals (such as gold) but at very low rates for others (such as lithium or tellurium) [4]. The political factor is often decisive as it ensures the availability of materials for recycling. For example, it plays a major role in the recycling of paper, with recycling rates varying significantly depending on the local situation. For example, Switzerland has a recycling rate of 82% [9], while Europe has a rate of 74% [10], and the worldwide rate is 59% [11]). It is worth noting that recycling rates vary dramatically based on the underlying acquisition and reporting method [12]. Despite the regional differences, the recycling sector is a major contributor to the economy, with a revenue of 1.9 bill. CHF in Switzerland in 2018, with an estimated growth rate of 3% per year [13]. In the United States, the Environmental Protection Agency (EPA) reported there were 681,000 employees in the recycling sector in 2020 [14].

Companies that are active in the recycling industry often have above-average performance [15], and the low insolvency rate among these companies indicates their resilience [16]. Recycling is an important process that, besides helping to reduce waste and to conserve resources, significantly contributes to the economy.

12.2 Analysis of a *Recycling* Case

Asphalted surfaces are a common feature in many towns and landscapes. These surfaces have a limited lifespan and require significant investments in maintenance, particularly from governments. When roads are repaired or renewed, large amounts of Reclaimed Asphalt (RA) are produced which must be replaced with new asphalt. In many highly developed economies, the development of new areas is rare, and the need for new roads is minimal. This creates excellent conditions for a closed material cycle, where RA can be used to repair or replace existing roads, rather than requiring the production of new asphalt. This helps to conserve resources and reduces waste in the road maintenance process.

12.2.1 Description of the Case

This chapter focuses on an application example of how the ideas of CE can be applied in the case of Recycled Asphalt Pavements (RAP). The company providing the use case is a Swiss construction services company with activities in building construction and civil engineering, headquartered in Switzerland with activities in the DACH region. The company generates a turnover of roughly 4 bill. CHF with around 8,000 employees, and belongs to the top 500 companies of Switzerland. The company is convinced that sustainability is an integral part of their way of doing business. There is a strong believe that only a sustainable company is fit for the future in terms of the environment, society, and the economy.

The company operates several asphalt plants, and makes continuous efforts to reduce its environmental impact. The main focus is reducing CO_2 emissions, but CE aspects are gaining importance. The company started paying more attention to aspects which involve designing products and processes with the goal of minimising waste and maximising the use of resources. This involves using reclaimed materials in the production process or designing products that can be easily recycled at the end of their lifespan. By considering both CO_2 emissions and CE aspects, the company is working to minimise its environmental impact and contribute to a more sustainable future.

The specific use case deals with an asphalt plant renewed in 2022, located in the French speaking part of Switzerland. The plant operates at an annually output of 70,000 t per year at a maximum capacity of 240 t/h. Asphalt mixtures are produced according to customer specifications in a batch process, with a RAP content of up to 50%. At present, it is the most modern asphalt plant in Switzerland, a lighthouse project that could serve as an example for further development of all asphalt plants in Switzerland. It has the latest technologies currently offered by asphalt mixing plant suppliers installed.

Lifecycle of Asphalt

Asphalt is a mixture of aggregates of different sizes which are mixed with a binder (bitumen) to form a homogeneous mass. Heating the mixture in a kiln is required to decrease the viscosity of the binder, resulting in a good adhesion of the aggregates. Asphalt mixtures are placed while hot, and form a stable road through the cooling process. A typical road consists of various layers of asphalt in different specifications and qualities. Larger aggregates and lower quality binder is typically used in base layers, and smaller aggregates and high performance binder for top layers.

During the aging of a road, the countless temperature cycles as well as the load of the vehicles deteriorate the bonding strength of the bitumen, which leads to small cracks. Water enters these cracks and the expansion of water when freezing causes asphalt pavements to crack, which results in the commonly known potholes. Additionally, the wear of a road pavement leads to fracturing of aggregates, the increasing fines content reduces the resistance of the pavement. This effect can be observed on old roads that appear to liquefy. For these reasons, the surface layer or sometimes even entire roads have to be renewed at regular intervals. Both aging effects can be reversed or compensated in the recycling process, which positively influences the suitability of RA for the recycling process.

Recycling Cycle of Asphalt

The recycling cycle of asphalt consists of three steps: Reclaiming of asphalt the building site, revaluation at the gravel plant, and production of new asphalt at the asphalt plant as indicated in Fig. 12.1.

In most cases, drill cores are taken from the most heavily stressed zones of a road (e.g. vehicle lanes) as preparation for the renewal of a road. In addition to condition monitoring of the road, this also enables an analysis of structure and condition of the various asphalt layers. This information is necessary in order to mill off the different

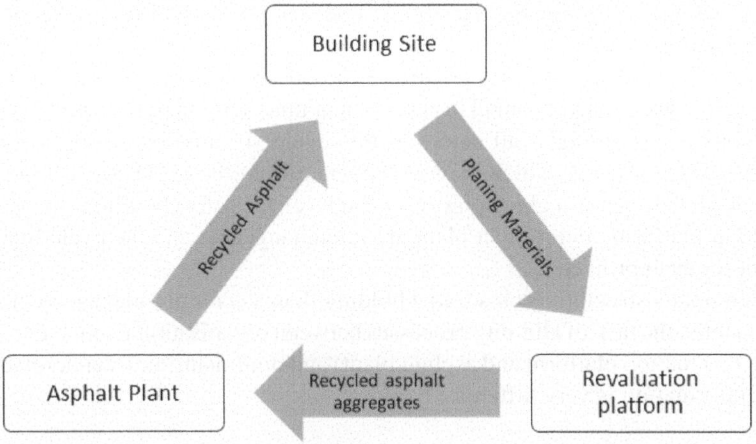

Fig. 12.1 Recycling cycle for asphalt

layers exactly according to their structure when removing a worn road surface. For each truck that transports the RA from the building site to the revaluation platform, it must be known from which layer the loaded RA was extracted. The planing material can then be processed in a revaluation platform with only little effort, as the aggregate size distribution is already close to specification. The asphalt lumps are broken up in the revaluation platform. This enables both the reduction of undesired fines as well as a sorting of RA fragments according to different size ranges. The challenge is to process agglomerates in such a way that they are as pure as possible without separating the valuable bitumen from the agglomerates.

The revaluation process of RA typically consists of screening, cleaning and optionally drying before it can be transported to the asphalt plant. Ideally, RA should be reclaimed in dry conditions and protected from moisture during transport and storage. Moisture in RA significantly increases the energy required for subsequent asphalt production.

The processed asphalt agglomerates are then transported to the asphalt plant for production of new asphalt. Here, different granulates of RA are mixed together with fresh gravel, so called white material, and new bitumen according to the customer's specification to produce new asphalt. In order to achieve good wetting of both new and recycled agglomerates, the mixture is heated in several steps to the required temperature of approx. 150 °C. Processing of RA requires temperatures approximately 15 °C higher than for new asphalt. This has a negative impact on energy consumption and CO_2 balance.

Organisation of the *Recycling* Cycle

In order to keep the closure of a road as short as possible, removal of old asphalt pavement and installation of a new pavement are in most cases carried out by the same company. In the use case under consideration, this company also owns the revaluation platform and the asphalt plant. This means the entire asphalt recycling cycle, including logistics, can be covered by one company. Accordingly, this company has the possibility to coordinate the whole value chain of RAP through an ecosystem of specialised subsidiaries.

However, it is also a specific feature of the construction sector that even large corporations have relatively small organisational units with their own profit and loss responsibility. Accordingly, all actors in this cycle are anxious to find economic solutions for themselves, which may well deviate from the cycle concept depicted. Due to the high decision-making power at local level, the actors building site operator, revaluation platform, and asphalt plant are interchangeable and the cycle may look different for each project.

The company structure is closer to a holding than a centrally managed company. Consequently, the task of aligning processes between organisational units (e.g. building site, revaluation platform and asphalt plant) and optimising and standardising the flow of information is very difficult.

Challenges

By far the biggest challenge in the use of RAP is the acceptance of the product by customers. Customers often pay little or no attention to this topic. Stalled processes, a lack of know-how, technical reservations and a lack of confidence in the quality of the product often hinder using asphalt with a higher recycled content. The peculiarities of public procurement, to which road construction is almost completely subject, massively limit the negotiating power of potential suppliers. Specifications are usually approved before a supplier gets informed about a project and accordingly, the suppliers' capability and desire to use an increased share of RAP is usually not addressed.

The presence of Polycyclic Aromatic Hydrocarbons (PAH), which were introduced in large quantities in road construction through the use of tar until the 1970s, is another challenges for asphalt recycling. These substances do not pose a hazard in bound form, but develop a carcinogenic effect during processing. Accordingly, very low thresholds apply for PAHs for newly manufactured asphalt. RA with a PAH content above the specified limit of 250 mg/kg cannot be recycled and must be landfilled or incinerated in a specialised plant. Currently, the only specialised plant capable of handling higher PAH levels is located in the Netherlands, which causes significant logistic costs and transportation carbon footprint.

Another challenge is the low level of standardisation of road surfaces. Almost every client's specifications deviate from each other. This leads to the fact that agglomerate mixtures from RA have to be separated at great expense and then mixtures have to be reassembled in order to achieve the required properties.

Availability of Raw Materials

Without taking into account the details of the different types of asphalt, it can be assumed that it consists of 95% aggregates and 5% bitumen as a binder. Different sizes of aggregates are considered to be readily and cheaply available in Switzerland, even though it is ultimately a finite resource. Many gravel pits have been intensively used in recent decades and are already exhausted, which leads to longer transport routes for alternative supply options. Typical prices for aggregates used for asphalt are between 30 to 40 CHF per metric ton excluding cost for logistics.

Bitumen makes up the largest part of the final price. Since it is a petroleum product, the price is subject to the dynamics of a spot market. Prices fluctuate, in 2022 the price range was 400 to 600 CHF per metric ton, but even peaks in the range of 750 CHF have already occurred. There are no natural deposits or refineries in Switzerland, which is why bitumen has to be imported from different sources worldwide.

In contrast, RA is available in abundance, and only about 50% of RA can be recycled, the rest has to be landfilled due to missing demand. 2,800,000 tonnes of RA are produced in Switzerland every year, this is equal to staggering 62,000 truckloads. Accordingly, the processing of RA represents a business opportunity for asphalt producers.

12.2.2 Drivers of the Case

Using RAP for producing new asphalt leads to a decrease of virgin resources used for production of asphalt equivalent to the content of RAP used in a specific recipe. Use of recycled asphalt also eliminates long transport routes, especially for bitumen, which is otherwise produced abroad. However, the local savings in CO_2 emissions are descreased due to higher temperatures required for processing of RAP. Other factors, such as the moisture content of agglomerates, or using of low-temperature asphalt, clearly outweigh the potential CO_2 savings of an increased RAP content on a local view. Consequently, in the context of the case, the environmental driver D.E1 (decrease the resources used in production) is rated with 0, indicating that despite the reduction in new asphalt production, the additional energy required for RAP processing negates the potential benefits. Accordingly, the driver D.E2 (decrease product carbon footprint) is assessed with negative relevance, reflecting the complex trade-offs in environmental impact.

Globally speaking, every 10% of RA used in asphalt reduce overall emissions in CO_2-equivalent by 8 to 10%. In case of a hot asphalt mixture with 60% RAP content, emissions can be reduced by 56% as shown in Fig. 12.2.

The use of RAP also introduces the driver D.E3 (resilience against resource-related risk) into the equation. Bitumen, a key component of asphalt, is obtained from crude oil, mainly by refinery technology through the distillation of crude oil. The use of RAP represents a secondary source of bitumen, which is available in large quantities at a very small price. Consequently, the integration of RAP as an alternative bitumen source reduces the dependence on crude oil, contributes to price stability, and reduces the cost of raw materials. Therefore, D.E3 is positively influenced.

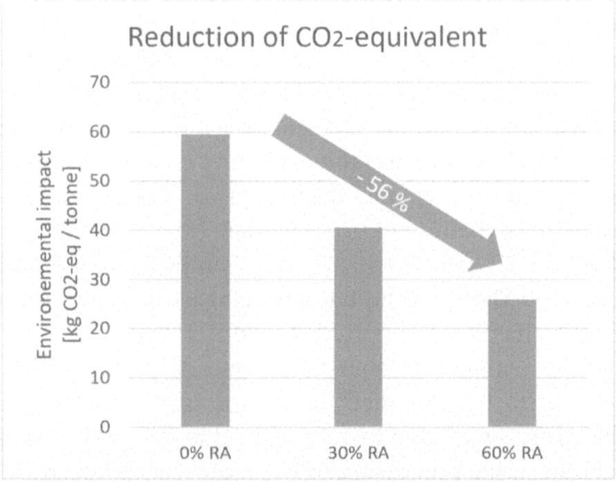

Fig. 12.2 Reduction of CO_2-equivalent (own representation derived from [17])

The high level of detail in the specifications of the asphalt mixtures and established standards massively limit the company's negotiation power to increase the share of RAP. Even if the technical equivalence of asphalt with a higher RAP content is mostly undisputed, this represents a significant trade barrier.

The permit for the construction and operation of an asphalt plant is in most cases tied to limitations with respect to emissions of pollutants and CO_2, as well as the allowed amount of stored material, giving D.I1 (compliant with legal requirements) extra weight. This leads to the dilemma that lower (local) emission values can be achieved with new asphalt due to lower processing temperatures. In contrast, the limitation of landfill space forces the company to use RAP. This trade-off requires the case company to continuously evaluate the optimal RAP content in its asphalt production. In doing so, they aim to strike a balance between meeting regulatory standards and maximising the use of RAP, thus giving D.I2 high relevance, as the company has to ensure compliance to avoid potential penalties associated with regulatory infractions.

Moreover, the development of the regulatory environment points towards an increased RAP share. It is a realistic scenario that the next revision of Swiss asphalt standards will prescribe a minimum RAP share in addition to the maximum RAP share. Accordingly, the company is already creating technological prerequisites to meet these future requirements (D.I3). Unfortunately, use of RAP is currently outside the public's field of vision, and there is no government support in the form of tax breaks or subsidy measures available; hence D.I4 is of no relevance.

Environment, Social and Governance (ESG) criteria are part of every business and investment decisions for clients, investors and society. Nevertheless, it cannot be expected to win a job solely due to the use of RAP. Besides meeting technical requirements, price is still the most important criterion for awarding contracts. Using RAP is only a competitive advantage due to its cost-reducing effects during the production process.

As a main actor in asphalt production in Switzerland, the company does not only want to meet industry standards, but define best practices and set new standards in the coming years. D.S3 (meet the standards of competitors) is therefore not relevant to the company's strategy. It is an official goal to organise specialist events and spread the word on how to increase the engagement in CE. The main driver towards a better ESG-rating is shareholder value and not competitiveness on the market. Although, driver D.S1 (image) and D.S2 (customer) are relevant, they were rated with '+' to reflect that other drivers carry greater significance for the company.

Prices of gravel as the primary raw material for asphalt production are characterised by strong temporal and regional fluctuations. However, this does not apply to the company concerned, as the supply of gravel can be covered by its own production sites. Nevertheless, opportunity cost do also apply to internal supply-chains, making it interesting to have a second source option for gravel by using RA, giving D.F1 (price volatility of raw material) some relevance.

The use of RAP offers significant advantages in terms of cost savings. It reduces binder procurement costs, which constitute a significant portion of the total cost of producing an asphalt mixture, and the procurement cost for aggregates is also reduced. On the other hand, using higher RAP contents increases processing costs

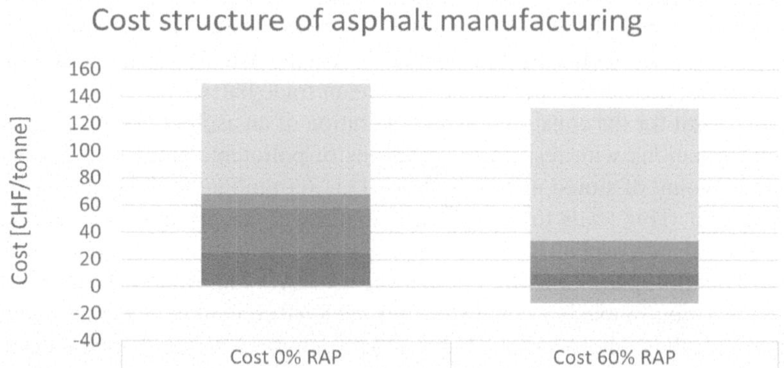

	Cost 0% RAP	Cost 60% RAP
Value Cration	82	98
RAP	0	-12
Energy	10	11
Gravel	33	13
Bitumen	25	10

Fig. 12.3 Cost structure of asphalt manufacturing

(storage, transport, preparation, quality control), leading to higher value creation of the asphalt plant, as shown in Fig. 12.3. In light of these factors, both D.F2 (cost saving opportunities) and D.F4 (create value from waste and production side streams) are highly relevant to the company. However, when considering the increased processing costs associated with RAP, these drivers, while important, are not as critical as institutional drivers. As a result, their relevance rating is '++'.

Production and delivery of a recycled mixture is more cost effective compared to a mixture without RA. Recycling of RA enables savings, but initial costs for equipment of mixing plants are high. However, with new technologies coming to the market, a greater financial added value is expected due to reduced energy and processing costs. It is important to emphasise that the the ecological aspect prevails in the use of RAP.

In the tender evaluation and grading of offers, the weighting of price is in most cases predominant (greater 50% of the overall grade). In contrast, environmental aspects, especially recycling and low-temperature asphalt, are weighted at maximum 15% of the overall grade. For this reason, the company cannot expect to achieve a higher market price through the use of RAP only. The lower cost structure of RAP leads to greater room for maneuvering, to gain market share with a low price strategy, or to increase margins. In this sense, the company rated D.F3 (competitiveness through innovation and business development) with a '+'.

Currently, only a fraction of the RAP generated can be recycled. As technology evolves, landfills full of RA may represent a future value to *Remine*, see Chap. 14.

Introducing new technologies and processes also creates development opportunities for employees. This can contribute to an attractive working environment. Beyond this, however, no social aspects of the business case are apparent.

As an industry leader towards becoming a more environmentally friendly company, the goals and achievements with respect to sustainability are actively promoted and communicated as ESG. However, the company does not consider D.R1 (social responsibility) and D.R2 (become a promoter of systemic change) as driving factors for asphalt recycling.

An overview of the relevance of all drivers is shown in Table 12.1.

Table 12.1 Drivers of the *Recycle* case

Cluster	Code	Driver	Context	Relevance
Environmental	D.E1	Resources	Decrease the resources used in production	0
	D.E2	Footprint	Decrease product carbon footprint	–
	D.E3	Risk	Resilience against resource-related risks (scarcity, supply challenges ect.	+
Institutional	D.I1	Legal	Comply with legal requirements and norms	+++
	D.I2	Punitive	Avoiding punitive measures	+++
	D.I3	Juridification	Anticipate or meet future legal obligations	+
	D.I4	Tax	Taxation benefits, access to supportive funds, and subsidies	0
Stakeholder	D.S1	Image	Improve corporate image, legitimacy, or credibility	+
	D.S2	Customer	Adapt to changing consumer behaviour, meet customer expectations, increase customer satisfaction	+
	D.S3	Practices	Meet the standards and follow the best practices of competitors	0
Financial	D.F1	Volatility	Price volatility of raw material	+
	D.F2	Savings	Cost saving opportunities	++
	D.F3	Competitiveness	Competitiveness through innovation and business development	+
	D.F4	Value	Create value from waste and production side streams	++
Responsibility	D.R1	CSR	Social and ethical responsibility	0
	D.R2	Promoter	Become a promoter of systemic change	0

0 representing insignificant importance, + indicating minor relevance, ++ reflecting notable relevance, and +++ signifying crucial importance

12.2.3 Enablers of the Case

The company's management is strongly committed to embed sustainability in all business operations. E.F1 (committed company culture) has been rated with '+++'. The company will only be fit for the future as a sustainable company, respecting the environment, society, and economy. It can therefore be assumed that sustainability aspects should be taken into account when investment decisions are made. However, the construction industry is typically very decentralised, with a high degree of decision-making power at the level of plant managers. Accordingly, there are great differences in the extent to which the sustainability concept of top management is also practised at a plant level.

As a listed company, the company has liquid assets of roughly 0.5 bill CHF and an equity ratio of 15 to 20%. The financing of interesting business opportunities with corresponding return prospects is usually not an obstacle. In this sense, E.F2 (available capital) is seen as an enabler that is a necessary prerequisite and has a relevance of '++'.

The company managed to bring the entire value chain in CE under its own control. Accordingly, it is significantly less dependent on external collaborations and relationships. Another advantage of the entire value chain loop is information flows within the company are easier to optimise. This results that by now, in some but not yet all cases, analysis results of drilling samples are available in the asphalt plant before starting the road's planning. In this respect, the enablers E.V1 (partnership) and E.V2 (transparency) do not have much relevance for the case company and have been rated with 0.

When it comes to enablers in the Technological cluster (E.T1 to E.T4), none of them have been considered as necessary for success. Producing asphalt with a recycled content of less than 50% is a mature process where equipment is readily available on the market. Asphalt with a recycled content higher than 50% currently requires higher temperatures in the production process that can damage recycled bitumen, leading to quality loss. However, technologies to overcome this limitation are at an advanced stage of development and should become established in the coming years.

However, there are still reservations about increasing the use of RAP in asphalt paving. One research task is to provide additional data for decisions on increased use of RAP. The company supports research both indirectly through industry organisations such as Asphalt Suisse, and directly through projects in cooperation with universities.

Furthermore, new cheaper and faster material testing methods are needed, because the validity of today's established testing methods (penetration, ring-ball, breaking point) is very limited at high RAP contents. Another weak point is that currently not all properties of new bitumen can be matched in the case high RAP contents.

Today, when road construction projects require the use of RAP with a high percentage of RA (above 50%), some asphalt producers and construction companies refuse to guarantee specifications. Such situations are not acceptable, because nowadays also mixtures with a high content of RA are considered a standard product. Standard product means a product that can be manufactured and installed with a limited and acceptable risk according to today's knowledge and technology (compared to conventional asphalt products). The use of RAP is no longer an innovation. Therefore, there is no reason for contractors having to cover additional risks (i. e., construction work without warranty) compared to mixes without RA.

Customers are becoming increasingly aware of the issue of RA. In Switzerland, various municipal, cantonal, and national stakeholders have adapted their tender documents so that the permissible RAP content may exceed the value specified in the standard [18]. Examples are listed in Table 12.2. This evolving awareness and adaptability among customers are key factors in enabling asphalt recycling initiatives, making E.C1 (increase customer engagement) an important enabler, which has been rated with a relevance of '++'.

Logistics used to supply the asphalt plants with RA would be the same to dispose of or landfill the RA. In reality an oversupply of RA is existing, and the restrictions on the amount of RA that can be stored on site is the obstacle compared to RA availability. For transporting RA to the asphalt plant, the company can use its own truck fleet, or subcontract the transportation. Given these circumstances and only E.M1 (available reverse logistics network) has relevance ('++'), while the other two

Table 12.2 Examples of customer-driven deviations from the standard [18]

Ashalt type	SN640 431-1-NA (%)	City of Berne (%)	Canton Berne (%)	Canton Argovia (%)	ASTRA (%)
Asphalt top layer (AC MR, PA, SDA)	0	0	0	0	0
Asphalt concrete for binder courses (AC B)	≤ 30	≤ 70	≤ 50	≤ 60	≤ 60
High modulus asphalt concrete (AC EME)	≤ 30	C1: ≤ 40 C2: ≤ 50	≤ 50	No requirement	C1: ≤ 40 C2: ≤ 50
Asphalt concrete for base layers (AC T)	≤ 60	≤ 70	≤ 80	≤ 60	≤ 60
Asphalt concrete for foundation layers (AC F)	≤ 70	≤ 90	≤ 90	≤ 80	≤ 90

Table 12.3 Enablers of the *Recycle* case

Cluster	Code	Enabler	Context	Relevance
Firm	E.F1	Culture	Committed company environment and culture	+++
	E.F2	Capital	Available capital or financial support to transition	++
Value chain	E.V1	Partnership	Build closer relationships/invest in collaboration and partnerships	0
	E.V2	Transparency	Enhance information sharing and transparency, make use of digital technologies and standardise information	0
Technological	E.T1	Research	Experiment and research	0
	E.T2	Mutuality	Joint decision-making (customers and suppliers)	0
	E.T3	R-technology	Availability of technical and technological know-how for a suitable value retention in-house	0
	E.T4	Outsourcing	Possibility to outsource the activity	0
Customer	E.C1	Engagement	Increase customer engagement and awareness, provide incentives	++
Material supply	E.M1	Logistics	Available reverse logistic network	++
	E.M2	Collection	Create simple collection/take-back systems	– – –
	E.M3	Symbiosis	Engage in industrial symbiosis	– –

0 representing insignificant importance, + indicating minor relevance, ++ reflecting notable relevance, and +++ signifying crucial importance

factors E.M2 (take back system) and E.M3 (industrial symbiosis) are seen rather negatively and their relevance has even been rated accordingly. A full overview of all enablers is shown in Table 12.3.

12.3 Outcome and Impact of *Recycle*

12.3.1 Ecological Results

The use of increased proportions of RA in asphalt mixtures has positive aspects on the environment and economy. Natural resources such as petroleum and aggregates can be saved, and the landfill space required for RA can be reduced.

The use of asphalt with an increased RAP content significantly reduces the emissions of greenhouse gases. The amount of CO_2 equivalent released corresponds closely to the share of RAP. In case of a hot asphalt mixture with 60% RAP content, the emissions can be reduced by 56%.

The total market size in Switzerland for reclaimed and fresh asphalt is approximately 2 mill. tonnes per year (as of 2018) [19].

Based on the assumption that only asphalt with a RAP content of 60% would be used in Switzerland, this results in theoretical saving potential of 67,000 tonnes of CO_2-eq emissions per year, compared to asphalt without RAP content for Switzerland.

Due to the higher temperatures in the kiln, which is required to increase the viscosity of the bitumen from the RA, the use of RA increases the local emission of CO_2 by approximately 10% compared to asphalt without an RAP content. This is a hurdle for the approval of new plants, although the global view clearly underlines the benefits.

12.3.2 Economic Results

Currently, the cost to dispose asphalt is about 20 CHF per tonne with a PAH content less than 250 mg/kg. Asphalt plants are currently in a situation where they can generate significant earnings by accepting RA. The use of RA further reduces the use of fresh bitumen, which has a positive effect on both the absolute cost of asphalt and the price fluctuations of the end product. Short-term price adjustments can only be passed on to the customer in exceptional cases, which is why the use of RA can significantly improve a producer's resilience.

On customer site cost savings of 5–6 CHF per tonne can be realised for every 10% of RA content used in new asphalt. Typical prices for asphalt are around 150 CHF per tonne.

The savings potential in the procurement of asphalt by the mostly state-owned customers is approx. 20% when using asphalt with a 60% RAP content.

12.3.3 Social Impacts

The specific use case is running in a Business-to-Business (B2B) environment. Touch points with the society are limited to the visible technical facilities in the landscape and the associated landfills and quarries. The recycling of asphalt can reduce the size of landfills and minimise the use of gravel quarries. This results in an enhanced environment and new habitats for people and animals.

12.4 *Recycle*: Summary and Learnings

The case study about the use of Reclaimed Asphalt (RA) in Switzerland showed the drivers and enablers around the increased use of RA in road construction and maintenance in Switzerland. Various consumers are pushing towards the use of a higher content of RAP for new asphalt pavements. Unfortunately, various regulations, contradictory requirements and established norms delay the process towards a higher RAP content. In recent years, various scientific studies have been published that have been able to refute or at least relativise existing quality concerns. Accordingly, nothing should be in the way of using increased RAP content.

The study showed, that the use of RAP has significant environmental benefits, such as reducing greenhouse gas emissions and saving natural resources, particularly in the production of new asphalt. Additionally, RAP can reduce costs by reducing the need for virgin materials and landfill.

However, the study also highlights several challenges in the use of RAP, including variability in quality and performance due to differences in the composition of the reclaimed material, and technical limitations in processing and mixing the material.

Therefore, to realise the environmental and economic benefits in road construction and maintenance requires careful planning and management of *Recycling* to ensure its effectiveness and sustainability.

The use case clearly showed that there is no one-size-fits-all approach to implementing a CE strategy with *Recycling*. A specific solution must be found for each use case, which takes into account the environment as well as the possibilities of the company concerned. A successful implementation of a CE project is a complex interplay between availability of (recycled) raw materials, availability of equipment, technical know-how of personnel involved on customer and supplier side, regulatory requirements for equipment and products, as well as a strong will to make a difference. This interplay is crucial for further development. In the case of asphalt recycling it can be assumed that the share of Reclaimed Asphalt (RA) will only increase slowly and in small steps.

More specifically, three learnings can be derived from this case study:

1. Use of *Recycling* is a way to significantly reduce emissions and the ecological footprint of a product. The reduction can be significant and becomes more transparent. In this case, asphalt producers move from buying a product with a high embedded energy content (bitumen), to the production of a product with direct energy consumption. This can lead to both internal conflicts of objectives and additional regulatory hurdles.
2. Knowledge about quality of the supplied raw material is a key factor in *Recycling*. The importance of knowledge about the reclaimed product has to be weighted higher than the purity of raw material, as this allows a more targeted use of the recycled material. In the described case this is a significant gap to be closed.
3. Only through collaboration of all stakeholders along the supply chain the implementation of a *Recycling* concept can be successful. A successful business model

relies on the cooperation of producers, customers, regulators and the bodies responsible for standardisation.

The study clearly showed that these factors have a positive effect on the economic efficiency of a *Recycling* solution. However, the case of asphalt recycling shows that opportunities are emerging even if not all conditions seem to be ideal.

References

1. Shimelmitz, R. (2014). *The recycling of flint throughout the Lower and Middle Paleothic sequence of Tabun Cave, Israel.*
2. Duckworth, C. N., & Wilson, A. (2020). *Recycling and reuse in the Roman economy.* (1st ed.). Oxford University Press. ISBN: 978-0-19-886084-6.
3. Schäfer, P. (2020). *Recycling - ein Mittel zu welchem Zweck?* Springer Spektrum. ISBN: 978-3-658-32924-2.
4. Worrell, E., & Reuter, M. A. (2014). *Handbook of recycling*. Elsevier.
5. Moro, J. L. (2021). *Baukonstruktion - vom Prinzip ins Details* 3. Auflage. Springer Vieweg, Stuttgart. ISBN: 978-3-662-64155-2.
6. Reike, D., Vermeulen, W. J., & Witjes, S. (2018). The circular economy: new or refurbished as CE 3.0? - exploring controversies in the conceptualization of the circular economy through a focus on history and resource value retention options. *Resources, Conservation and Recycling, 135*, 246–264. ISSN: 0921-3449.
7. *DIN EN 45557:2020 "General method for assessing the proportion of recycled material content in energy-related products"* (2020).
8. Ellen MacArthur Foundation. (2023). *Vision of a circular economy for fashion*. https://www.ellenmacarthurfoundation.org/our-vision-of-a-circular-economy-for-fashion.
9. *Kennzahlen & Quoten* Swiss Recycling. https://www.swissrecycling.ch/de/wertstoffe-wissen/kennzahlen-quoten (2022).
10. EPRC. (2020). *Monitoring report 2020; European Declaration on Paper Recycling 2016-2020.*
11. ICFPA. (2021). *ICFPA 2020-2021 Sustainability progress report.*
12. UNEP. (2011). *Recycling rates of metals - a status report.*
13. *Prognose: Umsatz der Branche Recycling bzw. Rückgewinnung von Werkstoffen in der Schweiz von 2012 bis 2018 und Prognose bis zum Jahr 2025 (in Millionen Euro)* Statista. https://de.statista.com/prognosen/415398/recycling-bzw-rueckgewinnung-von-werkstoffen-umsatz-in-der-schweiz (2022).
14. EPA, U. S. E. P. A. (2020). *Recycling Economic Information (REI) Report.*
15. Ahrend, K. -M. (2016). *Geschäftsmodell Nachhaltigkeit*. Springer. ISBN: 978-3-662-52880-8.
16. *Schweiz - Unternehmensinsolvenzen nach Branchen 2020* Statista. https://de.statista.com/statistik/daten/studie/958993/umfrage/unternehmensinsolvenzen-in-der-schweiz-nach-branchen/ (2022).
17. Kytzia, S., & Pohl, T. (2021). Ökobilanz der Herstellung von Asphaltbelägen. *Strasse und Autobahn, 2021*, 641–652. ISSN: 0039-2162.
18. *Wiederverwendung Ausbauasphalt und Einsatz Niedertemperaturasphalt - Best Practice Guideline* (Kies für Generationen KFG, Apr. 15, 2021). https://www.kiesfuergenerationen.ch/resources/20210416-Best-Practice-Guidline_d_def3.pdf.
19. Rubli, S. (2020). *Modellierung der Asphaltflüsse in der Schweiz* (Bundesamt für Umwelt, BAFU). https://www.aramis.admin.ch/Texte/?ProjectID=48398.

Chapter 13
R8 Recover

Reto Weber and André Podleisek

Abstract Although the *Recovery* of energy and materials is low in the value hierarchy, which means that not much of a product's value is retained at end of use and must be re-added to obtain a new, usable product, it is important for a circular economy. It poses the last chance to make use of remaining value of materials as much as possible before they end up in disposal sites as sunk costs and harm the environment in the process. Since the term *Recover* is vague and is very often used in an overlapping manner with other Rs, this chapter attempts to clearly define the term and set clear boundaries with respect to the other Rs. A company that is considered a pioneer in heat recover units for showers is presented as the use case of this chapter.

13.1 Introduction to *Recover*

By utilising the materials used in production or conserving the energy that remains at the end of a process, the value that has already been generated can be preserved. In a Circular Economy (CE), waste and by-products are seen as valuable resources to be used again. This approach can help close the loop on resource use, minimise pollution and waste, create economic opportunities, and enhance resilience to shocks such as volatile commodity prices. If the energy or resources were not to be extracted after the end of a process or a product's end of use, they would either be lost or need to be *Remined* from landfill again.

R. Weber
Kloten, Switzerland

A. Podleisek (✉)
Wettswil am Albis, Switzerland
e-mail: andre.podleisek@ost.ch

A. Podleisek et al. (eds.), *Circular Economy: The Next Level of Company Success*,
https://doi.org/10.1007/978-3-031-66406-9_13

Distinguishing *Recover* from Other Long Loop Rs

It is easy to distinguish Recover (R8) from short loop R (Refuse (R0), Reduce (R1)) and medium loop R (Reuse (R2), Repair (R3), Refurbish (R4), Remanufacture (R5), Repurpose (R6)) because in these, the use cycle of products and parts is extended, which is not the case for long loop R Recycle (R7), Recover (R8), and Remine (R9). An overview of all Retention Options (Rs) is shown in Sect. 3.1.3.

The distinction of the long loop Recycle (R7) from energy recovery is possible using the material perspective. In *Recycling*, the products' materials are processed to gain materials of the same or lower grade quality as used in the initial production, keeping the material as such. In *energy recovery*, only the energy value of the materials is retained.

The differentiation of *Recycle* from *Recover* is more difficult since materials can also be retained at the end of this process. In the European law, *Recycling* is defined as: 'any recovery operation by which waste materials are reprocessed into products, materials or substances whether for the original or other purposes. ...', whereas recovery is defined as: 'any operation the principal result of which is waste serving a useful purpose by replacing other materials which would otherwise have been used to fulfil a particular function, or waste being prepared to fulfil that function, in the plant or in the wider economy. ...' [1].

From this definition, it can be extracted that *Recycling* is a form of *Recovery*, but not all forms of *Recovery* are *Recycling*. Following example should illustrate the differentiation further: Let's say that a company produces a waste material as a by-product of their manufacturing process. The waste material has some value and can be used as a fuel source in another process.

- **Recycling**: The company can *Recycle* the waste material by reprocessing it into a new product, such as a raw material for their manufacturing process, or into a new product for another purpose.
- **Recovery**: The company can also *Recover* the waste material by using it as a fuel source in another process instead of using (potentially non-renewable) fuels.

So in this case, using the waste material as a fuel source is a form of *Recovery*, but it is not *Recycling* because the waste material is not being reprocessed into a new product.

The long loop R *Remine* (R9) is similar to *Recovery* with the intermediate step of the waste being stored in a landfill for an extended amount of time before being recovered. The storage of waste in landfills poses major environmental risks, so its *Remining* could help protect the environment while recovering resources [2]. In a CE, *Recycle* and *Recover* should take place to prevent any resources to end up unused in a landfill. *Remine* would only be used to clean up historical landfilled waste and is not a strategy applied to today's waste streams.

Categorisation of *Recover*

As indicated, the *Recover* can be split into energy recover and resource recover. The term energy recover can be differentiated into the following categories [3]:

Thermo-chemical recovery This is the most common conversion process. Energy is extracted from waste through high temperatures such as in combustion, gasification, pyrolysis, or liquefaction. Combustion is useful for recovering the chemical energy contained in materials and reducing the volume of solid waste destined for landfills. This reduces the need for fossil energy sources and methane generation from landfills [4]. Energy from combustion may be used for heating air or water, creating steam or electricity, or a combination of it [5]. Gasification is the process of turning a solid biomass into usable gases. For this, biomass is exposed to high temperature and brought into contact with a gasifying agent. The resulting gases can be used to provide energy through combustion, or for further catalytic synthesis into base chemicals or into storable energy carriers such as hydrogen [6]. Similar to gasification, pyrolysis turns solid biomass into usable liquids (bio-oil or pyrolysis oil – synthetic fuel which could substitute crude oil) and solids (biocharcoal – charcoal produced by pyrolysis of biomass). The latter process has been used for thousands of years for charcoal production and only in the last 30 years has it become relevant to produce bio-oil [6]. Liquefaction is a subcategory of pyrolysis for extraction of high-grade bio-oil [7].

Chemical recover The main chemical energy recovery process is called esterification. With it, biodiesel, which can be used as a direct substitute for diesel in engines without modifications, can be produced from oil crops, lipidic wastes, and algae. The oils or fats of these feed-stocks are brought to reaction with short-chained alcohols with or without a catalyst [8].

Bio-chemical recovery Energy can also be recovered by bio-decomposition of waste also known as composting. Biogas, which consists mostly of methane and CO_2, is produced by anaerobic digestion in landfills or biomass or through gasification as described above. Biogas from landfills has a high methane content and is dangerous if not properly contained due to its flammability and effects as a greenhouse gas [9]. Biomass can also be used in aerobic digestion to directly generate heat. No biogas is created with this, so the composting plant should be close to the energy consumer [10].

Kinetic and thermal recover During many industrial and daily life processes, kinetic or thermal energy is generated and often disposed of after the intended process. Following are two examples of energy recover from processes: Recovering kinetic energy by converting it into another form for immediate use or storage instead of disposing of it in the form of heat in a braking system. It can be found in vehicles such as cars, trams, etc. via the use of an energy storage such as springs, capacitors, or compressed air [11]. Recovering thermal energy from industrial processes in a seasonal heat storage for a foundry in Sweden [12]. In the example, waste heat from the foundry is recovered and captured by cooling water which is then pumped into the bedrock to heat it up. The heat from this bedrock storage can be used several months for heating and other appliances. The potential of waste heat recover is tremendous: The United States Environmental Protection Agency (EPA) and Department of Energy calculated that approximately 10% of all energy consumed in the United States is dissipated as waste heat [13]. Oppositely, it is also possible to store cold during winter months in the bedrock and then use it for cooling during summer months. In combination with the reuse of heat, such cold storage direct cooling can be 6–13 times more efficient than conventional air-cooled chillers [14].

The other type of *Recover*, resource recovery, is done mostly for sludges that remain after waste handling processes and industrial processes:

Chemicals recover Sewage sludge offers mostly nitrogen and phosphorous which can be extracted and used in other processes. Other recoverable resources are heavy metals, biopolymers, and enzymes [15]. This means that after the recovery it is still possible to Repurpose the remaining material.

Physical recover An example of physical waste recovery is the recovery of oil and water along with a dry solid from an oily sludge of a refinery [16]. According to the distinction to R7 *Recycle* made in Sect. 13.1 (Distinguishing *Recover* from Other Long Loop Rs), it is not seen as *Recycling* if the recovered resources are used in further processes where they substitute other (non-renewable) resources.

An example that would count as *Recycling* is the processing of sludges resulting after combustion of municipal solid waste for energy recover. To a large extend, metals are sorted out for recycling before combustion, but it is still possible to extract around 50 kg of metals per metric ton sludge [17]. This is material was missed during the initial recycling step, but the extraction process still counts as *Recycling* because the recovered resources will be used in new products. A similar situation is the use of dried sewage sludge to make bricks [15].

13.2 Analysis of a *Recover* Case

13.2.1 Description of the Case

Introduction

The following case was selected after interviewing different waste recovery plants managed by the government and a company of the private sector. It quickly became clear that the government managed facilities are not a good example of a use case as they are also supported by the government and thus do not face the same responsibilities as a small or medium enterprise. All information in the case was obtained through interviews with and documents provided by the company's CEO, unless otherwise stated.

Company

The selected company is a developer, manufacturer, and distributor of heat recover units for showers, its solution belonging to the thermal recover category. The current CTO came up with the idea for the product around the time of the 2007-2008 financial crisis when he was employed at an Innovation Factory. There, the employees had 100 hours at their disposal for their own projects, which were subsequently continued, sold, licensed, etc. by joint decision. Energy prices rose during the financial crisis, so the idea of recovering energy from wastewater, which is one of the biggest energy guzzlers in households [18], caught on with the other employees and it was decided to continue the idea in a spin-off. The company is located in Switzerland and currently has 8 permanent employees.

Products

Its products are drainage channels built into the floor for heat recover from wastewater when showering. An example can be found in Fig. 13.1. There are several types that differ in size/efficiency, types of installation and the scheme of the hot water connection of the shower.

The scheme shown in Fig. 13.2 is designed for households with a central boiler or storage tank. The warm part of the shower water, as can be seen in the left part of the figure, is initially fed completely by the hot water from the central heat source in order to achieve shower water with the exemplary temperature of 40 °C. During the shower, the wastewater flows through the heat recover unit, which is heating the cold water before it arrives at the mixing valve. The right part of the figure shows how the cold water has been preheated to 25 °C, so that the mixing valve can regulate the inflow of hot water. As a result, less hot water and therefore less heating energy is

Fig. 13.1 Example of a heat recovery drainage channel of the company (copyright by Joulia SA [19])

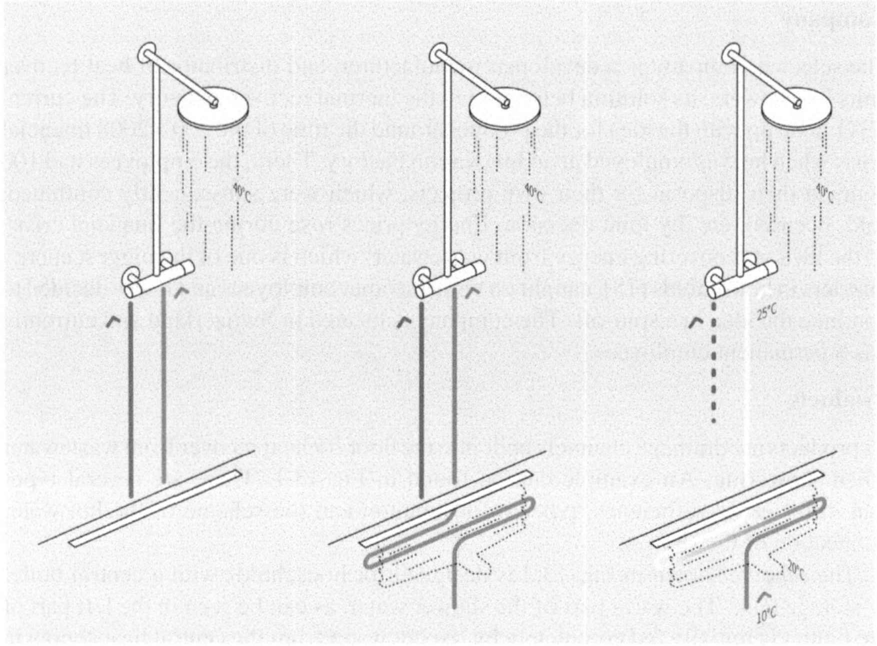

Fig. 13.2 Operating principle of the heat recovery units (copyright by Julia SA [19])

required at a constant shower temperature of 40 °C. The product thus fits perfectly into the thermal energy recovery category defined in Sect. 13.1 (Categorisation of *Recover*), since the waste heat is reused and substitutes part of the primary energy input.

While in Switzerland heat recover units are offered almost exclusively for installation in tiled floors, the company also sells the product in other EU countries as a version for use in shower trays.

There are three main target user groups:

- People who want to contribute to the fight against the climate crisis, but still like to take a warm shower. This can be achieved with the product, since less heating energy is required thanks to the heat recover.
- People who would like to take a warm shower but don't want to have a guilty conscience as a climate killer.
- People who want to save financially. This can be achieved with the product, since less heating energy is required thanks to the heat recover. In addition, the smaller infrastructure required for the (water) heating system results in lower investments.

There is a throwback to the environmental legitimacy of the second target group: If people are aware that the heat is being recovered, it is possible that they simply shower longer or more often because of a less guilty conscience. In that case, the total CO_2 emissions would increase despite higher savings. Nevertheless, the product also enables people who, for example, have a larger shower head with more flow volume, or people who like to shower longer or more often, to live more sustainably.

Comparable heat recover units as drainage channels or as units that run along the drainpipe are offered also by other companies.

Concept

The company sells these heat recover units indirectly via plumbers, as well as plumbing wholesalers and resellers abroad. The end customers can experience the product at trade fairs and either ask a plumber or wholesaler about it themselves, or the plumber or wholesaler suggests it to the end customer. The product is thus distributed via a three-tier distribution system, which is standard for residential plumbing fixtures.

Currently, the most active market are the Netherlands, where the reseller sells about 60% of all modules of the company. In addition, the heat recover units are supplied to manufacturers of plumbing products so that they can incorporate them into their own product lines.

The components are sourced in roughly equal parts from neighbouring EU countries and from Switzerland. The final assembly of the heat recover unit takes place at the company headquarters in Switzerland. If the customers have an adjustment request, this is done by partners nearby.

Development Process

The CTO was first able to build and test the idea privately at home in the shower tray. The effect wasn't very strong, but it was clearly detectable, so the idea was taken further.

Around 2011, a first version for a shower tray was successfully developed. At that time, however, the company noticed that the market trend was moving away from shower trays towards tiled shower cubicles. Thus, the development of the first version with a drainage channel design began.

Further Steps Taken

Marketing the shower channel after its development posed a challenge. Despite winning numerous innovation awards, the company struggled to garner the interest of building contractors, plumbing wholesalers, and plumbing product manufacturers. An important step was the drinking water certification from various authorities, which took up to 10 years to complete due to the lack of regulations and standards for heat recover units. Without drinking water certification, the product could not yet be sold in Europe, since shower water is categorised as drinking water here. Other manufacturers of heat recover units in Europe with support by the German Copper Institute pursued the same goal, which made the process of standardising the certification process faster.

Once the certificates were obtained, it was easier for the company to sell the product to wholesalers and manufacturers of plumbing products, since both credibility was proven, and the product was tested over time. In the case of the wholesalers, there was also the fact that their 'old guild' initially opposed heat recover and only in later years the top management stipulated CO_2 neutrality as a goal.

The company was able to break even and is currently profitable and has enough capital to continue to expand. Although the market is not yet highly competitive, it still needs to be further developed.

To date, the Innovation Factory still provides know-how and its network, and the company and Innovation Factory present together at trade fairs. The Innovation Factory benefits from this in that the company does marketing for them.

13.2.2 Drivers of the Case

A very strong driver of the innovation was the sharp rise of energy prices at the time the idea was conceived (D.E1). When studying where in households the most energy is lost, hot water going down the drain was identified as a big waste.

Carbon footprint was also a driver (D.E2). The company knew the target group of footprint-conscious people and wanted to provide them a product that would make it easier to achieve their goal of reducing energy use.

Image (D.S1) was a very important driver of the company. The company takes part in many innovation competitions and has already won around 20 innovation prizes to date. Building contractors and end customers, for example, can be convinced with those prizes. In contrast, plumbers, wholesalers, and manufacturers found it more important that the company achieved product certifications, especially as it is a young company. Therefore meeting legal requirements of the certifications was also an important driver (D.I1).

As already mentioned with regards to carbon footprint, customers were also a driver (D.S2). Footprint-conscious people, energy savers and people who don't want to have a bad conscience when showering can be reached.

As previously described, the volatility of the energy price during the financial crisis 2007–2008 was the main driver leading to the innovation (D.F1).

Also a very important driver for this innovation were the savings possible for the customer (D.F2). Heating costs for tenants or owners are reduced by the product. In addition, investment costs for owners for construction can be reduced due to a smaller heating infrastructure required.

The competitiveness (D.F3) achieved by differentiation through innovation was the strongest driver of the company, as this permitted the company to earn money.

For the company, this payed off as value (D.F4). Value is created from wastewater, so to speak. The company was able to become profitable and reach break-even.

An overview of identified drivers of the case is shown in Table 13.1.

Table 13.1 Drivers of the *Recover* case

Cluster	Code	Driver	Context	Relevance
Environmental	D.E1	Resources	Decrease the waste of heat to reduce energy consumption	+++
	D.E2	Footprint	Decrease product carbon footprint	++
	D.E3	Risk	Resilience against resource-related risks	0
Institutional	D.I1	Legal	Comply with legal requirements and norms	++
	D.I2	Punitive	Avoiding punitive measures	0
	D.I3	Juridification	Anticipate future legal obligations	0
	D.I4	Tax	Taxation benefits, access to supportive funds, and subsidies	0
Stakeholder	D.S1	Image	Improve corporate image, legitimacy, or credibility	+++
	D.S2	Customer	Adapt to changing consumer behaviour	+
	D.S3	Practices	Meet the standards and follow the best practices of competitors	0
Financial	D.F1	Volatility	Price volatility of energy	+
	D.F2	Savings	Cost saving opportunities for customers	+
	D.F3	Competitiveness	Competitiveness through innovation and business development	+++
	D.F4	Value	Less waste through recovering heat	+
Responsibility	D.R1	CSR	Social and ethical responsibility	0
	D.R2	Promoter	Become a promoter of systemic change	0

0 representing insignificant importance, + indicating minor relevance, ++ reflecting notable relevance, and +++ signifying crucial importance

13.2.3 Enablers of the Case

The most important enabler was the corporate culture of the CTO's original employer due to the provision of time for the employees to work on their own projects (E.F1).

Another important enabler was the available capital (E.F2). Since the development of the product took a very long time due to the changing market requirements and the complex situation with the standards, it was important that private investors did not leave in the meantime. It was probably also helpful that investors were connected to the Innovation Factory and therefore knew what they were getting themselves into. In addition, the Innovation Factory also provided manpower. Capital was also raised from three foundations focused on climate and technology innovation. Additional federal funding was considered, but this required a great deal of data and was not pursued. For example, test systems were required to collect data over longer periods of time. The company wanted to bring a product to the market and that would have slowed it down.

The ongoing partnership with the Innovation Factory is still important today (E.V1). They still provide know-how and their network. Establishing partnerships with wholesalers and manufacturers was very difficult at the beginning due to missing certifications, and the product was therefore considered immature. Now that the certifications are in place, but also because of society's greater drive towards sustainability, the company finds partners much easier. The company benefits greatly from partnering with a dealer in Holland, which accounts for about 60% of all products sold. Since wholesalers and plumbing product manufacturers are closer to the market, they give very exciting feedback to the company, which is closer to the technology and can implement feedback into the products. A sticking point for wholesalers is that advice is crucial when offering the product to customers. Although the product generates energy cost savings it also costs more, meaning customers might choose a cheaper product nonetheless. The sellers must therefore understand the functions and benefits of the product.

The sparse transparency in terms of demand and technical real-world data had a rather negative effect (E.V2). It is difficult for wholesalers to evaluate the technology because their customers do not ask for a solution they are not aware of, and if the wholesaler makes a targeted offer, the technology will first have to be explained to the customer in order to justify the additional costs. The difficulty with technical data is that while it can be collected on test benches, it is often not seen as real-world data because it was collected in a laboratory. New, real-world test concepts are currently being considered with technical colleges.

Research is important for the company (E.T1). By collecting data for a long time, certifications could be obtained, which were initially major blockers and today represent high technological market barriers for copy cats and competition. As a pioneer, the company had to help develop the standards, but today they have a stronger persuasive power. This enabler supports the drivers D.I1 and D.F3.

To focus on the development and distribution of the product, the company outsourced the manufacturing processes (E.T4). It is advantageous that manufacturers only need specifications and schemes and will define their production process themselves, so the company did not have to build capacity and competence for manufacturing. If a customer has special wishes for the cover of the shower channel, this can be adjusted externally by nearby partners of the company. Since products are sold to end customers via wholesalers and plumbers, the company depends on their understanding of the technology. For example, plumbers may have concerns about the more complex connection of the shower, the copper pipes, or the interrupted water lines, and thus potential for leaks.

Advertisement is only done by means of representing the company at trade fairs and explaining the technology to end customers there, making the product more tangible (E.C1). Customers can then contact wholesalers or plumbers about the product. Only the Dutch retailer does direct advertising for the heat recover unit.

Two additional enablers not present in the research framework (Sect. 3.2) were found: The company was influenced by legal conditions and the market. The legal circumstances, i. e., the initially missing test regulations and standards for certification and the time required to develop them, proved to be an enormous obstacle to the development of the company. Thanks to other small companies with similar products in the European Union (EU), it was possible to develop the European market, but it remained very small. Although the market is not highly competitive, it still needs to be developed further. This is much more expensive than product development and would also require more manpower.

An overview of identified enablers of the case is shown in Table 13.2.

Table 13.2 Enablers of the *Recover* case

Cluster	Code	Enabler	Context	Relevance
Firm	E.F1	Culture	Committed company environment and culture	+++
	E.F2	Capital	Available capital or financial support to transition	+++
Value chain	E.V1	Partnership	Build closer relationships/invest in collaboration and partnerships	+++
	E.V2	Transparency	Enhance information sharing and transparency, make use of digital technologies and standardise information	–
Technological	E.T1	Research	Experiment and research	++
	E.T2	Mutuality	Joint decision-making (customers and suppliers)	0
	E.T3	R-technology	Availability of technical and technological know-how for a suitable value retention in-house	0
	E.T4	Outsourcing	Possibility to outsource the activity	+
Customer	E.C1	Engagement	Increase customer engagement and awareness, provide incentives	+
Material supply	E.M1	Logistics	Available reverse logistic network	0
	E.M2	Collection	Create simple collection/take-back systems	0
	E.M3	Symbiosis	Engage in industrial symbiosis	0

0 representing insignificant importance, + indicating minor relevance, ++ reflecting notable relevance, and +++ signifying crucial importance

13.3 Outcome and Impact of *Recover*

13.3.1 Ecological Results

The ecological contribution of the company originates from the energy saving effect of the heat recover units. Therefore, it is calculated how much CO_2 the production of heat recover units generates and how much CO_2 can be saved with the use of heat recover units, as shown in Table 13.3. Only the components that make up more than 0.5% of the total weight of the product are considered [20].

If the smallest heat recover unit is operated with a shower head with a flow rate of 12 l/min, the heat recover efficiency is 21%. The efficiency of the heat recover units indicates how large the energy savings are for the treatment of domestic hot water. This was tested in stationary operation with cold water at 10 °C [21] and hot water at 40 °C. 'Real conditions' are approximated by a correction factor of 5% to simulate soiling, a shower duration-dependent correction factor for the startup time of the heat recover unit, and a correction factor for the hot water temperature to 38 °C.

Based on the survey it is assumed the average shower duration is 8.7 minutes, a person takes 6 showers per week [22], the wastewater has a temperature of 35 °C and the annual utilisation rate of the water heating unit is 70%. This results to potential savings with smallest heat recover unit in a single household of about 260 kWh per year.

With 128 g CO_2e/kWh [23], one person can save 33 kg CO_2 in one year. So, if a person uses the smallest heat recover unit, it will outweigh its product carbon footprint after about 2 years of use.

If the shower channel is used for a service life of 30 years [24], a person with the smallest heat recover unit could save a total of 934 kg CO_2 over its lifetime or 31 kg of CO_2 per year.

Table 13.3 Components of smallest and largest heat recovery unit (data according to [20])

Material	Smallest unit (12.76 kg)		Largest unit (15.73 kg)	
	Weight (kg)	CO_2e (kg)	Weight (kg)	CO_2e (kg)
Brass	0.95	3	2.26	7
Chrome steel	8.68	55.8	8.22	52.9
Steel	–	–	0.09	0.4
Copper	0.83	4.9	2.75	16.3
Polypropylene	–	–	0.1	0.4
Hemp fiber board	0.25	0.2	0.29	0.2
Cardboard boxes	1.61	0.5	1.14	0.3
Paper, printed	0.09	0.3	0.355	1
Total of considered materials	12.4	64.5	14.52	78.5

A Swiss person with annual emissions of about 4,000 kg CO_2 [25] could save close to 1% of their annual CO_2 emissions by using a heat recover unit. With the largest version, ca. 600 kWh could be saved due to the better efficiency of 45%, which corresponds to 77 kg of CO_2 per year. The largest version has a carbon footprint equivalent of 78 kg and would amortise after just one year. It would reduce the annual CO_2 emissions of a Swiss person by nearly 2%.

Considering additional variable factors such as the smaller required heating infrastructure, a consulting company [26] calculated the lifetime carbon footprint for an apartment building of 16.2 people and different types of heat recover units. They found an approximated 18 to 82-fold saving in CO_2 emissions during the lifetime of a heat recover unit compared to its production. The 82-fold saving was calculated with the largest heat recover unit. This corresponds to a saving of 6,445 kg CO_2 over 30 years and an average of 2.7 people per shower channel (6,366 kg excluding production emissions). This results in slightly higher results of 78 kg CO_2 saving per person per year and thus approx. 2% of the CO_2 footprint of a Swiss person.

With regards to the service life of the heat recover unit, it must be noted that this was determined on the basis of the individual components. However, the heat recover unit is a complex product and may therefore have a shorter lifespan than the individual components and need repairs which were not included in the calculation.

13.3.2 Economic Results

No data was found regarding the *economic potential* of wastewater heat recover in residential homes in Switzerland. The company would not disclose sales figures or balance sheet data and could only say that it was operating profitably and had reached break even. To create a rough picture the following assumptions were made: Based on the current number of 8 employees and average salaries without accounting for other factors, more than 240 of the smallest heat recover units or 180 of the largest would need to be sold for break-even. Comparing these (potentially too small) sales volumes to the roughly 49,000 newly built [27] and 46,000 renovated apartments [28] in Switzerland per year, it becomes clear that a huge market potential is currently only scratched. A multi-million CHF market for wastewater heat recover can be imagined.

To assess the *economic impact for users*, a comparison between the heating costs saved from using a heat recover unit over thirty years and the additional investment costs is made. The amortisation period was then calculated for households of various sizes, given the lack of financial data from the company.

If the investment costs of heat recover units are compared to those of a conventional drainage channel, the smallest version including additional installation costs is around 1,300 CHF more expensive and the largest version around 2,400 CHF more. Table 13.4 and Fig. 13.3 shows the possible savings and amortisation times under consideration of different energy prices and household sizes.

Table 13.4 Possible savings and amortisation of heat recovery units

Energy price (CHF per kWh)	Houshold size	Smallest unit		Largest unit	
		Savings (CHF)	Amortisation period (yrs)	Savings (CHF)	Amortisation (yrs)
0.10	Single	−520	50	−582	39.6
0.10	2-person	260	25	1,236	19.8
0.10	Average (2.7-person)	806	18.5	2,509	14.7
0.10	3-person	1,040	16.7	3,054	13.2
0.15	Single	−130	33.3	327	26.4
0.15	2-person	1,040	16.7	3,054	13.2
0.15	Average (2.7-person)	1,859	12.3	4,963	9.8
0.15	3-person	2,210	11.1	5,781	8.8

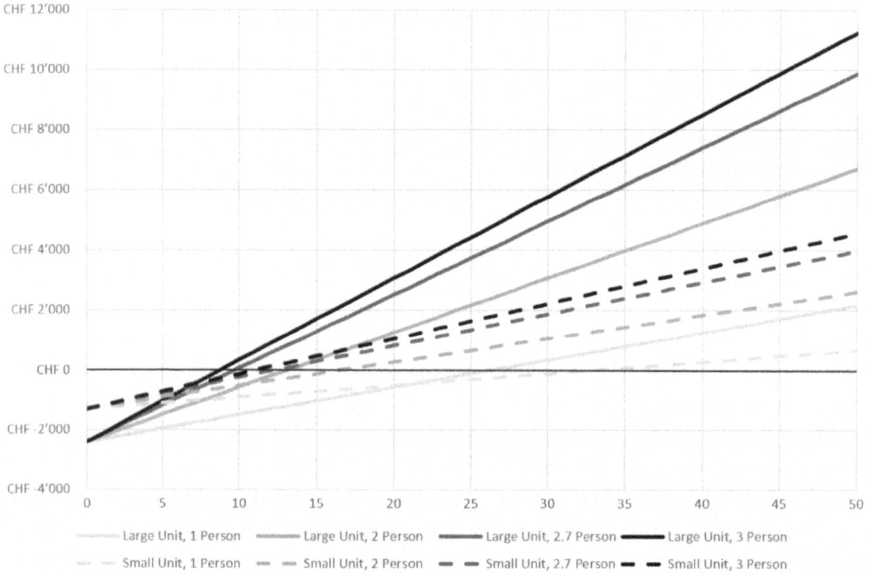

Fig. 13.3 Comparison of possible savings of the heat recovery units at 0.15 CHF per kWh over 50 years

No subsidies were taken into account in the calculations. In Switzerland, the product is recognised by the Minergie standard for low energy housing. A reduction of energy requirement of the entire hot water demand of 13–46% is credited [29]. Looking at the heating energy prices, it should also be noted that consumer prices have been on an upward trend for the last 50 years [30].

In the calculations, the installation scheme for hot water storage tanks or electric boilers in the basement was analysed. The efficiency of the installation is based on measurements made by the company. If other systems are chosen to generate hot water, this also leads to different total efficiency values and thus to varying savings.

The manufacturer grants a guarantee period of five years on the heat recover unit and two years on the cover. If the heat recover unit has a defect after the warranty period has expired, money has to be invested in repairs and the situation can differ from the calculation above.

13.3.3 Social Results

For the social impact, the effect of a heat recover unit for low-income tenants was considered regarding the lower cost for heating, and what proportion of wage this accounts for. Heat recovery unit could relieve tenants or social welfare if installed in social housing.

Based on the data above and an average multi-year heating energy price of 0.10 CHF per kWh over the last few years (using pellets, heating oil or natural gas) [30]), individual tenants could save 26 CHF per year with the smallest version and 60 CHF per year with the largest version of the heat recover device. Recently, prices have increased significantly, so if today's heating energy price of some 0.15 CHF per kWh is applied, 50% more could be saved, 39 CHF and 90 CHF respectively.

Comparing these savings with the current heating energy price to a net salary of 4,000 CHF, which is the 10% quantile in Switzerland for a full-time job [31], this accounts for 1% and 2.3% respectively. For the Swiss average household (2.7 people), the savings are 105 CHF and 243 CHF respectively, so 2.6 and 6% in case of a single source of income. Based on this calculation it can be said that a heat recover unit has a noticeable effect on the heating energy bills of low-income tenants and thus improving their disposable income and mitigating effects of rising energy prices.

13.4 *Recover*: Summary and Learnings

The use case shows that a young company was able to enter the market with an energy recover product. It fulfils the criterion defined in Sect. 13.1 (Categorisation of *Recover*) of substituting a primary energy input by reusing the waste heat. The product can appeal not only to people who care about sustainability, but also to others thanks to financial savings. The product has opened up a new market segment previously considered unattractive by manufacturers of plumbing products, partly due to missing standards and testing procedures for certification. The company was aware of the lack of this enabler previously unidentified (see Sect. 13.2.3), but it still slowed the company down in market expansion. It was crucial that investors and foundations were aware of both the consequences of the enabler and the product's potential while the market and certifications were being developed, to avoid losing interest and dropping out. Despite successful certifications and achievements in the market, the market is still small but also not very competitive. However, this should change as society becomes more aware of sustainability issues and saving energy becomes inevitable. With the certification process now being standardised [32], it will be easier for potential competitors to develop their own heat recover units in the future. As a result, market and competition are both expected to grow.

The ecological impact in Sect. 13.3.1 showed that if the product is used regularly, it will be CO_2 negative within a few years and, over its estimated service life of 30 years, save a multiple of the CO_2 emissions that occurred during production. This effect increases with the number of people showering in the shower equipped with the heat recover unit. Due to the increasing efficiency with a higher number of people using the heat recover unit, it would make sense to install it in large facilities where many people shower in the same shower room, such as gymnasiums or barracks. Real-world data would be advantageous for a more precise analysis of the efficiency values. The company is clarifying projects with technical universities to collect values that are more closely based on actual use, as mentioned for enabler E.V2 in Sect. 13.2.3.

When calculating the social impact in Sect. 13.3.3, it was visible that heat recover units would be worthwhile for tenants, based on the assumption heating costs are charged in full to the tenants. If heating costs were subsidised, the social effect of the heat recover unit i slower or no longer exists. Still, there would be an economic effect in favour of the owners of the social housing or social welfare bearing the costs.

With regards to economic impact in Sect. 13.3.2, it became clear that the product contributes more quickly to CO_2 savings than to financial savings when compared to conventional drainage channels. While the calculation was simplified by excluding possible unplanned repairs, it also excluded potential grants and subsidies. If subsidies were included, the savings calculation would be much more positive. Similar to the ecological impact, a higher number of users results in higher economic impact/savings.

The company's future looks secure due to the market potential of almost 100,000 units per year in Switzerland alone. The company has positioned itself in a growing market, yet in its infancy but increasing due to energy and climate concerns. In order to expand the market volume and maintain a high market share, the company needs to develop public awareness about the technology and its product.

The use case showed that presumably unattractive niche markets can be successfully entered by new companies. It is important that investors are brought on board as early as the innovation phase. These investors have to understand the technology and its potential, and have to maintain a resilient partnership and investment. This is extremely important in a market constricted by standards. Either the product is subject to extensive testing for certification or the standardisation must be expanded.

In the latter case, partnering with competitors facing the same problem and forming an alliance is worthwhile. Even though a portion of the market volume could thus be lost, developing the newly opened niche market requires attracting attention and educating wholesalers or installers – a task better shared with others to gain momentum.

References

1. *Directive 2008/98/EC of the European Parliament and of the Council of 19 November 2008 on waste and repealing certain Directives (Text with EEA relevance).* http://data.europa.eu/eli/dir/2008/98/oj/eng.
2. *ReMining – A new business.* https://re-mine.eu/ (2022).
3. *Energy recovery.* Circular Economy Guide. https://www.ceguide.org/Strategies-and-examples/Dispose/Energy-recovery (2022).
4. US EPA, O. *Energy recovery from the combustion of Municipal Solid Waste (MSW).* https://www.epa.gov/smm/energy-recovery-combustion-municipal-solid-waste-msw (2022).
5. *Introduction to biomass combustion – Farm energy.* https://farm-energy.extension.org/introduction-to-biomass-combustion/ (2023).
6. Rosendahl, L. (Ed.) (2013). *Biomass combustion science, technology and engineering.* Oxford; Philadelphia: Woodhead Publishing. ISBN: 978-0-85709-131-4.
7. *Lecture 13: Pyrolysis & Liquefaction.* College of Agricultural Sciences. https://agsci.oregonstate.edu/bioenergy-education/lecture-13-pyrolysis-liquefaction (2022).
8. Abomohra, A. E.-F., Wang, Q. & Huang, J. (2022). *Waste-to-energy: Recent developments and future perspectives towards circular economy.* Cham: Springer International Publishing. ISBN: 978-3-030-91570-4.
9. *Biogas-Renewable natural gas - U.S. Energy Information Administration (EIA).* https://www.eia.gov/energyexplained/biomass/landfill-gas-and-biogas.php (2022).
10. Smith, M. & Aber, J. (2022). *Recover energy from composting to heat water on farms - Progressive Dairy — Ag Proud.* https://www.agproud.com/articles/31492-recover-energy-from-composting-to-heat-water-on-farms.
11. Homutescu, A. (2017). *German pat.* 102015221552A1. https://patents.google.com/patent/DE102015221552A1/en.
12. Andersson, O. & Hägg, M. (2008). Preliminary design of a seasonal heat storage for ITT Flygt, Emmaboda, Sweden. http://ubeg.de/IGEIA/IGEIA-D10---feasibility-study-factory-Sweden.pdf.
13. Waste heat to power systems. https://www.epa.gov/sites/default/files/2015-07/documents/waste_heat_to_power_systems.pdf
14. Willemsen, G. Open-loop geothermal heat pump systems in the USA and aquifer cold storage in the Netherlands - similarities and differences. https://www.researchgate.net/publication/251300833_OPEN-LOOP_GEOTHERMAL_HEAT_PUMP_SYSTEMS_IN_THE_USA_AND_AQUIFER_COLD_STORAGE_IN_THE_NETHERLANDS_-_SIMILARITIES_AND_DIFFERENCES.
15. Stamatelatou, K. & Tsagarakis, K. P. (Eds.) (2015). *Sewage treatment plants: Economic evaluation of innovative technologies for energy efficiency.* London: IWA Publishing. ISBN: 978-1-78040-501-8.
16. *Resource recovery from Oily Sludge for a Midwestern Refinery.* Veolia North America. https://www.veolianorthamerica.com/case-studies/resource-recovery-oily-sludge-midwestern-refinery (2022).
17. Kahle, K., et al. (2015). *Bottom ash from WTE plants - Metal recovery and utilization.* ISWA – International Solid Waste Association.

18. Schweiz, M. (2023). *Anwendungshilfe zu den Gebäudestandards MINERGIE ®/ MINERGIE-P ®/MINERGIE-A ®*. https://www.minergie.ch/media/230109_anwendungshilfe_minergie_p_a_v2023.1_de.pdf.
19. *Joulia*. https://joulia.com/ (2022).
20. *ecoQuery*. https://v31.ecoquery.ecoinvent.org/Account/LogOn?ReturnUrl=%2fHome%2fIndex (2022).
21. Schweizerischer Ingenieur- und Architektenverein. (2015). *SIA 385/2:2015 - Anlagen für Trinkwarmwasser in Gebäuden - Warmwasserbedarf, Gesamtanforderungen und Auslegung*.
22. Dichter Institut Marketing und Forschung. (2015). *Umfrage zum Dusch- und Badeverhalten der SchweizerInnen*.
23. Krebs, L. & Frischknecht, R. (2018). Umweltbilanz Strommixe Schweiz 2018.
24. *Lebensdauertabelle - HEV Schweiz*. https://www.hev-schweiz.ch/vermieten/verwalten/lebensdauertabelle/.
25. Crippa, M., et al. (2021). GHG emissions of all world: 2021 report. Luxembourg: Publications Office of the European Union. ISBN: 978-92-76-41546-6.
26. Kuhn, A. & Knecht, K. (2022). *Warmwasser-WRG mit Joulia: Synergie- und Integrationsstudie mit Bilanzierung Schlussbericht*. Dübendorf: s3 GmbH.
27. Bundesamt für Statistik. *Bau- und Wohnungswesen*. https://www.bfs.admin.ch/bfs/de/home/statistiken/bau-wohnungswesen.html (2023).
28. Drzimalla, P. (2023). *MuKEn 2014 – was machen die Kantone 2021?* Energie-Experten. https://www.energie-experten.ch/de/wohnen/detail/muken-2014-was-machen-die-kantone-2021.html.
29. Joulia, A. G. (2019). *Die Dusche im Minergie Standard*. https://joulia.com/wp-content/uploads/2020/10/Joulia_Minergie_D_Nov19_web.pdf.
30. Bundesamt für Statistik. *LIK, Durchschnittspreise für Energie und Treibstoffe, Monatswerte (ab 1993) und Jahresdurchschnitte (ab 1966). - 1.9.1966-30.11.2022 — Tabelle*. Bundesamt für Statistik. https://www.bfs.admin.ch/asset/de/23772757 (2022).
31. *Quantilsgrenzen (monatlicher Nettolohn), Voll- und Teilzeitbeschäftigte nach Beschäftigungsgrad - Privater Sektor undöffentlicher Sektor zusammen - Schweiz [T2 _netto] - 2006, 2008, 2010, 2012, 2014, 2016, 2018, 2020 — Tabelle — Bundesamt für Statistik*. https://www.bfs.admin.ch/bfs/de/home/statistiken/arbeit-erwerb/loehne-erwerbseinkommen-arbeitskosten/lohnniveau-schweiz/verteilung-nettoloehne.assetdetail.21224964.html (2022).
32. *Dusche mit Wärmerückgewinnung*. Retrieved March 18, 2020 from https://www.dvgw-cert.com/medien/leistungen/download__antrag-go-zp..._/zp_650.pdf (2020).

Chapter 14
R9 Remine

Nicolas Hofer, André Podleisek, Katharina Luban, and Sylvan Mächler

Abstract Our research journey revealed that *Remine* is a highly specific case within the spectrum of resource retention strategies, often overlooked yet critically important due to the increasing scarcity of resources. Despite an extensive search, we were unable to partner with a case company for in-depth research on this topic. Consequently, our analysis in this chapter adopts a more generic viewpoint. We will draw insights from publicly accessible contributions, rather than focusing on a singular, in-depth case study. The implementation and execution of *Remine* are complex processes that often span several years to decades, are influenced by a range of drivers, and often include significant government involvement. This extended time frame and the intricate nature of *Remine* processes are explored through two examples to clarify the concept and distinguish it from *Recycling*. The first example delves into the extraction of metals, illustrating how *Remine* operates in this context. The second case study examines land as a resource and its relationship with *Remine*. Various drivers and enablers associated with these scenarios are highlighted. Ultimately, the chapter aims to shed light on the significant impact *Remine* has on global resource management and sustainability.

14.1 Introduction to *Remine*

So far, many different Retention Options (Rs) have been discussed in the area of Circular Economy (CE). The last R to be considered is Remine (R9). It represents

N. Hofer
Zürich, Switzerland

A. Podleisek (✉)
Wettswil am Albis, Switzerland
e-mail: andre.podleisek@ost.ch

K. Luban
Sevelen, Switzerland

S. Mächler
Vorderthal, Switzerland

A. Podleisek et al. (eds.), *Circular Economy: The Next Level of Company Success*,
https://doi.org/10.1007/978-3-031-66406-9_14

not just the last option of the long loop Rs but also the final one in the CE model, and stands as the final intervention in a CE framework to reduce the accumulation of waste (see Fig. 14.1). *Remine* specifically targets materials that have been consigned to landfills. In this context, 'landfill' extends beyond traditional waste disposal sites on land to encompass any locations where waste is inadvertently discarded, including rivers, seas, and oceans.

According to Reike et al. [1, p. 257], *Remine*, or also known as 'Urban Mining' or 'Landfill Mining', is the most forgotten or ignored point of CE. However, there were people who recognised the potential of 'Urban Mining' early on. Already in 1972 Jane Jacobs mentioned the term 'Urban Mining.' She stated that in the future cities and landfills will be another form of mines [2]. The role of *Remine* itself is becoming more significant in terms of resource scarcity. Various studies declare that by 2050 the demand for metallic resources will already exceed the reserves in the earth many times over.

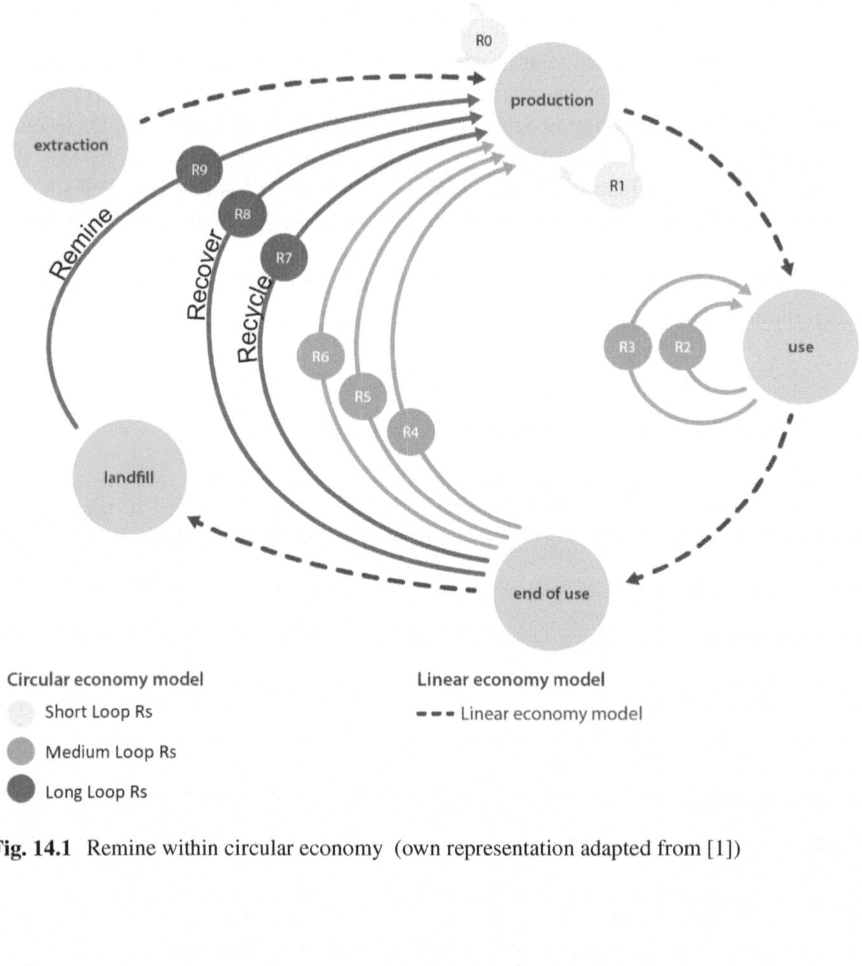

Fig. 14.1 Remine within circular economy (own representation adapted from [1])

Hence, in the context of metal resources, two distinct terms are commonly used. 'Primary Stock,' refers to the resources still available in the ground, yet to be extracted by mining companies. Conversely, 'Secondary Stock' describes all metals that have already been extracted and utilised. Both these concepts are crucial in understanding the availability and lifecycle of metal resources [3, p. 7–9].

In addition to strategic considerations, there are economic and technological drivers. The technology must be advanced enough to enable the extraction of metals from waste. Rogers [4] also describes in his book that the technological aspect determines the feasibility. This determines which useful resources are extracted or not.

Closely connected to this is also the economic efficiency. There can be many costs such as time, prospecting, site preparation, or labor, to name a few. Therefore, it should be possible to roughly estimate the output and yield of such a venture. Further, aspects such as new technologies or rising land prices can also be important enablers. A new technology can make the previously expensive extraction of metals from slag, for example, now inexpensive and effective. Thus, it may also make sense to store certain wastes with metallic components for a longer period of time. Until such time as there are technological opportunities to extract them profitably. Another driver may be the price of land. For example, the landfill is located on a piece of land that has increased enormously in value in recent years. Then it becomes economically interesting to move or dissolve the landfill [5].

Thus, it is obvious that many different factors have to be considered if the principle of *Remine* is to be applied.

14.1.1 Distinguishing Remine from Other Long Loop Rs

As previously mentioned, various terms are used interchangeably for *Remine*. This reflects the lack of a fixed definition and the dependency of the term on specific applications. Furthermore, the interpretation of each term often overlaps, adding to the complexity.

Distinguishing *Remine* from other Retention Options (Rs) can be more difficult than anticipated. For example, Recycle (R7) and *Remine* share several similarities, making it sometimes difficult to define a clear difference. A key factor in differentiating these Rs is the aspect of time. *Recycling* is a process that occurs in 'real time', dealing with waste materials shortly after they are produced. In contrast, *Remine* usually takes place many years later.

As an example, the metal extraction from slag can be named. When slag is first deposited, it may not be economically feasible to recycle it due to various constraints. As a result, the slag is relegated to a landfill. These landfills, initially intended as 'final sinks', are increasingly being viewed as temporary storage sites. It may take several decades before market conditions or technological advancements make it viable to excavate the slag for *Recycling*. This process of extracting valuable materials from long-term storage or waste deposits is what we define as *Remining*.

Triggered by factors such as rising market prices for raw materials or more cost-effective extraction technologies, *Remining* involves the retrieval of previously discarded materials. This concept is further explored in our first use case in Sect. 14.2, where the 'digging up' of slag for its valuable content is a classic instance of remining.

Occasionally a strategic aspect is pursued by *Remine*, in order to create a certain independence from mine-rich countries. With the creation of landfills and the recovery by means of *Remine*, the attempt is made to manage own 'mines' and to contain the dependence in the area of metals. Also, politics and society take influence on the *Recycle* aspect, whereas *Remine* is rather only economically/technologically driven [6].

Cossu and Williams [7] also agree here that the concepts of *Remine* and *Recycle* can be interpreted in the same way, as both principles pursue the processing of waste to extract resources for human benefit. Recover (R8) also shares some similarities with *Remine*. As an example, the principle of incineration technologies is mentioned, which are considered rather negatively, since possible resources are burned for heat, while exhaust gases are produced. In the field of *Remine*, however, this is a standard mechanism for enhancing the concentration and recovery of metals from combustion residues.

Furthermore, *Remine* can, in certain respects, be considered a subset of *Recycle*. This is primarily because *Remine* focuses on extracting specific valuable raw materials from previously discarded waste. Unlike the broader ambit of *Recycle*, which aims to process a wider range of materials, *Remine* is typically more targeted.

Now that the theoretical foundation has been laid, the subsequent sections will dive into practical applications. We will explore two distinct real-world scenarios where *Remine* is applied. The first case focuses on remining valuable resources, the second case delves into remining as a strategy for recovering land for development.

14.2 Analysis of a *Remine* Case for Valuable Resources

The Elbisgraben landfill, situated in Switzerland, plays a crucial role in the realm of waste management and disposal. Each year, this facility becomes the final sink for approximately 50,000 tonnes of non-combustible waste. At the same time, other landfills for other wastes are also maintained [8].

There are three disposal compartments at the Elbisgraben landfill: reactor waste landfill, residuals landfill, and a slag landfill [9].

At the Elbisgraben landfill, not only is non-combustible waste deposited, but slag from waste incineration plants is also processed. Using a specialised metal recovery machine, the facility effectively extracts various metals from the incineration slag. This embodies the principle of *Remine*.

The project's achievements have been acknowledged in numerous reports, highlighting its success across several dimensions. Politically, the facility operates within the strict parameters set by the Waste Ordinance. Since 2015, this legislation has stipulated stringent limits on the metal content allowed in landfill material, specifically capping non-ferrous metals to 1%. The Elbisgraben plant demonstrates exceptional performance in this regard, with a recorded non-ferrous metal content of a mere 0.13%, which is well below the legal threshold [10]. This not only illustrates compliance with national waste management policies but also showcases the facility's role in setting new industry benchmarks.

Another stringent requirement for landfill operation is the extraction of at least 70% of metals from the slag – a target that the Elbisgraben facility not only meets but surpasses. With the installation of a new, cutting-edge plant, the Elbisgraben landfill has reported metal extraction rates from slag of up to 95%. This significantly reduces the metal content in the landfilled slag to a mere 0.08%.

The benefits of this operation extend far beyond regulatory compliance. The process of recovering metals from slag at Elbisgraben is markedly more eco-friendly compared to traditional mining methods. Every year, around 40,000 tonnes of slag, comprising of approximately 10% metals, are processed at the site. Notably, two-thirds of these metals are valuable non-ferrous types, including copper, aluminium, silver, and gold. By extracting and reintroducing these metals into the economic loop, the Elbisgraben landfill plays a great role in fostering a CE. This practice not only fulfils political mandates but also delivers economic benefits, as the metals reclaimed are sold back to businesses and industries [11].

14.2.1 Impacts Through the Case

The Elbisgraben landfill's metal remine system has established a precedent that extends beyond its economic success, influencing both political and environmental strategies. Its model has been adopted by several international landfills, reflecting the global impact of this innovative approach. The replication of this principle in other countries attests to the effectiveness and appeal of the system. Domestically, the financial impact is significant as well; the Elbisgraben landfill has realised 750,000 CHF in revenue from the sale of remined metals. This not only showcases the economic viability of the project but also underscores its potential for widespread adoption and influence it can have on waste management practices.

14.2.2 Drivers of the Case

Reflecting on that case study, we can infer the potential drivers. Although the actual drivers are not explicitly stated, the circumstances suggest a few likely candidates.

Institutional drivers likely played a central role in driving that *Remine* case. This is particularly evident in the operation's adherence to and exceeding of legal landfill waste content standards. The commitment to regulatory compliance (D.I1) not only ensured adherence to current legal requirements but also positioned the operation favourably in terms of potential future legislation (D.I3). By staying below the mandated thresholds, the initiative demonstrates proactive measures that can provide a buffer against future regulatory changes or increased environmental scrutiny. Furthermore, by performing beyond the minimum legal standards, the operation effectively minimised the risk of governmental sanctions (D.I2), thereby safeguarding against possible punitive measures.

Economic drivers undoubtedly also plays a significant role. The facility's ability to profit from reselling remined metals to the industry has not only opened up a new market but also contributed to a sustainable revenue stream. This strategy exemplifies the driver D.F2, as it realises cost savings by transforming waste into a valuable commodity. Furthermore, the initiative enhances the company's competitiveness (D.F3) by tapping into innovation and business development, creating a competitive edge in the market for precious metals.

The project also adeptly addresses D.E3, resilience to resource-related risks, by providing a more stable, locally sourced supply of metals. By extracting resources from slag, a by-product of their own processes and other incineration plants, the company prevents resource wastage and actively extracts embedded value, which would otherwise remain untapped in discarded electronic parts. *Remine* efficiency is the key here to ensure that the highest possible amount of metals is reclaimed and reintroduced into the production cycle.

14.2.3 Enablers of the Case

The enabler E.T1, experimenting with or developing new technologies, can be considered as instrumental in making *Remine* so successful in that case. The introduction of innovative and advanced technologies has broken new ground in how slag can be processed to extract precious metals. This progress underscores enabler E.T3, availability of technical and technological know-how in-house, reflecting the internal capabilities that drive such innovative recycling practices forward.

Additionally, the initiative has leveraged E.M3, engaging in industrial symbiosis, by ensuring that the by-product of waste incineration, i.e. slag, is not wasted but instead used as a source for metals.

14.2.4 Threats of the Case

In concluding this case, let us also reflect on the broader implications of *Remine* initiatives, such as those undertaken by the Elbisgraben landfill. Although there are economic benefits, we must also weigh the potential environmental costs that could arise. The concern of 'replacing one evil with another' serves as a cautionary reminder that the benefits of remining should not be overshadowed by unintended consequences.

Before embarking on similar initiatives, careful consideration must be given to the full environmental impact. The transportation of slag over long distances, for example, could present a significant environmental setback due to the increase in emissions from logistics operations. There must be a balance between the benefits of *Remine* and the economy of scale. It is imperative to minimise carbon emissions and other forms of pollution. Therefore, it is essential to integrate strategic environmental management practices such as optimising logistics, using greener transportation methods, or even establishing localised remining facilities to reduce transit distances.

14.3 Analysis of a *Remine* Case for Land Development

In the industrial district of Zurich, Switzerland, the construction company Eberhard embarked on a landmark development project. Central to this initiative was the transformation of a service facility into a structure stretching 425 m in length. This expansion was carried out on top of the Herdern landfill. A site that has served the city until 1950. Over the years, this landfill, initially a clay pit filled with household sweepings, cinders, foundry sands, and residual excavated materials, had transformed into an ecological compensation area. As Zurich grew and the demand for land intensified, the economic impetus for redevelopment became undeniable. Despite the ecological value that the closed landfill had cultivated over time, the city's expansion and the increasing value of urban land led to the decision to dismantle the landfill. The project spanned over a year, systematically emptying the landfill to reclaim the land for new development, thus underlining the economic feasibility and desirability of reclaiming land in a growing urban context. The challenge was not just the sheer volume but the complexity of contamination of the excavated material – each layer presented a different degree of pollution that necessitated a carefully strategised approach for remediation [12].

14.3.1 Impacts Through the Case

In this scenario, the resource in question extends beyond various metals, as seen in prior cases, to a much larger scale: land itself. Previously, the landfill was a blockade, preventing the space from being utilised for new development purposes. The redevelopment project left a significant mark both financially and environmentally. The construction costs for the venture reached a total of 10.2 mill. CHF. The financial outcomes, however, extended well beyond this initial investment. The reclamation and subsequent redevelopment of the land resulted in new returns that surpassed the initial costs. This showcases a successful transformation process of the previously dormant landfill area into a profitable and valuable development. Through Remine, the land has been liberated and rendered anew, a feat that recycling processes do not encompass. In the grand scheme of CE, remining reintroduces the resource, i. e.

land, back into the usable spectrum, thereby fostering economic growth, benefiting the environment, and improving the well-being of the community.

14.3.2 Drivers of the Case

This use case adeptly addresses the drivers of D.S2, as it embodies the adaptation to changing consumer behaviour and meets expectations of a significant customer – the city itself. It also embodies D.E2, by significantly reducing the carbon footprint, with the 'product' in this scenario being the once landfilled area that has undergone a transformative cleanup. Additionally, it touches upon D.E1, as the repurposing of land effectively decreases the need for using new uncontaminated land. It also resonates with D.R1, showcasing a commitment to social responsibility by rehabilitating a space that was once a source of environmental concern.

14.3.3 Enablers of the Case

The project's success could be significantly attributed to the following two enablers. Firstly, the collaborative relationships established between various stakeholders (E.V1) were crucial. Partnerships with local government entities, environmental agencies, and development firms allowed for a cohesive approach to redesigning the land's use. Secondly, a Reverse Logistics (RL) network (E.M1) was established to manage the flow of materials out of the landfill, ensuring that any contaminated waste was extracted and appropriately redirected, reducing the environmental impact of the cleanup.

14.3.4 Threats of the Case

While the project presented several innovative solutions, it was not without its concerns. One such concern was the potential for 'Ethical Conflict,' particularly in relation to the ecological compensation area. The dismantling of this area raised concerns about the displacement of local species that had come to rely on this habitat, including wild bees, lizards, and other thermophilic animals and insects. The careful planning of the construction schedule to accommodate hibernation patterns mitigated some of these concerns, allowing for the relocation of these species with minimal disruption.

Another concern was the 'Resource Conflict,' which manifested in the dual perception of the land as both a landfill and a habitat. The eventual repurposing of the land needed to be evaluated to ensure that it did not simply exchange one set of ecological challenges for another. The long-term implications of the land's use had to be considered and a clear vision for its sustainable development been demonstrated.

14.4 *Remine*: Summary and Learnings

Remine emerges as a concept, unlocking potential across various sectors that may not be immediately evident. It is a process enabled by a confluence of factors, each integral to realising its full potential. The cases discussed illustrate this complexity. Initially, the implementation of *Remine* can seem impractical due to technological limitations or the absence of economic incentives. Unlike *Recycle*, which reintroduces resources into the cycle shortly after use, *Remine* often requires a considerable temporal investment, spanning decades, before the conditions align to make it viable and market-driven.

In the instances examined, this maturation period was crucial. At first, the proposition of *Remine* was not profitable or necessary, but as years passed, drivers and enablers converged to activate the market mechanisms needed for its success. The first case also highlights the significance of *Remine* in the realm of material recovery, releasing materials previously deemed less critical or abundantly available.

Despite the overlapping terminology and the occasional oversight in discussions about the CE, the role of *Remine* is indispensable. It may operate on a delayed timeline, but its function is essential to complete and closed the resource loop with a final option of material processing strategy when other long loop Rs can not.

References

1. Reike, D., Vermeulen, W. J., & Witjes, S. (2018). The circular economy: new or refurbished as CE 3.0? – exploring controversies in the conceptualization of the circular economy through a focus on history and resource value retention options. *Resources Conservation and Recycling, 135,* 246–264. ISSN: 0921-3449.
2. Jacobs, J. (1972). *The economy of cities*. Harmondsworth: Penguin Books. ISBN: 0–14-021473-9
3. Nakamura, T., & Halada, K. (2015). *Urban mining systems* (1st ed.). Springer Japan, Tokyo. ISBN: 978-4-431-55075-4
4. Rogers, E. M. (1995). *Diffusion of innovations* (4th ed.). Free Press, New York, NY. ISBN: 0-02-926671-8.
5. Krook, J., & Baas, L. (2013). Getting serious about mining the technosphere: a review of recent landfill mining and urban mining research. *Journal of Cleaner Production, 55,* 1–9. ISSN: 0959-6526.
6. Bunge, R. (2021). *Unterschied recycle und remine: Gespräch: Personal communication November 14, 2021.*
7. Cossu, R., & Williams, I. D. (2015). Urban mining: Concepts, terminology, challenges. *Waste Management, 45,* 1–3. ISSN: 0956-053X.
8. Kanton Basel-Landschaft. (2022). *Deponie Elbisgraben* baselland.ch. https://baselland.ch/politik-und-behorden/direktionen/bau-und-umweltschutzdirektion/industrielle-betriebe/abfallanlagen.
9. Kanton Basel-Landschaft. (2022). *Abfälle*. https://baselland.ch/politik-und-behorden/direktionen/bau-und-umweltschutzdirektion/industrielle-betriebe/abfallanlagen/dienstleistungen/abfalle.
10. Wieland, B. (2021). Baselland: Neue Metallrückgewinnungsanlage spürt sogar Münz auf. *Aargauer Zeitung AG.* https://www.bzbasel.ch/basel/basel-stadt/ein-monster-das-sogar-muenzen-aufspuert-ld.2207520.

11. Maier, S. (2021). Recyling: Effiziente Metallrückgewinnung auf der Deponie Elbisgraben. Baublatt. https://www.baublatt.ch/kommunal/recyling-effiziente-metallrueckgewinnung-auf-der-deponie-elbisgraben-31861.
12. Serviceanlage SBB Herdern. Eberhard Unternehmungen. (2022). https://eberhard.ch/referenzen/inhalt/Serviceanlage%20SBB%20Herdern%20-%20Z%C3%BCrich-Altstetten

Part III
Practical Implications of Circular Economy

Chapter 15
Introduction to Part III

Katharina Luban

Abstract In Part III of this book, we will translate our research findings into practical applications and present the synthesis of the Book. The chapters are dedicated to the practical aspects of implementing Circular Economy successfully. First, a case of the retention option *Reuse* was chosen to demonstrate the use of drivers and enablers when evaluating circular ideas or business models. Plastics *Recycling* and an in-depth analysis of digitalisation as a main enabler of *Remanufacture* round the picture. All findings of the research of the book are summarised in the Rapperswil Circular Economy Model.

About Part III of This Book

In Part I of the book, the framework of drivers and enablers of Circular Economy (CE) was developed and the drivers and enablers were linked to the different Retention Options (Rs). A survey revealed the current state of CE application in companies and the financial relevance of these initiatives. In Part II, a case study methodology was applied to investigate each R in detail. Successful cases were presented for each R.

The third part of this book is dedicated to the practical aspects of successfully implementing Circular Economy. First, a case of the R *Reuse* was chosen to demonstrate the use of drivers and enablers when evaluating circular ideas or business models. Plastics *Recycling* and an in-depth analysis of digitalisation as a main enabler of *Remanufacture* round the picture. All findings of the research of the book are summarised in the Rapperswil Circular Economy Model.

Chapter 16: Using the Framework of Drivers and Enablers to Evaluate Circular Business Ideas

In Chap. 16, the example of reusable food packaging served as a demonstration of how to bring transparency to the evaluation process of circular ideas. When using a similar procedure, businesses will gain a deeper understanding of how to best leverage the drivers and enablers of their ideas, and they can also use the procedure to compare any CE idea or business model with the solutions of competitors.

K. Luban (✉)
Sevelen, Switzerland
e-mail: katharina.luban@ost.ch

A. Podleisek et al. (eds.), *Circular Economy: The Next Level of Company Success*,
https://doi.org/10.1007/978-3-031-66406-9_15

Chapter 17: Rethinking and Positioning Plastics for Circular Economy

Despite legislative efforts, a significant proportion of plastic materials continue to contribute to waste accumulation. In Chap. 17, Professor Daniel Schwendemann, a leading figure in the domain of plastics processing, provides insights on key considerations and challenges inherent to plastics. He suggests moving from a general rejection of plastics to fostering a culture of responsible plastic use and effective *Recycling*.

Chapter 18: Digital Transformation of Remanufacturing

Chapter 18 describes a Digital Product Lifecycle (PLC) Approach to *Remanufacturing*. Professor Dr. Ramesh Subramoniam has been working in the field of supply chain, manufacturing, and data since many years and has used his experience to develop a framework for a digitised PLC integrated with business phases including remanufacturing stages. He points out that future success in remanufacturing will hinge on companies adopting a mindset open to change, backed by a robust technology infrastructure to facilitate the transition.

Chapter 19: The Rapperswil Circular Economy Model – A Guide to Successful Implementation

All findings and learnings of our research are summarised in Chap. 19. The summary concludes in the Rapperswil Circular Economy Model, that provides guidance on how to address the implementation of circularity in a specific company. The basis is a 6-step management model starting with understanding the current situation, then setting vision and goals, setting ambition levels, defining possible actions including prioritisation, developing a roadmap, and finally measuring results.

Chapter 16
Using the Framework of Drivers and Enablers to Evaluate Circular Business Ideas

Katharina Luban, Mladen Petraskovic, and Raphael Meyer

Abstract In the contemporary dynamic business environment, the imperative for a robust analysis and comparison of business ideas has grown significantly. In this chapter, we demonstrate how the framework of drivers and enablers of circular economy can be integrated in the assessment and comparison of various business models dedicated to realising a *Reuse* idea. Although the assessment of an existing circular business model is demonstrated, the procedure can also be applied to the evaluation of new ideas or for comparing ideas to those of the competition.

16.1 Goal and Methodology of Assessing Circular Concepts

Technological advancements and evolving consumer preferences drive companies to innovate and adapt. Therefore, robust analyses and comparison of various business strategies have become increasingly essential. This is especially true when evaluating business cases in the field of Circular Economy (CE), where new players, preferences, and values must be considered and where experience with similar cases is scarce. In this chapter, we will demonstrate how circular aspects can be integrated into the assessment of business models or ideas.

Assessments help prioritise viable concepts and design options [1]. They allow the rejection of less promising ideas, guide development processes, aid in decision-making regarding the introduction of new business models, and persuade investors of the potential success of such models.

K. Luban (✉)
Sevelen, Switzerland
e-mail: katharina.luban@ost.ch

M. Petraskovic
Schaffhausen, Switzerland

R. Meyer
Freienbach, Switzerland

A. Podleisek et al. (eds.), *Circular Economy: The Next Level of Company Success*,
https://doi.org/10.1007/978-3-031-66406-9_16

Numerous authors have made efforts to articulate the concept of a business model [2]. Taking into account these detailed analyses, this chapter adopts the perspective that a business model is understood as the logic of how a business creates value and delivers it to customers [3]. Therefore, although the assessment of a completely new circular business model is demonstrated, the general ideas can be applied to evaluate general business ideas in the field of CE.

To formulate a set of appropriate criteria to assess the strategic potential for the success of business models, Höse et al. [4] propose to combine the resource-based view (e.g., [5]) with the market-based view (e.g., [6, 7]). These theoretical concepts are originally focused on an economic analysis, and, therefore, the evaluation criteria equally do so. This is why Bocken et al. suggest improving the criteria or incorporating additional criteria when the number of stakeholders to be taken into account expands and ecological aspects must be incorporated [8].

Following this suggestion, the analysis of the circular business models should contain the following criteria:

- ecological aspects (emissions, energy balances, etc.),
- organisational aspects (handling, logistics, cleaning, etc.),
- stakeholders (needs, willingness, etc.) and
- economic efficiency (costs, competition, etc.).

The framework of Drivers and Enablers proposed in Sect. 3.2 reflects these criteria. We will put it into action and demonstrate its suitability to evaluate and compare Circular Economy Business Models (CEBMs) and ideas. The underlying circular idea we evaluated was the reduction of the environmental impact of food packaging. The challenge was to define the best way to retain the value or at least some value of packaging for takeaway food, which is currently disposed of without further thought.

According to the findings of our previous research, the first step was to decide on an appropriate Retention Option (R). Initial considerations led to the assumption that the strategy *Reuse* would be a good option to pursue. The decision for this option was followed by the identification of three realised business models that had already been implemented in this field.

To verify the R, the drivers of CE, listed in Sect. 3.2.2 were analysed with respect to the idea of reusing food packaging. All identified business models face the same situation and therefore all drivers apply for their situation. The findings were compared to those of the Case Study on *Reuse* in Chap. 7. From this analysis, we knew that some drivers are more relevant to this Retention Option (R) than others. Reassuring these important drivers for the idea of reusing food packaging confirmed the decision for *Reuse* as the most appropriate CE option.

Once the driving forces of CE were confirmed for the *Reuse* case, the enablers of CE implementation were used to compare the different business models. The qualitative evaluation was transferred to figures to allow for a quantified evaluation. According to our findings in earlier chapters, certain Enablers are more relevant than others, depending on the R. According to their relevance to the R *Reuse*, the enablers were weighted. Multiplying the evaluation values by weighting factors resulted in a utility value for each business model.

On top of this, special attention was paid to the financial model of the business models, calculating the break-even for the customers who might decide to implement the ideas. We can show that finance models add a valuable dimension to the utility value calculated by means of the enablers.

16.2 Choosing the Best Retention Option for Food Packaging

When developing a circular business model or idea, the first point to clarify is the potential R. In the field of food packaging, the collection and recycling of disposable pizza boxes, coffee cups, and other cardboard and paper food packaging would be an intuitive way to reduce resource consumption and waste disposal. However, coatings, oil, and grease residues make the circular approach of *Recycle* (R7) difficult or impossible.

Within the concept of value retention, the next-best options would be allocated higher in the Value Hill (see Fig. 3.2). A brainstorm session showed that no good and especially long-lasting ideas to Repurpose (R6) soiled food packaging could be developed. The Medium loop R from Reuse (R2) to Repurpose (R6) all have in common that they require a defined collection system and therefore some sort of Reverse Logistics (RL). This factor proves to cause high implementation efforts and costs. Therefore, the option Reuse (R2) was chosen as the preferred mode of value retention because it has the highest retention value (see Sect. 3.1). Sure enough, the option *Refuse/Reduce* would have even a higher impact on value retention, but this option is not in the hands of the providers, but rather must be driven by the customers who refrain from ordering or transporting food. Since this is not a viable scenario, the circular option of *Reuse* was chosen as the best R to pursue.

These considerations are confirmed by the fact that several startups decided to take advantage of the *Reuse* option for food packaging to define their circular business models. Introducing reusable packaging results in less packaging waste. Cities and municipalities can use these concepts to reduce littering on the one hand and reduce waste disposal costs on the other. By reusing takeaway packaging instead of throwing it away, numerous resources, as well as CO_2 are saved [9]. In the following chapter, we will take a closer look at these concepts. To do so, we identified distinct company cases in the context of reusable packaging that subsequently were analysed in detail.

16.3 Three Business Models for Reusable Food Packaging

Pizza boxes, coffee cups, and packaging in general create tonnes of waste, but fortunately, awareness of local and ecological production and regional cycles to reduce the environmental impact is increasing. In this context, several startups and organisations

Fig. 16.1 Concept of renting reusable food packaging

offer reusable systems to replace disposable packaging to provide an alternative to the linear model of take-make-dispose. Good examples can be found in the catering sector and will be used in this chapter although reusable systems to package other products do exist as well, such as postal deliveries with a 'dispobox' [10] or small load carriers in the automotive industry [11].

The providers of reusable food packaging follow the underlying logic of 'renting' the packaging to the consumers (see Fig. 16.1). The rental process usually starts with downloading an app and registering for the service. Then the food is ordered or bought at a service-partner, who offers the reusable packaging service. When receiving the food, the customer has to register the reusable packaging, usually by scanning a bar code. After food consumption, the packaging has to be returned and the product is 'checked out' from the customer account.

The concepts for reusable packaging differ in details such as cost of rental or deposit, way of returning the packaging, etc. Three solutions will be introduced to illustrate these differences. The three examples were chosen according to the availability of information, either through personal contacts or publicly available data. The final choice fell on the three companies reCIRCLE, Vytal, and GreenToGo.

16.3.1 reCIRCLE

Description

The company reCIRCLE sells and rents various types of food boxes, pizza boxes, and drinking containers. It distributes its packaging products in the gastronomy sector, which can be reused several hundred times before being recycled. In this way, the company wants to promote CE and minimise disposable packaging. The goal of reCIRCLE is to replace disposable packaging with reusable, high-quality, affordable, and more ecological alternatives. The company advertises local production, short transport distances, and high-quality, durable products. The special feature of reCIRCLE is that they even considered *Recycling* in the concept from the start: Although being a supplier of *Reuse* packaging, their products are also designed for recyclability.

Fig. 16.2 Concept of reCIRCLE (own presentment based on [9])

Concept

reCIRCLE's food packaging reuse-concept is visualised in Fig. 16.2. All reCIRCLE partners take back the used boxes and refund the deposit of 10 CHF for food boxes or 5 CHF for drink containers. With the app, it is even possible to borrow the boxes free of charge for 7 days. A prerequisite for return or exchange is that the boxes, which consist of a tray and lid, are returned completely. Defective boxes can be exchanged for new ones at partner restaurants for free. The company reCIRCLE takes care of the recycling. The partner businesses regularly sort out heavily discoloured, scratched, or cut containers. These are collected twice a year and replaced with usable ones. This ensures that flawless products always circulate in the network. The exchange usually takes place in spring and fall. The Life Cycle Assessment (LCA) shows that, despite cleaning, boxes are more environmentally friendly than disposable packaging after only 15 uses.

Product

reCIRCLE offers packaging in different sizes. All products are manufactured in Switzerland. The box shell is made of Polybutylene Terephthalate and glass fibres, the cup is made of Tritan and the lids of the boxes and cups are made of Polypropylene. The boxes can be washed several hundred times and are also dishwasher and microwave safe. For drinks, reCIRCLE offers cups with a lid, which they call ISY.

16.3.2 Vytal

Description

Vytal is a company in Germany that offers reusable alternatives for food and beverages. The company is market leader in Europa, has more than 4,500 partners and more than 330,000 private users. It offers a solution for the requirements resulting from the reusability law in Germany that came into force on January 1, 2023. This

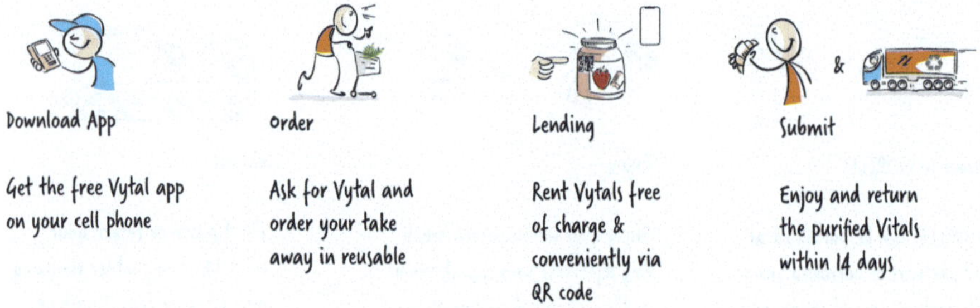

Fig. 16.3 Concept of Vytal (own presentment based on [12])

law entails a mandatory obligation for restaurants to offer a reusable solution for food and beverages depending on the size and number of employees in the establishment. Restaurants pay a filling fee for each box that a customer takes and registers. The filling fee is intended to be less than the cost for a single-use packaging. Vytal operates without fixed monthly costs and there is no minimum contract period. The customer pays a deposit to ensure the return of the box [12].

Concept

The concept is shown in Fig. 16.3. To use the solution, the Vytal app must be installed. It allows the consumer to find partner restaurants that have reusable packaging in their range. Then, orders can be placed via the app or by phone, adding Vytal containers for takeaway. By QR code, the Vytal box is registered at pickup or delivery and the account is debited with a deposit of 10 EUR per bowl and 4 EUR per cup. After a 14-day return period, the reusable packaging must be returned; otherwise, the deposit will be retained by Vytal. The use of Vytal containers can also be handled offline, that is, without using the app but instead using a credit card. However, with the offline method, only two packages can be borrowed at a time, while with the app method there is no limit.

Product

Vytal containers are made of bisphenol-A (BPA) free Polypropylene (PP) and thermoplastic elastomer (TPE). These plastics are easily recyclable and microwavable. The containers were developed according to European Union (EU) standards and are manufactured locally, in part with the participation of workshops for disabled people. According to the manufacturer, they can be used at least 200 times. From just ten fillings, they are more environmentally friendly than disposable packaging and save up to 30 kg of CO_2 over the course of their lifecycle compared to disposable packaging made of styrofoam or aluminium.

16.3.3 GreenToGo

Description

GreenToGo is one of several sustainability and waste reduction initiatives managed by 'Don't Waste Durham', a local non-profit organisation. The programme works with restaurants in the city of Durham, situated in North Carolina (U.S.) to provide reusable containers that members can use instead of disposable packaging. The initiative supplies restaurants with clean GreenToGo containers and monitors inventory and restocks as needed. Operators conduct in-person training for partner restaurants, provide signage for customers, and offer demonstrations at restaurants to explain to customers how the concept works. They offer a full service, so customers do not have to worry about dirty containers or the deposit, shown in Fig. 16.4. Restaurants have an on-site return station, which can be placed indoors or outdoors, where customers can return soiled GreenToGo containers [13].

Concept

Figure 16.4 illustrates the GreenToGo programme's framework. To use the programme, customers purchase an annual subscription to GreenToGo, which allows them to borrow between one and four boxes. The price is a minimum of 25 USD and increases by 5 USD for each additional box. Once customers have used their reusable containers, they can drop them off at drop-off stations near participating restaurants and stores built by volunteers from recycled wood.

The participating restaurants will be stocked with reusable containers. When a GreenToGo member eats in the restaurant or orders takeout, they can request a GreenToGo container for their leftovers or for their meal by confirming their active membership with the cashier on their mobile device. The person then uses a clean GreenToGo container for the food they ordered, and the customer checks it out on their mobile device using the QR code provided. The supply of containers in restaurants is electronically monitored based on check-out data and replenished by

Fig. 16.4 Concept of GreenToGo (own presentment based on [13])

GreenToGo staff as needed. The employees collect the dirty bins weekly from drop-off stations by bicycle to avoid an additional impact on the environment.

Product

The GreenToGo container is a reusable takeout box, beyond no other products are offered. In order to use this container, a membership is required, which is linked to the associated app. There is no offline method, that is, the usage of GreenToGo services without an app.

16.4 Analysis of Drivers to Understand the Business Situation

Drivers are factors that push concepts of CE and their existence will be an important factor for successful CE implementation. In the case of reusable food packaging, all service providers address the same CE concept. Therefore, the driver cannot be used to differentiate business models and naturally they match the analysis in Chap. 7. The assessment of the driver nevertheless serves to understand the current business situation of the three companies, leading to either questioning or confirming the possible success of the case.

Within the group of driver that arise from environmental issues, the main driver of reusable packaging is clearly the reduction of resource consumption (D.E1). The whole idea of reusable packaging is based on the ambition to decrease the resources needed for disposable cups in takeaway restaurants and cafes. To prevent Green washing, this idea of 'less resource consumption' must be considered in the overall environmental footprint. That is why the concepts of reCIRCLE, Vytal and GreenToGo all show a comprehensive consideration of this topic, e.g., by defining their logistics (using e-bikes instead of combustion engines) and by performing efficient cleaning of the returned packaging.

A second relevant driver within the cluster of environmental driver considers the reduction of the products's environmental footprint (D.E2). All of the companies work with materials that are durable as well as recyclable at the end of the use cycles. Resource-related risks (D.E3) could not be identified as drivers for the case.

Looking at Institutional factors, the law on reusable packaging posed by the EU is clearly a driver for reusable packaging [14–16]. This law requires all catering establishments to offer reusable packaging (D.I1). In some cities punitive measures for disposable food packaging are already in place, making this another important driver for the businesses (D.I2). Even if in Switzerland there are no legal requirements yet in place, businesses expect future legal obligations (D.I3), so this is another institutional factor driving reusable food packaging.

A minor driver is the monetary incentive from the government (D.I4) in the form of tax refunds or financial contributions for partners, but of course, this will help to make the cases successful.

The analysis of driver within the cluster Stakeholder is based on the assumption that restaurants offering well-known and functional concepts of reusable food packaging will benefit from increased reputation levels (D.S1), and that these new services are necessary to adapt to changing customer behaviour and meet new customer requirements (D.S2). However, this driver is less strong than the legal drivers.

All three companies are first-movers in the field of reusable packaging. Therefore, the need to follow best practices from competitors or catch up with existing standards is not considered as a driver (D.S3).

The volatility of product prices of current food packaging (D.F1) is not a driver, neither is the opportunity to save costs (D.F2) a reason to engage in reuse concepts in the first place. Especially for startups, financial success is vital, so they have to make money. But since they will keep it for themselves, their argumentation in the direction of customers is 'doing good without paying more'. This means that although money is not a driver for customers to accept alternatives for disposable packaging, the implementation will be supported by the long-term opportunity to save money. These calculations highly depend on the business models and will be therefore discussed in a separate chapter, comparing all three cases. The remaining Financial driver *increasing competitiveness* (D.F3) or *creating value from waste* (D.F4) are not recognised as drivers for the cases of reCIRCLE, Vytal and GreenToGo either. Overall, we can state that Financial drivers are not overly relevant factors to implement CE for food packaging.

When analysing the driver of the cluster Responsibility, we see highly relevant drivers again. Actually, these drivers are considered key drivers of the concepts, because all startups in the field of reusable food packaging take on social and ethical responsibility (D.R1) and express the desire to become a promoter of systemic change (D.R2) by increasing awareness of waste reduction in society and offering solutions to do so.

Table 16.1 shows the summary of the analysis of the driver of the CE case of reusable food packaging. Drivers that are considered strong receive +++, those that are clearly present are rated with ++, slight relevance is expressed with + and the absence of relevance is marked by 0.

16.5 Enablers of Implementing Reusable Food Packaging Systems

As the discussion of the *Financial* drivers showed: From customer perspective, the introduction of a reuse alternative to disposable food packaging is not primarily based on financial considerations. For customers, in some cases, the improvement in value retention might even be worth it to pay more or to undergo changes in behaviour and to accept inconveniences that did not exist before. For a long-term success of CE implementation we assume that customers of reusable food packaging (both restaurants and final customers) are willing to undergo changes and/or to make

Table 16.1 Drivers for the *Reuse* case

Cluster	Code	Driver	Relevance
Environmental	D.E1	Resources	+++
	D.E2	Footprint	++
	D.E3	Risk	0
Institutional	D.I1	Legal	++
	D.I2	Punitive	++
	D.I3	Future Juridification	++
	D.I4	Tax	+
Stakeholder	D.S1	Image	+
	D.S2	Customer	+
	D.S3	Practices	0
Financial	D.F1	Volatility	0
	D.F2	Savings	+
	D.F3	Competitiveness	0
	D.F4	Value	0
Responsibility	D.R1	CSR	+++
	D.R2	Promoter	+++

a monetary contribution. Depending on the loss of comfort and the strength of other drivers, they will be more or less willing to accept higher prices. The required changes in behaviour as well as the resulting cost are a differentiation point of the three business models on reusable food packaging and they will be crucial for the success of the implementation. In this chapter we will first discuss monetary contributions and will later use the established framework of enablers of CE to assess the different approaches of reCIRCLE, Vytal (Vytal) and GreenToGo.

16.5.1 Financial Considerations of the Business Models

In a business environment, it would be naïve to ignore monetary issues. All three providers of reusable food packaging will have to earn money in the end, so neither of them claims to reduce the cost of their customers, neither the restaurants nor the final consumers. They all rather claim to 'do good' at more or less the same price. We looked at the different business models to compare cost and effort.

reCIRCLE

The Basic subscription at reCIRCLE is shown in Table 16.2. It costs 210 CHF for a restaurant, a price composed of 150 CHF partnership fee and 60 CHF subscription fee. The delivery of new boxes and cups costs 18.50 CHF every time they are ordered.

Table 16.2 reCIRCLE subscription models based on [9]

Subscription type	Basic	Flat XXS	Flat M
Cost per Year	210.- CHF	330.- CHF	930.- CHF
Purchase and Buyback of reCIRCLEs	Box 10.- CHF / ISY 5.- CHF	10% Discount	10% Discount
Replacement of Worn reCIRCLEs	Additional Fee	23 Pieces	135 Pieces
Shipping Fee	18.50 CHF	Free	Free
reCIRCLE Cutlery Fee	1.- CHF / Set	Free	Free
Activation of Multiple Branches	×	✓	✓

The investment in reCIRCLE products will cost 10 CHF per food box and 5 CHF per cup. The higher the proportion of reCIRCLE products in a restaurant, the higher the investment will be. However, the products that are in circulation will be financed by the deposits of the customers, which they have to pay when buying the food in reusable packaging. When deciding on more advanced subscription models, the purchasing conditions become even better; that is, if a restaurant wants to exclusively offer reCIRCLE products (100% share), a larger subscription is worthwhile because 10% discount is granted on the purchase price of the boxes and, for example, in the case of the Flat M subscription, 135 packages are replaced free of charge in the event of a defect. To reduce the risk that comes with the investment, reCIRCLE also offers the option of returning undamaged packaging at any time at the purchased price. The overall investment can be calculated as follows:

*Purchase price reusable packaging = (meals sold per day * price reusable packaging + drinks sold per day * price reusable cups) * return cycle in days * proportion of reCIRCLE products*

To calculate a break even, we need to know the current cost, which is calculated using the following formula:

*Current Cost = (meals sold per day * price disposable packaging + drinks sold per day * price disposable cups) * number of open days per year * (1-proportion of reCIRCLE products)*

As a calculation example, we analysed a restaurant which sells 50 meals and 100 drinks a day. The restaurant is open 220 days per year and has a proportion of 50% reusable packaging. The price for disposable packaging is 0.2 CHF per meal and 0.08 CHF per drink. With an average return cycle of 7 days and assuming the prices quoted for the Basic subscription, the break-even point of this sample restaurant is reached after only 2 years (see Fig. 16.5).

It should be noted that the calculations are based on the boxes being returned after seven days *at the latest* and in *undamaged* condition, which cannot be assumed in practice, but this will not postpone the break-even point much, as the cost of damage

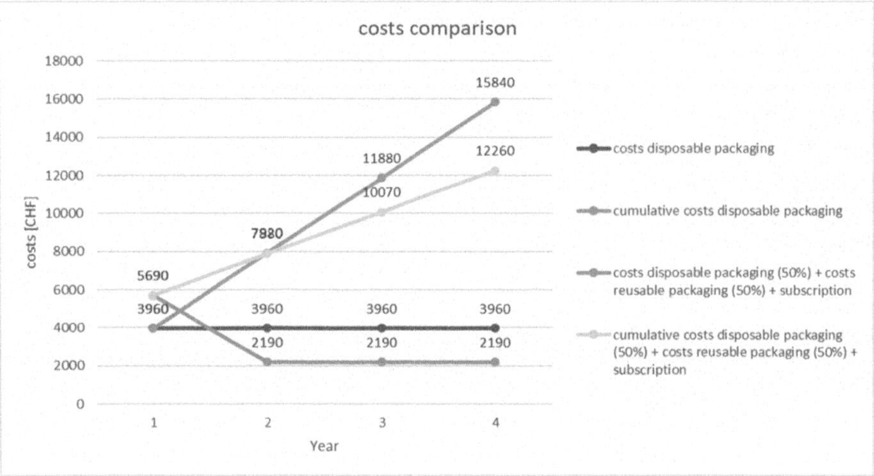

Fig. 16.5 Break-even point for restaurants offering reCIRCLE packaging

and unreturned packaging stays with the end customers. The business model of reCIRCLE seems attractive to restaurants because a participating restaurant saves costs once they reach the break-even point. However, the inconveniences of this model lie with the restaurants, who independently have to collect and clean the reusable food packaging, but the cost and effort for the end consumers are low and they might value the restaurants' engagement as active contributors to environmental protection.

Vytal

Vytal does not offer a subscription system, so restaurants do not have to face fixed monthly or yearly costs. Vytal offers reusable packaging free of charge and generates its revenue through filling costs. Thus, the revenue of reusable packaging is derived from the number of times it is used. That is, every time a customer orders food in a reusable Vytal box, the partner restaurant has to pay 30 cents to Vytal. This cost model is interesting for restaurants that want to pay per use and do not want to take any risks in investments. On the other hand, there are also no potential cost savings for restaurants as long as their current disposable packaging does not exceed 30 cents per piece.

Thus, while the model is attractive to Vytal, the resulting inconveniences of recollecting and cleaning are not compensated monetarily for the restaurants. We assume that this model will most likely be successful where restaurants are forced to offer reusable packaging solutions without considering this new option as a business case.

GreenToGo

GreenToGo requires a restaurant that wishes to participate in the GreenToGo system to pay 500 USD per year. This amount includes all types of packaging that the restaurant wishes to use. Additionally, a return station will be placed in or near the restaurant unless there is already a station nearby. In addition to the costs associated with participation, the programme aims to reduce any inconveniences resulting for restaurant operators and their staff. The annual fee of 500 USD covers the complete service: GreenToGo collects and cleans reusable packaging and constantly supplies participating restaurants with a sufficient number of clean packaging.

The downside is the fact that end consumers must pay a monetary contribution to the system. This might be an entry barrier, but only when the system is used scarcely. The membership costs 25 USD per year for a 1-box GreenToGo membership. Membership includes the GreenToGo mobile app and GreenToGo check-out/check-in privileges for one year at participating restaurants in Durham. If more than one GreenToGo box is needed, the membership can easily be upgraded to a 2-, 3- or 4-box membership using the GreenToGo mobile app, each of which costs 5 USD extra per box.

With a yearly fee of 500 USD, no collection and cleaning effort, and no investment risk, GreenToGo is a good model for restaurants wishing to reduce their environmental footprint. However, this will only work if the customers are willing to pay the monetary contribution of 25 USD per year.

16.5.2 Using the Enablers of CE to Compare Business Models

Enablers are influencing factors that support the successful implementation of an R. Their absence may be perceived as barrier, but if they do not exist, they can be intentionally built up. The evaluation of the three business models of reusable food packaging is based on the general enablers of CE defined in Sect. 4.2.1. According to the findings in the use case analysis of a specific *Reuse* use case in Sect. 7.2.3, most of these general enablers are important for implementing *Reuse* cases. Two of these enablers are especially vital for every startup: A committed company environment and culture is important (E.F1) and they all have to build close relationships and/or invest in collaboration and partnerships (E.V1). However, these two enablers do not serve as distinguishing factors when comparing three different startups and will therefore not be considered when comparing the three food packaging cases.

When analysing the *Reuse* use case, we did not find evidence for *Technological* enablers (E.T1, E.T2, E.T3). However, we assume that technology-based features, such as a functional and user-friendly app, recognised payment methods and – where applicable – well-designed collection systems, are relevant to all providers of reuse food packaging. Additionally, all three startups need to engage in industrial symbiosis (E.M3) to develop and realise their products. We will not follow up on these enablers as they do not apply to the respective cases, but it must be stated that they might be relevant, especially, when technology and partnerships can be identified as differentiation options for the business models.

All other general enablers will be considered in accordance with the discussions of the *Reuse* use case in Sect. 7.2.3. We state a slight relevance to the enablers *Transparency by making use of technologies* (E.V2) to consider user-friendliness in the procedures that depend heavily on happy customers. We also decided to evaluate the enablers *Outsourcing* (E.T4) because all three business models rely on activities that are not within their own premises and outsourcing activities have a significant influence on the availability of *Capital* (E.F2).

The results of these initial considerations as to the relevance of enablers and their suitability for the comparison of the three business models are summarised in Table 16.3.

Table 16.3 Enablers for the *Reuse* case

Cluster	Code	Enabler	Context	Relevance
Firm	E.F1	Culture	Important, but no feature for differentiation	0
	E.F2	Capital	Investments will have impact on the financial situation of a startup	+++
Value chain	E.V1	Partnership	Important, but no feature for differentiation	0
	E.V2	Transparency	Making use of technologies to get user buy-in	+
Technological	E.T1	Research	Not considered a feature for differentiation	0
	E.T2	Mutuality	Not considered a feature for differentiation	0
	E.T3	R-technology	Not considered a feature for differentiation	0
	E.T4	Outsourcing	Outsourcing activities have an influence on financing requirements	+
Customer	E.C1	Engagement	Customer engagement is a key aspect of success	+++
Material supply	E.M1	Logistics	Integral part of all three business models	+
	E.M2	Collection	A simple collection or take-back system is an integral part of the business models and has a big impact on user acceptance	+++
	E.M3	Symbiosis	Important, but no feature for differentiation	0

0 representing insignificant importance, + indicating minor relevance, ++ reflecting notable relevance, and +++ signifying crucial importance

Quantifying the Criteria

To objectively compare the business models of reCIRCLE, Vytal and GreenToGo, the method of utility analysis was applied. In the process of utility analysis, information will be collected and quantified to calculate utility scores for each alternative. This involves two steps: scoring and weighting criteria to arrive at an overall measure of utility. Each criterion is assigned a weight based on its relative importance in achieving the decision objectives. The weighting reflects the significance of each criterion in contributing to the overall success of the decision. This was reached by converting the current weighting of the enabler into numbers. The importance of the various enablers was reflected by the multiplier: For an applicable enabler, one point was assigned; enablers that do not apply did not receive a point. Table 16.4 shows the conversion of the relevance of enablers to multipliers.

In a second step, the score for each enabler had to be defined. To stay as objectively as possible, the enablers were divided into different criteria. These criteria were chosen in accordance with the requirements and goals of the concept. Each criterion was evaluated in a binary way: It was either evident (1) or not (0). In the end, the assigned points were summed up to arrive at a fulfilment score of 1 to 5 points. Table 16.5 shows the conversion of qualitative evaluations into scores from 0 to 5. The combination of the scores of each enabler and their respective weights is executed by multiplying the quantitative result of each enabler by its assigned weights. This process ensures that each enabler contributes proportionally to the overall utility score of the idea. The summation of these weighted scores provides a comprehensive and weighted evaluation of each alternative. The resulting utility scores offer a quantitative representation of how well each alternative aligns with the established criteria, considering both the specific data points and the importance assigned to each criterion. This systematic approach allows objective comparison and prioritisation of alternatives, ultimately aiding in the selection of the most optimal course of action based on the calculated utility scores.

Table 16.4 Converting the relevance of the drivers and enablers into a multiplier

Relevance	Multiplier
0	0
+	2
++	3
+++	4

Table 16.5 Description of the scores in the cost-utility analysis

Points	Description
1	Not fulfilled
2	Sufficient-satisfactory
3	Satisfactory-good
4	Very good
5	Outstanding

Assessing the Three Business Models

To quantify the enablers that were identified to be relevant for the business model analyses and that were weighted with numbers from 1 to 5, they were scored by breaking them down into criteria. Those who met many criteria scored higher than those who did not. First, the *Transparency* of the three use case companies was analysed (E.V2). Transparency refers mainly to transparency of data by means of digitalisation. Table 16.6 shows the evaluation of this enabler based on six different criteria, all of which refer to the functionality of the apps of the three solution providers. It becomes obvious that Vytal's app offers all functions, and thus receives the full score of 6. The reCIRCLE app receives one point less because it does not offer an impact function. The impact function in Vytal's app shows how much disposable packaging an individual has saved and the impact this has on the environment. This feature is considered to be an incentive to continue and extend the usage of reusable packaging. The GreenToGo app scores poorly, as it does not offer an impact function or allow for orders and delivery services via the app. The app can only be used to scan the packaging, create a payment method, and locate partner restaurants and collection points.

Outsourcing (E.T4) is an enabling activity for companies. Since all startups need to concentrate their activities on their core business, the score for this enabler is considered to be high in case the companies do not complete certain activities themselves. This is evaluated by means of considering five activities, which form the criteria shown in Table 16.7. Points are awarded in case the relevant activity has been outsourced, because this leaves potential for scaling up the business easily. All service providers have outsourced the manufacturing of the products (E.T4.1). Transport, i.e., the delivery of the products to the partner restaurants (E.T4.2), is carried out by all of them. GreenToGo does the cleaning in-house (E.T4.3) and does not offer the possibility to refund the deposit offline, so GreenToGo does not get a point for these two criteria. This leads to a total score of 4 for reCIRCLE and Vytal and a score of 1 for GreenToGo.

Table 16.6 Evaluation criteria for enabler E.V2 transparency

E.V2	Transparency	reCIRCLE	Vytal	GreenToGo
E.V2.1	Ordering possible (pick-up)	1	1	0
E.V2.2	Check-in and check-out function (via QR codes or scanning)	1	1	1
E.V2.3	Google Maps/Map	1	1	1
E.V2.4	Create means of payment (for deposit-free lending)	1	1	1
E.V2.5	Delivery service	1	1	0
E.V2.6	Impact function	0	1	0

Table 16.7 Evaluation criteria for enabler E.T4 outsourcing

E.T4	Outsourcing	reCIRCLE	Vytal	GreenToGo
E.T4.1	Production of packaging	1	1	1
E.T4.2	Transportation of packaging	0	0	0
E.T4.3	Packaging cleaning	1	1	0
E.T4.4	Deposit refund offline	1	1	0
E.T4.5	Inventory control	1	1	0

Table 16.8 Evaluation criteria for enabler E.C1 engagement

E.C1	Engagement	reCIRCLE	Vytal	GreenToGo
E.C1.1	Deposit-free lending possible	1	1	1
E.C1.2	Lifetime guarantee (deposit refund)	1	0	1
E.C1.3	Free rental (deposit excluded)	1	1	0
E.C1.4	Lending of more than one box possible	1	1	1
E.C1.5	Return by third person possible	1	1	0
E.C1.6	Offline use possible (without app)	1	1	0

Customer commitment has a significant influence on the success of the concept. From Table 16.8 it can be seen that reCIRCLE meets all criteria of E.C1 and therefore receives the full score. Vytal does not provide a lifetime guarantee of deposit refund. A full deposit refund is only possible within 14 days, after which only half of the deposit amount is refunded. Therefore, Vytal is assigned 5 points. GreenToGo scores below average on the E.C1 commitment because no third-party return is possible. Customers must use their own cell phones to scan the box or QR code at the collection point. In addition, offline use is not possible to check out food packaging. This can be a significant barrier to entry for users, as they must first install the app and create a payment method to use the system. Therefore, GreenToGo receives 3 points.

Reverse logistics (RL) (E.M1) is another important enabler for the successful implementation of a reuse option because it is the prerequisite to close the cycle. When returning products to the original distributor, several barrier can hinder RL. The potential to overcome these hurdles is evaluated with the criteria shown in Table 16.9. One barrier is – once again – the usability of the app, because it shapes the willingness to join the system. The system does not work if the application is not accepted by the customers (E.M1.1). All providers offer the possibility to recycle the packaging and thus close another cycle (E.M1.2) at the end of the product lifetime. Partner restaurants can return the reusable packaging at any time (E.M1.3). For customers, open systems would be easier to implement leaving the option to choose the best provider in a certain case. It would be an advantage for end consumers if,

Table 16.9 Evaluation criteria for enabler E.M1 logistics

E.M1	Logistics	reCIRCLE	Vytal	GreenToGo
E.M1.1	App usability	1	1	0
E.M1.2	Recycling option	1	1	1
E.M1.3	Return of packaging possible for partner restaurants	1	1	1
E.M1.4	System openness for other suppliers	0	0	0

Table 16.10 Evaluation criteria for enabler E.M2 collection

E.M2	Collection	reCIRCLE	Vytal	GreenToGo
E.M2.1	Return at various partner restaurants	1	1	1
E.M2.2	Return at delivery service	0	0	0
E.M2.3	Collection points with public access, partner restaurants excluded	0	0	1

for example, Vytal packaging could be returned to reCIRCLE partner restaurants. Unfortunately, none of the software is compatible with one of the others (E.M1.4). In the end, Vytal and reCIRCLE score 3 points for criterion E.M1, GreenToGo scores 2.

The last enabler under assessment is the realisation of the take-back. E.M2 assesses the option(s) for collecting used food packaging. Generally, we distinguish three forms of collection: Table 16.10 provides an overview. All providers allow packaging to be returned to various partner restaurants (E.M2.1). None of the providers opens up the possibility of returning packaging to a delivery service, i.e., handing the delivery person a used reusable box at the door when making a new delivery (E.M2.2). Only GreenToGo sets up collection stations in public, mostly in industrial areas, making the system independent of opening times and thereby enhancing the accessibility of the collection points. Therefore, GreenToGo receives the highest score for this criterion (E.M2.3) and for the overall enabler.

Weighted Score

Finally, the normalised scores for all enablers were multiplied by their assigned weights. This way it is possible to emphasise the importance of each enabler in the overall decision. The result is a set of weighted scores for each alternative. We summed up the weighted scores for each alternative to obtain an overall utility score. This score reflects how well each alternative aligns with the decision objectives, considering the assigned weights and individual criterion performances.

Table 16.11 depicts our evaluation (E), the weighted evaluation (WE), and the aggregated scores. It becomes obvious that reCIRCLE received the highest overall

Table 16.11 Utility analysis

Enabler	Multiplier	reCIRCLE		Vytal		GreenToGo	
		E	WE	E	WE	E	WE
E.V2 Transparency	2	5	10	6	12	3	6
E.T4 Outsourcing	2	4	8	4	8	1	2
E.C1 Engagement	4	6	24	5	20	3	12
E.M1 Logistics	2	3	6	3	6	2	4
E.M2 Collection	4	1	4	1	4	2	8
Aggregate score			**52**		**50**		**32**

score with 52 points. It is followed by Vytal with 50 points and GreenToGo with 32 points.

16.6 Summary and Learnings

In this chapter, a systematic evaluation method emerged from the framework of the drivers and enablers of CE developed in Sect. 4.2.1. The framework provided a structure for comparing the drivers and enablers associated with three business models addressing the same idea to reduce the footprint of disposable food packaging. First, a discussion of the business understanding and the analysis of the drivers for this idea led to the definition of *Reuse* as the most promising and compatible value retention option.

In addition to environmental, institutional, stakeholder, and responsibility drivers, potential savings could drive successful CE initiatives. When developing new business models, cost savings is, however, not a significant driver. This is because the business ideas are shaped around the fact that implementing the circular economy is driven by forces, and business want to earn money while doing good at the same time. However, they will convince their customers only if the inconveniences are not too high and the implementation is not too costly. This we analysed as 'enablers'. We differentiated the three startups according to their products, their concepts, and the cost customers have to face when choosing the respective startups as partners.

As for money matters, reCIRCLE requires an initial investment that will pay off after 2–3 years depending on the number of reCIRCLE boxes in circulation. Vytal charges per use, making the entry into their system free of investments. However, when scaling, there are no possibilities to reduce cost. GreenToGo includes monetary contributions from restaurants as well as end-users. In return, they provide collecting, cleaning, and redistribution services, thereby reducing inconveniences to a minimum.

To make a systematic and unbiased assessment of the different use cases, we applied a utility analysis on the basis of the general enablers of CE. Through its application, we shed light on key components, concepts and benefits of the individual cases, assigning scores and weightings according to the importance of the enablers to the case and their fulfilment. The quantification of evaluation criteria is not an objective method and depends heavily on the needs and understanding of the business, but it is a base for unbiased comparison and discussions. The utility value analysis resulted in the highest overall score for the concept of reCIRCLE. However, in some categories other concepts score higher. For example, the GreenToGo collection system is considered superior to the reCIRCLE and the Vytal collection system.

In the end, the utility analysis based on enablers of CE generates transparency. Depending on the actual realisation of the case, the weightings can be adjusted and scores can be adapted. This shows its practicality and utility in real-world scenarios. When using a similar procedure, businesses can gain a deeper understanding of how to best leverage driver and enablers for a CE during the evaluation of circular business ideas, but they can also use the resulting procedure to compare any CE idea or business model with existing competitors' solutions.

References

1. Schallmo, D. (2018). *Geschäftsmodelle erfolgreich entwickeln und implementieren: mit Aufgaben, Kontrollfragen und Templates* 2., überarbeitete und erweiterte Auflage. 282 pp. ISBN: 978-3-662-57604-5. Springer Gabler, Berlin, Heidelberg.
2. Schallmo, D., & Brecht, L. (2013). *Geschäftsmodell-Innovation: Grundlagen, bestehende Ansätze, methodisches Vorgehen und B2B-Geschäftsmodelle* 335 pp. ISBN: 978-3-658-00244-2. Springer Gabler, Wiesbaden.
3. Teece, D. J. (2010). Business models, business strategy and innovation. *Long Range Planning, 43*, 172–194. ISSN: 00246301.
4. Höse, K., Süß, A., & Götze, U. (2022). Sustainability-related strategic evaluation of business models. *Sustainability, 14*, 7285. ISSN: 2071-1050.
5. Barney, J. B. (2014). *Gaining and sustaining competitive advantage* 4. ed., Pearson new internat. ed. 508 pp. ISBN: 978-1-292-02145-4. Pearson, Harlow.
6. Porter, M. E. (2008). *Competitive Advantage: Creating and Sustaining Superior Performance* (2nd ed.). ISBN: 978-1-4165-9584-7. Free Press, Riverside.
7. Porter, M. E. (2014). *Competitive Strategy: Techniques for Analyzing Industries and Competitors*. ISBN: 978-1-4165-9035-4. Free Press, Riverside.
8. Bocken, N., Short, S., Rana, P., & Evans, S. (2013). A value mapping tool for sustainable business modelling. *Corporate Governance: The international journal of business in society, 13*, 482–497. ISSN: 1472-0701. https://www.emerald.com/insight/content/doi/10.1108/CG-06-2013-0078/full/html.
9. *reCIRCLE Wir gehen den Mehrweg!* reCIRCLE. https://www.recircle.ch/en/.
10. Die Schweizerische Post. (2024). *Dispobox–Die ökologische Mehrwegverpackung* Die Post. https://www.post.ch/de/pakete-versenden/verpacken-und-adressieren/dispobox.
11. DE-PACK SA GmbH & Co. KG. *Small Load Carriers* KLT VDA. https://klt-vda.com/. (2024).
12. *Vytal - Das nachhaltige Mehrwegsystem*. https://www.vytal.org/. (2023).
13. *GreenToGo — Take-out without the trash*. https://durhamgreentogo.com/. (2023).
14. European Union. (2019). *Directive (EU) 2019/904 of the European Parliament and of the Council of 5 June 2019 on the reduction of the impact of certain plastic products on the environment (Text with EEA relevance)* June 5, 2019. https://eur-lex.europa.eu/eli/dir/2019/904/oj.
15. Bundesministerium der Justiz. (2023). *Verpackungsgesetz - VerpackG*. https://www.gesetze-im-internet.de/verpackg/__33.html.

16. Plasteurope.com. (2023). *PACKAGING TAX GERMANY: Judgment by Federal Administration Court: Tübingen waste levy is lawful / Municipalities should follow the example–Nabu*. https://www.plasteurope.com/news/PACKAGING_TAX_GERMANY_t252867/.

Chapter 17
Rethinking and Positioning Plastics for Circular Economy

Daniel Schwendemann, André Podleisek, and Nicolas Hofer

Abstract The European Union has enacted a range of directives with the objective of elevating recycling rates. Despite these legislative efforts, a significant proportion of plastic materials continue to contribute to waste accumulation. This scenario not only poses environmental challenges, but also represents untapped economic opportunities in the realm of plastics recycling. The intricacies of plastic types and the corresponding recycling methods are multifaceted. Gaining a better understanding of these elements can provide invaluable insights that can reshape the perception and use of plastics. In this chapter, we will delve into these complexities, drawing on the authoritative expertise of Professor Daniel Schwendemann, a leading figure in the domain of plastics processing. He provides insights on key considerations and challenges inherent to plastics recycling.

17.1 Introduction to Circular Economy in Plastics

Plastics have become integral to our daily lives, even though their large-scale production only started back in the 1950s. Since then, plastic has found applications in various sectors, such as packaging, construction, automotive, electronics, and household, to name a few. The demand for plastics has been growing steadily, with an average annual production growth rate of 8.5% [1].

Currently, the annual production of plastics exceeds 390 mill. metric tonnes. The overwhelming majority, 352 mill. metric tonnes [2], are primary fossil-based plastics – manufactured from virgin fossil fuels like crude oil or natural gas. This stands in contrast to the relatively modest quantity of 32 mill. metric tonnes of secondary

D. Schwendemann (✉)
Pfullendorf, Germany
e-mail: daniel.schwendemann@ost.ch

A. Podleisek
Wettswil am Albis, Switzerland

N. Hofer
Zürich, Switzerland

A. Podleisek et al. (eds.), *Circular Economy: The Next Level of Company Success*,
https://doi.org/10.1007/978-3-031-66406-9_17

plastics – reclaimed through *Recycling* – and 6 mill. metric tonnes of bio-based plastics – derived from renewable resources like corn or wheat.

By 2050, the demand for plastics is expected to double compared to today [3]. According to OECD estimates, plastic consumption will even triple by 2060 without government measures [4].

This is not surprising when we consider the global context: as the world's population increases and levels of prosperity rise, there is an inevitable surge in product demand. In addition, a significant portion of the components of products are made of plastics. Due to their low density – around 1 to 8 compared to steel – plastics have already overtaken metals in terms of volume for several years (see Fig. 17.1).

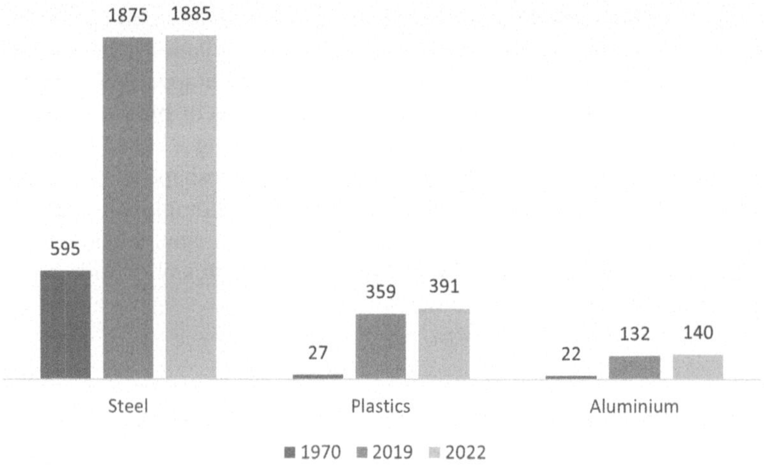

Fig. 17.1 Global production of materials in mill. tonnes/year in 1970, 2019 and 2022 (own representation based on [5, 6])

The growing demand carries with it significant environmental challenges, including ecosystem degradation, increased emissions of greenhouse gases, and health risks linked to microplastics and their additives [7]. According to the OECD, in 2019 plastics contributed to roughly 3.7% of global CO_2 emissions, amounting to 1.8 bill. tonnes of which 90% are due to production processes [4]. It is obvious that considering plastics in a Circular Economy (CE) is a must.

17.2 Trends in Plastics Use

Media outlets and environmental campaigns frequently highlight the detrimental effects of plastic pollution, from the Great Pacific Garbage Patch to the ingestion of microplastics by marine life (e. g., [8, 9]). The extensive media coverage on plastic pollution exerts a significant influence on public opinion, policy decisions, and corporate behaviour [10]. Surrounded by negative narratives, many companies are steering towards a complete abandonment of plastics, a trend that Professor Schwendemann criticises. Instead, the focus should shift toward the necessity to critically differentiate between the use and misuse of plastics.

The material offers a variety of beneficial properties, such as durability, lightness, and impermeability, which can be harnessed for sustainable purposes. For example, plastic can significantly extend the shelf life of perishable food items and thereby mitigate food waste. A mere 1.5g of plastic film can extend the shelf life of cucumbers from 3 to 14 days [11, 12]. An examination of environmental trade-offs revealed that the ecological advantages stemming from reduced food waste are markedly higher, by a factor of 4.9, compared to the environmental drawbacks linked to the plastic packaging [13]. Therefore, the outright abandonment of plastics is an oversimplified and potentially counterproductive reaction to the complex environmental challenges at hand. Instead, with the institution of rigorous collection and recycling procedures, plastics can manifest a myriad of benefits in the realms of efficiency and resource stewardship. The discourse should move from a wholesale rejection of plastics towards fostering a culture of responsible plastic usage and effective recycling.

17.2.1 Plastics Recycling Today

Until recently, the litmus test for plastics recycling was straightforward: it had to be cheap. For secondary plastics to penetrate the market, they had to be more cost-effective than their virgin predecessors. Otherwise, they would not have a market besides some niche applications. The oscillations in oil prices largely determined the attractiveness of plastics recycling; when prices soared, recycling became an economically viable alternative. Conversely, a decrease in oil prices often led to diminished interest.

In retrospect, four key periods stand out for heightened interest in recycling. The 1980s initiated the first wave, driven by an energy crisis and high oil prices, that promoted recycling as an alternative to primary plastics production. The early 1990s saw a second wave, triggered by Iraq's invasion of Kuwait and subsequent oil price spikes. The early 2000s marked a third wave, fuelled by another surge in oil prices and met with advances in recycling technology. The fourth wave occurred in 2008, when all-time record-high oil prices led to increased interest in recycled plastics. Stabilisation in oil prices after that period reduced the urgency around recycling activities.

However, the paradigm has undergone a significant shift, entering what is now dubbed the '5th Recycling Wave.' This contemporary wave is not solely dictated by economic factors; rather, it is increasingly propelled by societal expectations and progressive political landscapes [14]. In the European Union (EU), rigorous regulatory measures are being put in place. Among these are the mandates for a 55% recycling rate for plastics by 2030, along with a directive to recycle a minimum of

70% of all packaging waste [15]. Concurrently, the Single Use Plastics Directive (2019/904) [16] and the updated Circular Economy Action Plan (COM/2020/98) [17] have outlined a target to achieve a recycling rate of 90% for plastic bottles by the year 2029, and have prescribed that these bottles should encompass at least 30% recycled plastic by 2030. Failure to comply with these targets incurs financial penalties. In addition, the reinvigorated demand for recycled plastics has largely been stimulated by taxation on non-recycled plastic packaging.

In this changing landscape, companies are being incentivised to act, not just due to legal imperatives like the taxation of Carbon footprint but also through additional economic levers like carbon credits for reduced CO_2 emissions – a market that has generated more than 152 bill. EUR in revenues since 2013 [18]. Hence, recycled material has a higher attributed value beyond its material price, making it attractive even if it is priced higher than primary plastics. Companies seeking to increase revenue or decrease their tax burden will, after having explored the usual cost-saving approaches, look for new concepts and optimise their Total Cost of Ownership (TCO) of materials in a lifecycle perspective.

Recycling has a big potential in this regard. Specifically in packaging applications, for which nearly 165 mill. metric tonnes are used, representing approximately 44% of global plastic consumption [2]. The short use phase and overall lifespan of these materials make them ideal candidates for recycling. Successfully giving these plastics a second lifecycle could result in an annual saving of up to 165 mill. metric tonnes of plastics. And this if it is only recycled once! The harsh reality is that in 2021 recycled plastic comprised a mere 8.3% of the world's total plastic production [2].

17.3 Plastics Types and Their Recyclability

Plastics can be fundamentally divided into three categories based on their response to heat and the reversibility of the chemical reactions in their formation: thermoplastics, thermosets, and elastomers.

Thermoplastics have the unique characteristic of being able to be melted and reshaped multiple times without undergoing significant chemical alterations [19]. This unique characteristic is attributed to their molecular structure, where polymer chains are held together by weak intermolecular forces, allowing for easy reversibility between solid and liquid states. Within the EU including Norway, Switzerland, and the United Kingdom, thermoplastics such as Polypropylene (PP) and Polyethylene (PE) lead the market, followed by Polyvinylchloride (PVC) and Polyethylene Terephthalate (PET) [20]. Together, these thermoplastics account for more than two-thirds of the European plastic demand, as illustrated in Fig. 17.2. Given their characteristics and market dominance, these thermoplastics are prime candidates for recycling initiatives, offering significant opportunities for resource conservation and waste reduction.

Thermosets, on the other hand, undergo an irreversible chemical process during their initial formation, resulting in a permanently set three-dimensional network

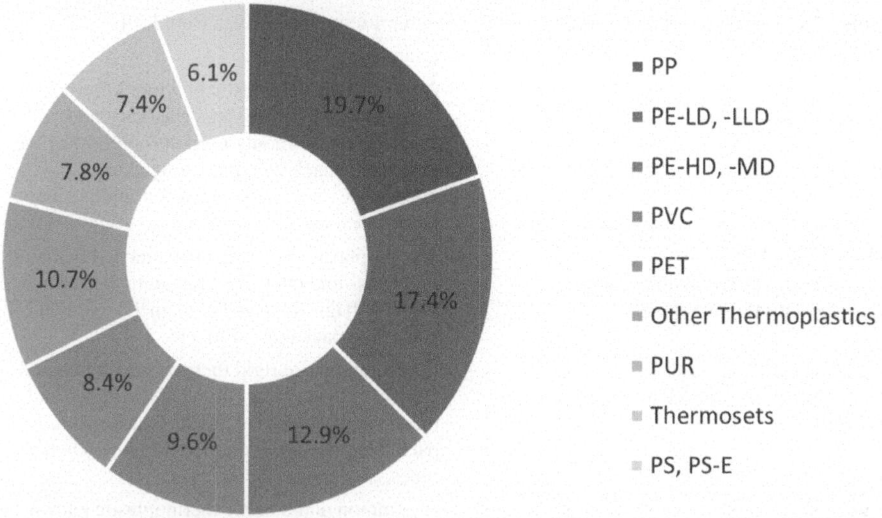

Fig. 17.2 Distribution of primary plastics by polymer type (EU27+3) (own representation derived from [20])

structure [19]. This characteristic endows them with high resistance to heat, chemicals, and mechanical stress, making them ideal for applications that require durability and stability, such as in the automotive, aerospace, and electronics industries (more applications of different plastics can be found in Table 17.1). However, this very attribute of irreversible cross-linking also means that they are generally not recyclable in the same manner as thermoplastics. Once set, thermosets cannot be remelted or reprocessed, limiting their potential for traditional recycling methods and posing challenges for waste management.

Elastomers are distinguished by their elastic properties, primarily because of the presence of weak intermolecular forces within their partially cross-linked structure [21]. This unique feature enables elastomers to recover their original form after being stretched, a characteristic essential for applications demanding flexibility and resilience. Such applications include the manufacturing of automotive tyres and various industrial seals, where the ability to withstand repetitive stress and deformation is crucial. Among the most recognised elastomers is natural rubber, although the field also encompasses a range of synthetic alternatives like styrene-butadiene rubber (SBR) and silicone rubber, each offering distinct properties suitable for specific industrial needs. However, the recycling of elastomers presents a challenge, as their complex molecular structure makes them less amenable to conventional recycling processes, thus posing a limitation in the context of a CE [22].

Table 17.1 Plastic types and examples of use [own representation derived from [20]]

Plastic type	Examples
PP Polypropylene	A versatile thermoplastic polymer known for its resistance to chemicals, elasticity, and toughness. Commonly used in food packaging, sweet and snack wrappers, hinged caps, microwave containers, pipes, automotive parts, bank notes, etc.
PE-LD/PE-LLD Low-Density Polyethylene/Linear Low-Density Polyethylene	Thermoplastic polymers characterised by their toughness and ductility. Frequently used for reusable bags, trays and containers, agricultural film, food packaging film, etc.
PE-HD/PE-MD High-Density Polyethylene/Medium-Density Polyethylene	Denser and more rigid thermoplastics than PE-LD with superior chemical resistance and mechanical properties. Used for cosmetics packaging, toys, milk bottles, shampoo bottles, pipes, houseware, etc.
PVC Polyvinyl chloride	A common and durable thermoplastic known for its strength and versatility. Used for window frames, profiles, floor and wall covering, pipes, cable insulation, garden hoses, inflatable pools, etc.
PET Polyethylene terephthalate	A clear, lightweight thermoplastic often used in packaging such as bottles for water, soft drinks, juices, cleaners, etc.
PUR Polyurethane	A thermoplastic known for its exceptional resilience, toughness, and durability. Is used for building insulation, pillows and mattresses, insulating foams for fridges, etc.
PS/PS-E Polystyrene/Expanded Polystyrene	Are very versatile thermoplastics which can be rigid or foamed. Food packaging (dairy, fishery), building insulation, electrical & electronic equipment, inner liner for fridges, eyeglasses frames, etc.
Other thermoplastics	Such as LEGO bricks (ABS); optical fibres (PBT); eyeglasses lenses, roofing sheets (PC); touch screens (PMMA); cable coating in telecommunications (PTFE); and many others in aerospace, medical implants, surgical devices, membranes, valves & seals, protective coatings, etc.
Thermosets	Includes thermosets such as phenolic resins, epoxide resins, melamine resins, urea resins and others. Used for applications where resistance to heat, corrosion, and electrical insulation is pivotal.

17.3.1 Plastics Recycling Methods

Several methods are available to recycle plastic, each with its unique advantages and limitations. These methods include mechanical, chemical, and thermal recycling. An overview is given in Fig. 17.3.

Mechanical *Recycling*

Mechanical *Recycling* refers to the reprocessing of polymers through a series of physical methods, including shredding, washing, drying, and extrusion, without altering the fundamental structure of the polymer chains. It is the most widely used method for processing plastic waste. The mechanical recycling process faces several challenges, including complications arising from coatings and paints on the plastics. Additionally, some contaminants may not be entirely soluble, which can adversely affect the mechanical properties of the material. The end result is often a decrease in the quality of the recycled product. Mechanical recycling of mixed plastic waste can also produce blended polymers, which compromises the material's original properties [23].

Especially with PET a very high quality can be achieved with mechanical recycling if the input streams are very clean and homogeneous. The viscosity (i.e. the length of the molecular chains) can be increased again by means of physical processes. This plastic is therefore ideally suited for entry into the CE [24].

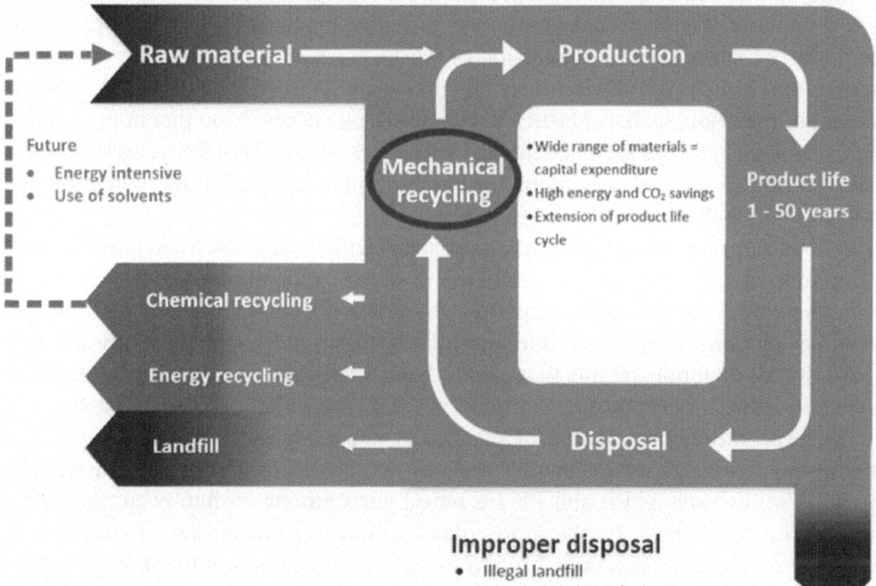

Fig. 17.3 Schematic overview of the end-of-life scenarios for plastics

Chemical *Recycling*

Chemical recycling is a suite of reprocessing technologies that decompose polymer chains, converting them into high added-value materials such as oligomers, monomers, base chemicals, and various forms of hydrocarbons (solid, liquid, or gas) [23]. This method is particularly promising for dealing with mixed and contaminated plastic waste when separation is neither economically viable nor technically straightforward. For that reason chemical recycling techniques are being strongly advocated. However, the energy requirement for this process is quite significant, even exceeding that of standard plastic production [25].

It is also important to note that there is not only one chemical recycling process. There are a variety of processes, each of which naturally has advantages and disadvantages and also different energies and emissions. The various processes were examined in the study by the Nova Institute in Hürth. In the area of depolymerisation to monomers, there are three subgroups: enzymolysis – a biochemical process, solvolysis – a chemical process and the thermochemical process – for depolymerisation to monomers. In a broader sense, conversion processes such as pyrolysis to naphtha, gasification to syngas and Combustion to CO_2 are also included [26].

Thermal *Recycling*

Efficient mechanical and chemical recycling processes are based on effective sorting mechanisms, which can be manual or automated. However, the complexity and diversity of plastics often make them difficult to segregate. Thermal recycling stands as a pragmatic approach and alternative in waste management. This approach is particularly focused on the conversion of waste, especially plastics, into energy by incinerating it at high temperatures. The process effectively transforms the caloric content of materials such as plastics into usable heat energy. Note that in the context of this book this process is considered to be Recover (R8) and not Recycle (R7). Thermal recycling is also advantageous for handling plastics that are too contaminated or degraded for the other recycling methods.

In countries like Switzerland, this approach is widely adopted. Most plastics, with the notable exception of PET, are directed to waste incineration plants.

Such plants, of which there are many in Switzerland, fulfil the function of thermal recycling by converting waste into energy. Thermal recovery is currently the best option for all material streams that cannot be mechanically or chemically recycled, especially as sensible recycling solutions are not yet available for many products.

However, the calorific value of plastics is not recognised in energy recovery, emphasises Schwendemann. You have to pay for the disposal of plastic waste, even though plastics such as PE and PP are pure hydrocarbons – often even purer than petrol at filling stations. Furthermore, plastics emit large quantities of CO_2 during combustion. The combustion of 1 ton of plastics produces approximatley 2.7 tonnes of CO_2, which is often presented in a more positive light, as it is compared with the combustion of coal to generate energy, which produces 3.67 tonnes of CO_2 [27]. The new approaches to Carbon Capture Utilisation (CCU) and the subsequent merging of CO_2 and green hydrogen into plastics are opening up new possibilities and merging chemical and thermal recycling [26].

17.4 Plastic *Recycling* Challenges

Despite the growing emphasis on recycling, significant challenges remain. One of the main obstacles is the efficient collection of plastic waste, a concern exacerbated by the fact that a considerable volume of plastic still ends up in landfills, even in EU [28, 29]. The perceived negligible economic value of plastics at the end of their lifecycle is shown in the lack of voluntary collection.

Plastic waste can be categorised into two primary types: pre-consumer and post-consumer waste. Pre-consumer, or post-industrial waste, consists of materials generated during the manufacturing phase before the product is sold to the end-user. This category includes items like manufacturing off-cuts, defective products, and surplus materials. Once a product has been used and discarded by the consumer, it is classified as post-consumer waste.

In Switzerland, the efficacy of recycling varies significantly depending on the type of plastic. For example, PET Recycling Switzerland reports that the recycling rates for PET used in plastic bottles are impressively high, approximately 82% [30].

Recycling rates for other types of plastic are significantly lower. This is partly due to the fact that Switzerland currently has no large-scale sorting and recycling infrastructure for plastics other than PET. The 'collection 2025' strategy aims to collect all packaging plastic waste throughout Switzerland [31].

However, there is a downside to disposal monopolies such as PET Recycling Switzerland. They hinder the collection of certain plastics for reuse by other organisations. These monopolies currently exist mainly for PET, especially for water and soft drink bottles. Other similar collection systems could be set up for other types of plastic, although country-specific regulations must be taken into account. The waste monopoly is often in the hands of the municipalities, which means that many individual solutions have to be developed and secured in a compatible manner.

It is also important to understand that not all plastics are recyclable indefinitely. For many types, repeated recycling cycles can degrade their polymer chains, affecting the plastics' viscosity and overall performance. PET, however, is an exception in this regard. During the recycling process, the polymer chains of PET can recover, allowing them to withstand multiple recycling cycles without significant degradation. Despite the high recycling potential of PET, not all PET products are recycled equally. The 'bottle-to-bottle' recycling process, currently works efficiently predominantly only for water and soft drink bottles. This method is less effective for other PET products, such as shampoo bottles, which are often sorted out and not recycled. While this process shows great success with PET bottles, the development of effective recycling solutions for other types of packaging materials is still in its infancy.

The handling of post-consumer waste varies by country. As we have mentioned before, in Switzerland PET is collected separately, while all other types of plastic waste are aggregated. Consequently, post-consumer waste often comprises a mixture of plastics of varying compositions, which can be contaminated with organic matter such as food residues or inorganic materials such as paper. This complexity makes post-consumer waste more challenging to recycle compared to post-industrial waste, which is less contaminated and can bypasses the need for sorting and washing.

17.4.1 Odour Challenge

Another serious obstacle to increasing recycling rates and, in particular, to increasing the use of recycled plastics is the smell. During the service life of packaging, odours or even low-molecular components of food can diffuse into plastic and these substances can lead to an entire recycling batch with a bad odour during thermal processing. This reduces customer acceptance of the use of recyclates. The industry is currently introducing processes for odour reduction through advanced sorting, hot washing and vacuum treatment. New processes such as stripping are being tested on extruders to treat the material gently [32].

17.4.2 Contamination Issue

A further challenge in the domain of plastic recycling pertains to the potential introduction of hazardous substances into the recycling stream. This issue arises e.g. when plastics that have been in use for extended periods are reclycled. They can contain now outdated and potentially harmful additives. Common contaminants include obsolete flame retardants or plasticisers, which were not initially recognised as hazardous but have since been identified as problematic.

The presence of such contaminants in the recycling stream is not just a minor concern; it poses significant health and safety risks. These risks are not limited to the recycling process itself but extend to the future use of recycled materials. For instance, products manufactured from contaminated secondary plastics can later be subject to recalls, posing threats to consumer safety as well as company reputation. Measures must be implemented to ensure that these harmful substances do not perpetuate within the recycling cycle [33].

17.4.3 Economic Hurdle

The economic aspects of plastic waste management form a complex and challenging landscape, mainly due to the competing interests of *Recycling* and *Energy Recovery* methods [34]. One of the most significant hurdles in plastic recycling is the high cost involved in the collection, sorting, and processing of waste materials. These expenses significantly influence the final market price of recycled (secondary) plastics, often rendering them less economically viable compared to virgin materials [34]. Given these economic challenges, it becomes essential to identify and implement strategies aimed at reducing the costs associated with plastic recycling. Advancements in technology, particularly in the areas of sorting and processing, hold considerable promise. Such innovations have the potential to reduce operating costs, making recycled plastics more competitive in the market. The future of sustainable plastic waste management may well depend on our ability to balance economic viability with environmental responsibility, ensuring that recycled plastics are not only an ecologically sound choice but also an economically rational one.

17.4.4 Logistics Debate

Professor Schwendemann emphasises the need to critically assess the ecological balance in the collection of plastics for recycling. The logistics involved can significantly impact the environment, potentially offsetting the benefits of recycling itself.

The discourse on plastic waste logistics, particularly its environmental impact, is vital in understanding the complexities of plastic recycling. The contrast between local and global collection logistics adds to this complexity, with each approach having its own merits and limitations.

Local collection systems are commended for reducing transportation distances and consequently CO_2 emissions, aligning with efforts to mitigate climate change. However, these systems often lack the necessary infrastructure and technology for efficient processing, leading to higher costs per unit of recycled material. Furthermore, the limited scale of local operations often does not justify the investment in advanced sorting and processing technologies, affecting the quality and range of plastics that can be recycled.

On the other hand, global collection logistics benefit from economies of scale, enhancing overall efficiency. Large-scale operations can invest in more sophisticated technology, improving both the yield and quality of recycled plastics. Yet, the environmental cost of transporting plastics over long distances to centralised facilities is a significant concern. The CO_2 emissions from transportation can sometimes outweigh the ecological advantages of recycling.

17.4.5 Bioplastic Dilemmas

Professor Schwendemann emphasises that the move toward alternative plastics such as bioplastics is not as straightforward as it may appear. A bioplastic is in essence a material that possesses at least one of the following properties:

Is made from renewable resources like corn or wheat, i.e. bio-based
and/or
is biologically degradable.

This results in a taxonomy that includes: plastics that are bio-based but not biodegradable; plastics that are fossil-based and biodegradable; and plastics that are both bio-based and biodegradable. This highlights a significant ambiguity in the public perception and labelling of bioplastics [35].

Bio-based plastics, heralded for their potential to reduce reliance on fossil fuels, are not necessarily less impactful on the environment. It involves agricultural inputs and processes that have their own ecological footprints, such as land use changes and resource consumption. For example a Life Cycle Assessment (LCA) comparison for polylactic acid (PLA) and PET shows that PLA generally has a lower environmental impact in terms of Global Warming Potential (GWP). However, PET has advantages in terms of acidification and eutrophication potential. Moreover, PLA production can have significant impact on ecosystem quality. The production process of PLA drinking bottles, while eco-friendlier in some respects, can adversely affect biodiversity and the health of ecosystems [36].

A critical issue in plastic waste management, as highlighted by Professor Schwendemann, is the sorting and disposal of biodegradable plastics. Presently, these plastics must still be discarded with regular waste due to the difficulties in distinguishing biodegradable from non-biodegradable plastics. The primary limitation is that existing sorting technologies cannot reliably ensure that only biodegradable plastics are directed towards composting processes. Overcoming this obstacle would enable the effective segregation of biodegradable plastics from other waste streams, facilitating their reintroduction into the natural cycle. This, in turn, positions biodegradable plastics as a renewable alternative in the production process, a concept further illustrated in Fig. 17.4.

Furthermore, the decomposition rate of biodegradable plastic is currently too slow to be compatible with industrial composting cycles [37]. However, it is important to emphasise that defossilisation targets cannot be achieved without the use of bio-based plastics. It is therefore important to tackle the circular use and/or the use of carbon in the cycle (CCU) in addition to bio-based plastics [38].

While biodegradable plastics offer an alternative to traditional plastics, their application is somewhat restricted by their limited mechanical strength. This limitation affects their viability as a replacement in various uses where durability and robustness are essential. However, they are an excellent alternative for areas where biodegradation is desired, such as disposable cutlery, plant pots, or growth and bite protection covers, where the strength of these materials can be used [39, 40].

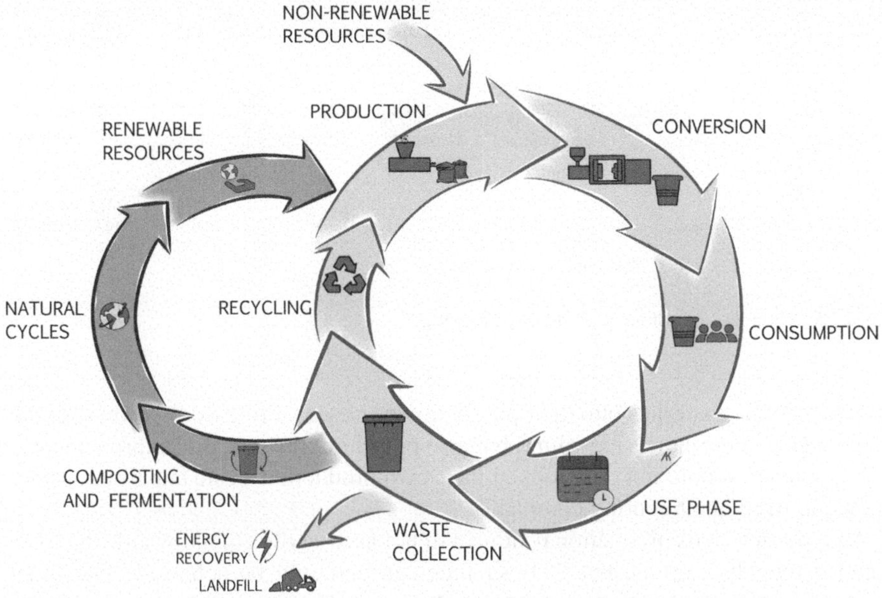

Fig. 17.4 Plastics recycling future

17.5 Opportunities and the Path Forward

In the realm of plastic recycling, the approach of companies towards collection needs a fundamental shift. The prevailing practice of collecting plastics should not be an end in itself; rather, it is crucial to frontload the recycling process with a strategic and purpose-driven mindset. This approach involves a critical reassessment of product design and manufacturing processes: identifying which components or products could be intentionally designed and specified to incorporate recycled materials.

A major aspect of this shift involves the rethinking of the specifications of the materials. Often, alternative materials, such as recycled plastics, are available but fail to meet the stringent criteria set for conventional materials. This situation necessitates a reevaluation of whether such narrowly defined specifications are always necessary or if there is room for more flexible approaches. Specifically, parts of products that are less visible or subject to less stringent quality or aesthetic standards present an opportunity to use recycled plastics effectively without compromising functionality or quality. The end-of-life scenario must already be considered during the development phase, preferably with a further field of application; if this is not possible, the quality target, i.e. the place of use, must at least be defined during sorting. High-quality recycling is then possible, see Fig. 17.5.

Industrial waste, which is characterised by higher purity and quality than post-consumer waste, is another valuable opportunity for companies to recycle materials

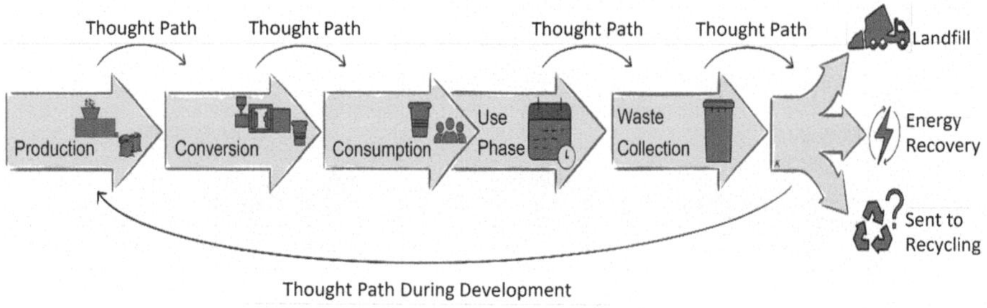

Fig. 17.5 Thought of path for high quality recycling

and incorporate materials into their production processes. Professor Schwendemann sees progress in in-house recycling, but also points out the need for improvements.

Companies should not only look at their own production waste but also consider the waste from neighbouring companies.

Nevertheless, this production residue will not necessarily count towards the EU's planned recycling quota, but CO_2 savings can certainly be achieved. The exact interpretation of the legislation still has to be awaited [41].

17.5.1 Successful Recycling Cases

A notable success story from Spain demonstrates the great potential of innovative recycling practices. In this case, unsold pizza packaging, made from PET (transparent PET shell), became a focus. Annually, around 2,000 tonnes of this plastic, primarily due to manufacturing errors, printing mistakes, or simply unsold products, were generated. Traditionally, this material was discarded. However, a transformative approach was adopted by a local household goods manufacturer who discovered a method to recycle these PET shells into broom handles. This example serves as a powerful demonstration of how rerouting waste streams can create new product opportunities. The pizza company benefits from reduced disposal costs, while the household goods manufacturer obtains free raw materials. This local solution, crafted through collaborative efforts, also achieves minimised logistics.

While this approach is promising and shows tangible benefits, it is important to acknowledge that such strategies can be demanding and can face scalability challenges. Solutions effective in one region, like Spain, might not directly apply to others, such as Switzerland, due to different regulatory environments, consumer behaviours, and available raw materials. Often, it is the medium-sized enterprises that are pioneering and implementing these innovative recycling solutions.

Another very interesting example from Switzerland is ski boot Recycling. At the ARGO workshop for handicapped people in Davos, ski boots are collected, sorted by colour, and roughly cut up. Each part is then scanned for its plastic type

Fig. 17.6 Recycling ski boots to plastic granulates

using a Fourier transformation infrared spectrometer (FT-IR) and the thermoplastic polyurethane (TPU) is extracted. This is then processed on a twin-screw extruder, homogenised, the melt filtered and regranulated. Cell phone cases are produced in the injection molding process [42]. The Freitag AG Circ-Case is made entirely from ski boot recyclate (see Fig. 17.6) and can itself be recycled.

A third very interesting project links Southeast Asia with Switzerland. In a social project in the Andaman Sea, Thailand, plastic waste is collected in the water, on the beach, and near the beach and then sorted and cleaned. The old PET bottles are then turned into high-quality plastic granules for the manufacture of watch cases in a specially designed process. This example shows very impressively what can be achieved when specifications are opened up. Today, watch cases can be manufactured from Ocean Bound materials by specifically modifying the characteristics [43, 44]. At the start of the project, PET plastic was not included in any of the specifications.

17.6 Rethinking the Plastics Industry

Amid the climate crisis, the extraction of resources is increasingly scrutinised. Companies already investing in recycling and exploring its potential applications are better positioned for a sustainable future. Insights from the Innosuisse project 'Towards a NetZero Plastics Industry' reveal significant circular opportunities in rewiring value chains for sustainability [45]. New approaches need to be thought of and tackled. The path to a CE is not easy, but it is worth it, as new local value chains are often created. This aspect is also crucial for success; in future, we must decouple the increase in added value from the use of raw materials. It is crucial for the success of the CE for plastics that we move away from the end-of-life view in the sense of waste. At the end of its lifecycle, the material must still be seen as a raw material, in the sense of a resource. This can be achieved because with plastics, at least with thermoplastics, we are dealing with materials that can be easily recycled, and if this is not possible, we can at least use new approaches to keep the carbon in the cycle utilise the energy content. Giving the material a value as waste is therefore not a bold demand but a reality.

References

1. Chalmin, P. (2019). The history of plastics: from the Capitol to the Tarpeian Rock. *Field Actions Science Reports, Special Issue 19*. ISSN: 1867-8521.
2. Plastics Europe. (2022). *Plastics - the Facts 2022*. https://plasticseurope.org/wp-content/uploads/2022/12/PE-PLASTICS-THE-FACTS_FINAL_DIGITAL.pdf.
3. Van Velzen, T. (2023). *The road to sustainable plastics by 2050 (part 1)* WUR. https://www.wur.nl/en/show-longread/the-road-to-sustainable-plastics-by-2050-part-1.htm.
4. OECD. (2022). *Global Plastics Outlook: Policy Scenarios to 2060*. ISBN: 978-92-64-97364-0. OECD.
5. Material Magazin. (2023). *Welches sind die wichtigsten Werkstoffe?* https://www.materialmagazin.com/index.php/gute-frage/welche-sind-die-wichtigsten-werkstoffe.
6. Wirtschaftsvereinigung Stahl. (2023). *Welterzeugung 1970/2022* Wirtschaftsvereinigung Stahl – Informationen des Verbandes der stahlerzeugenden Unternehmen in Deutschland. https://www.stahl-online.de/startseite/stahl-in-deutschland/zahlen-und-fakten/.
7. Saleem, J., Tahir, F., Baig, M. Z. K., Al-Ansari, T., & McKay, G. (2023). Assessing the environmental footprint of recycled plastic pellets: A life-cycle assessment perspective. *Environmental Technology & Innovation, 32*, 103289. ISSN: 23521864.
8. Kottasová, I. (2023). *The Great Pacific Garbage Patch is now so huge and permanent that a coastal ecosystem is thriving on it, scientists say* CNN. https://www.cnn.com/2023/04/17/world/plastic-pollution-ocean-ecosystems-intl-climate/index.html.
9. United Nations Environment Programme. (2017). *Marine Litter: Socio-economic Study*. https://wedocs.unep.org/20.500.11822/26014.
10. Bailey, I. (2022). Media coverage, attention cycles and the governance of plastics pollution. *Environmental Policy and Governance, 32*, 377–389. ISSN: 1756-932X.
11. Plastics Europe. (2008). *The Compelling Facts About Plastics*. https://plasticseurope.org/wp-content/uploads/2021/10/2006-Compelling-facts.pdf.
12. Benz, M. Die verpackte Gurke ist ökologisch sinnvoll – Konsumenten sollten sich weniger über Plastik aufregen. *Neue Zürcher Zeitung*. ISSN: 0376-6829. https://www.nzz.ch/wirtschaft/die-verpackte-gurke-ist-oekologisch-sinnvoll-warum-sich-konsumenten-weniger-ueber-plastik-aufregen-sollten-ld.1742745.
13. Shrivastava, C., et al. (2022). To Wrap Or to Not Wrap Cucumbers? *Frontiers in Sustainable Food Systems, 6*, 750199. ISSN: 2571-581X.
14. Shamsuyeva, M., & Endres, H. -J. (2021). Plastics in the context of the circular economy and sustainable plastics recycling: Comprehensive review on research development, standardization and market. *Composites Part C: Open Access, 6*, 100168. ISSN: 26666820.
15. European Union. (2018). *Directive (EU) 2018/852 of the European Parliament and of the Council of 30 May 2018 amending Directive 94/62/EC on packaging and packaging waste* May 30, 2018. https://eur-lex.europa.eu/eli/dir/2018/852/oj.
16. European Union. (2019). *Directive (EU) 2019/904 of the European Parliament and of the Council of 5 June 2019 on the reduction of the impact of certain plastic products on the environment (Text with EEA relevance)* June 5, 2019. https://eur-lex.europa.eu/eli/dir/2019/904/oj.
17. European Commission. (2020). *A new Circular Economy Action Plan For a cleaner and more competitive Europe (COM/2020/98)* 2020. https://eur-lex.europa.eu/legal-content/EN/TXT/?uri=COM%3A2020%3A98%3AFIN.
18. European Commission. (2023). *What is the EU ETS?* https://climate.ec.europa.eu/eu-action/eu-emissions-trading-system-eu-ets/what-eu-ets_en.
19. Rennie, A. R. (1999). In *Mechanical properties and testing of polymers* (ed Swallowe, G. M.) red. by Brewis, D. & Briggs, D. Series Title: Polymer Science and Technology. Springer Netherlands, Dordrecht. ISBN: 978-94-015-9231-4.
20. Plastics Europe. (2021). *Plastics - the Facts 2021: An analysis of European plastics production, demand and waste data*. https://plasticseurope.org/wp-content/uploads/2021/12/Plastics-the-Facts-2021-web-final.pdf.

21. BOP Team. (2022). *Elastomer Properties and Where to Use Them* BOP Products. https://www.bop-products.com/blog/elastomers/elastomer-properties-and-where-to-use-them/.
22. Alarifi, I. M. (2023). A comprehensive review on advancements of elastomers for engineering applications. *Advanced Industrial and Engineering Polymer Research, 6*, 451–464. ISSN: 2542-5048.
23. Ragaert, K., Delva, L., & Van Geem, K. (2017). Mechanical and chemical recycling of solid plastic waste. *Waste Management, 69*, 24–58. ISSN: 0956053X. https://linkinghub.elsevier.com/retrieve/pii/S0956053X17305354.
24. Walker, J. *Kreislaufwirtschaft setzt einen Schulterschluss vieler Partner voraus, Keynote Swiss Plastics Expo Forum* in. Swiss Plastics Expo (Luzern, Jan. 18, 2023). https://www.swissplasticsplatform.com/de/e/kreislaufwirtschaft-setzt-einen-schulterschluss-vieler-partner-voraus.23387.
25. Quantis. (2020). *Life Cycle Assessment of plastic energy for the chemical recycling of mixed plastic waste* (Plastic Energy). https://plasticenergy.com/wp-content/uploads/2020/10/Plastic-Energy-LCA-Executive-Summary.pdf.
26. Krause, L., Carus, M., Raschka, A., & Plum, N. (2022). *Mapping of advanced recycling–Providers, technologies, and partnerships* (nova-Institut GmbH). https://renewable-carbon.eu/publications/product/mapping-of-advanced-recycling-providers-technologies-and-partnerships/.
27. DIW Berlin. (n.d.). *Auf dem Weg zur Klimaneutralität: Plastikrecycling muss stärker in den Fokus rücken* Publisher: DIW Berlin. Retrieved November 30, 2023 from https://www.diw.de/de/diw_01.c.820601.de/auf_dem_weg_zur_klimaneutralitaet__plastikrecycling_muss_staerker_in_den_fokus_ruecken.html.
28. European Parliament. (2023). *Plastic waste and recycling in the EU: facts and figures*. https://www.europarl.europa.eu/news/en/headlines/society/20181212STO21610/plastic-waste-and-recycling-in-the-eu-facts-and-figures.
29. Baran, B. (2020). Plastic waste as a challenge for sustainable development and circularity in the European Union. *Ekonomia i Prawo. Economics and Law, 19*, 7–20. ISSN: 18982255.
30. Federal Office for the Environment. (2023). *PET beverage containers*. https://www.bafu.admin.ch/bafu/en/home/themen/thema-abfall/abfallwegweiser--stichworte-a--z/pet-getraenkeflaschen.html.
31. Swiss Recycle. (n.d.). *Wichtige Gesetzestexte zur Separatsammlung und Kreislaufwirtschaft* Rechtsnormen. Retrieved November 29, 2023 from https://swissrecycle.ch/de/wertstoffe-wissen/rechtsnormen.
32. Daniel Schwendemann. (2023). *Geruchsminimierung von PP-Rezyklat auf dem Compounder* in. 18. Rapperswiler Kunststoffforum.
33. Jandric, A., et al. (2020). Bromierte Flammschutzmittel in Elektroaltgeräten: Untersuchung der Brom-Konzentration nach Kunststofftypen und Gerätekategorien mittels Röntgenfluoreszenzanalyse. *Österr Wasser- und Abfallw, 72*, 68–76. ISSN: 0945-358X.
34. Paletta, A., Leal, W., Balogun, A., Foschi, E., & Bonoli, A. (2019). Barriers and challenges to plastics valorisation in the context of a circular economy: Case studies from Italy. *Journal of Cleaner Production, 241*. ISSN: 0959-6526.
35. Endres, H. -J., & Siebert-Raths, A. (2009). *Technische Biopolymere: Rahmenbedingungen, Marktsituation, Herstellung, Aufbau und Eigenschaften*. ISBN: 978-3-446-41683-3. Hanser, München.
36. Gironi, F. & Piemonte, V. (2011). Life cycle assessment of polylactic acid and polyethylene terephthalate bottles for drinking water. *Environmental Progress & Sustainable Energy, 30*, 459–468. ISSN: 1944-7442.
37. Rahman, M. H. & Bhoi, P. R. (2021). An overview of non-biodegradable bioplastics. *Journal of Cleaner Production, 294*, 126218. ISSN: 09596526. https://linkinghub.elsevier.com/retrieve/pii/S0959652621004388.
38. Plastics Europe. (2023). *Plastics Europe veröffentlicht Plastics Transitions Roadmap* Plastics Europe DE. https://plasticseurope.org/de/2023/10/26/plastics-transition-roadmap/.

39. Bachmann Plantec AG. (2023). *NETPOT: Topflösung aus min. 50% weniger (Bio)-Kunststoff* BACHMANN PLANTEC AG. https://www.bachmann-pflanzentrays.ch/produkte/netpot/.
40. Fluidsolids. (2023). *Food Grade–Fluidsolids Biocomposites.* https://www.fluidsolids.com/loesungen/fluidsolids-food-grade/.
41. European Commission. *Guidance on waste definitions* Sept. 21, 2021. https://ec.europa.eu/docsroom/documents/46954/attachments/8/translations/en/renditions/pdf.
42. Flury, M. (2023). *Recycling ist nicht einfacher, aber nachhaltiger* KunststoffXtra. https://kunststoffxtra.com/recycling-ist-nicht-einfacher-aber-nachhaltiger/.
43. Flury, M. (2020). Dem Abfall einen Wert geben. *KunststoffXtra, 7*, 26–27. ISSN: 1664-3933.
44. Tide. (2022). *Maurice Lacroix launches the Aikon #tide.* https://www.tide.earth/en/news/maurice-lacroix-aikon-tide/.
45. University of St. Gallen. (2023). *Innosuisse flagship project: How plastics can be produced and used sustainably* HSG - Uni SG. https://www.unisg.ch/en/newsdetail/news/innosuisse-flagship-project-how-plastics-can-be-produced-and-used-sustainably/.

Chapter 18
Digital Transformation of Remanufacturing

Ramesh Subramoniam

Abstract Companies today make data-driven decisions and *Remanufacturing* business is no exception to that evolving trend. The networks are evolving with data being collected from all phases of the product lifecycle from design, prototype, manufacturing, aftermarket, returns remanufacturing and recycling. This chapter provides a framework for a digitized product lifecycle integrated with business phases including *Remanufacturing* and validates them with case studies based on the prior experience of the author working for major companies in various industries.

18.1 Digital *Remanufacturing* – Accelerator for the Circular Economy

Remanufacturing is a comprehensive and rigorous industrial process by which a previously sold, leased, used, worn, remanufactured, or non-functional product or part is returned to a like-new condition, same-as-when-new or better than when in new condition, both from a quality and performance perspective, through a controlled, reproducible, and sustainable process [1]. A remanufactured product is 30 to 40% less expensive, uses 85% less energy and contributes to product recycling. This makes the production system both environmentally friendly and economically beneficial. Additionally, since remanufactured products are handled in the local geographic regions where customers reside, the system also has social benefits, including job creation for the local community.

Remanufacturing was for the longest time ignored by the Original Equipment Manufacturers (OEMs) that originally designed and manufactured the product. As a result, many independent remanufacturers handled the product, such as Cardone Industries in Florida, USA. The OEMs, however, have a significant advantage in *Remanufacturing* products. Since it is their product design, they can *Remanufacture* the products according to their own quality standards. In collaboration with service

R. Subramoniam (✉)
Professor of Practice, Operations Management, The University of Texas, Dallas, 969 Redwing Drive, Coppell, USA
e-mail: ramesh.subramoniam@utdallas.edu

A. Podleisek et al. (eds.), *Circular Economy: The Next Level of Company Success*,
https://doi.org/10.1007/978-3-031-66406-9_18

providers and retailers, OEMs also have access to product return data, which they can use for future improvements.

Harvard Professor Michael Porter believes that Internet-of-Things (IoT) will change the way companies do business in the future [2, 3]. The rapid increase in smart, connected products has resulted in a growing volume and variety of real-time data. Companies also became more aware of the need to protect their own data more than 12 years ago and have since taken control of the data in most cases, for example, from third-party logistics providers, distributors, and the like. The new trend is to monetize the data, where possible, and also to help support data-driven decision making at all levels of the organization. We are also living in an age where our economic activities are threatening our existence on planet Earth. With the world's population increasing by 83 mill. every year, there is a strong need to move away from a "throw-away" society to a "take-back" society and establish more social and environmental guidelines and processes [4].

Companies have been operating in their silos for a very long time. In many cases, they have missed the importance of working in a collaborative manner and sharing data with different stakeholders in the reverse supply chain. Adopting a more collaborative approach can lead to improved efficiency and effectiveness in both production and consumption systems. Specifically, adopting such measures can contribute to achieving the United Nations Sustainable Development Goal, specifically goal number 12, which focuses on responsible production and consumption. The goal sets benchmarks to be met by 2030, emphasising that current material demands should not result in over-extraction or degradation of environmental resources [5]. We also live in a world today with a massive digital transformation in progress. With technologies such as cloud computing, 3D printing, blockchain, IoT and Artificial Intelligence (AI) providing solutions, we will discuss how they can add value to Reverse Logistics (RL) and *Remanufacturing* operations.

18.2 Drivers for *Remanufacturing* Digital Transformation

There are several risks associated with the implementation of circular strategies in *Remanufacturing* such as the mismatch between fluctuating demand, supply, and the value of used components that cause uncertainties regarding costs and returns on investment [6]. Specific challenges include gaps in information flow concerning the condition, availability, and location of assets currently in service [7–9]. Industry 4.0 (I4.0) addresses these issues by leveraging digital technologies to collect data, thereby mitigating some of these challenges [6]. Currently, companies are engaged in digital remanufacturing transformation for the following three reasons:

Improve Customer Experience OEMs will benefit from knowing why their products were returned and which design features need immediate improvement to prevent customer attrition. This is a top-line impact for OEMs and their supply chain partners.

Reduce Costs Traditionally, *Remanufacturing* has served as a low-cost solution alternative for obtaining replacement parts during the warranty period, leading to a cost reduction of 30–40%. Also, having access to product tracking and tracing data allows the remanufacturer to pre-arrange labor and other resources before the product arrives at their facility. The condition of the product is equally important to the remanufacturer, as it helps to better define the routing of the product. This could potentially result in cost savings with their third-party logistics (3PL) providers by minimizing the number of times the device is handled, based on prior information about the product's condition from retail centers.

New Business Models A good understanding of product returns data, when collected by retailers, can provide valuable information to the OEM and other stakeholders. By analyzing product returns data, retailers can identify monetization opportunities that can lead to the development of new business models. For example, insights from the returns data can be used to create specialized approaches like a "replacement-as-a-service" model or leasing programs for products. Real-time data can be a significant competitive advantage in the aftermarket for maintenance and replacement opportunities.

Let us take the automotive industry as an example; the average lifetime for automobiles is 10–12 years in both the EU and the U.S. In 2010, 40.2 mill. automotives in the U.S. and 7.8 mill. in the EU reached their End-of-Life (EOL), accounting for 5% of global industrial waste [9]. This results in an increased interest for automotive OEMs to *Remanufacture* and *Recycle* as part of their long-term sustainability efforts. I4.0 [10] digitizes and integrates processes vertically across the entire organization, as well as horizontal integration that stretches beyond the internal operations from suppliers to customers and all key value chain partners. It includes technologies from track and trace devices to real-time integrated planning with execution. This broader technology integration opportunities will provide OEMs with an even better way to provide solutions in the remanufacturing sector if they take a PLC approach, as outlined in this chapter.

18.3 *Remanufacturing* Product Lifecycle and Pre-pandemic Research

The PLC depicted in Fig. 18.1 illustrates the typical trajectory of a product, marked by an initial increase in production volume, reaching a maturity stage, and eventually experiencing a decline. The various lifecycle stages for Original Equipment (OE) production, Original Equipment Service (OES) during the warranty period, and Independent Aftermarket (IAM) post-warranty, as would be typical for a major automotive supplier, are illustrated in Fig. 18.2.

For many companies, *Remanufacturing* was often an afterthought, overlooking the chance to get involved earlier in the stages of business pursuit, product development, and process development before any of the aforementioned PLCs were

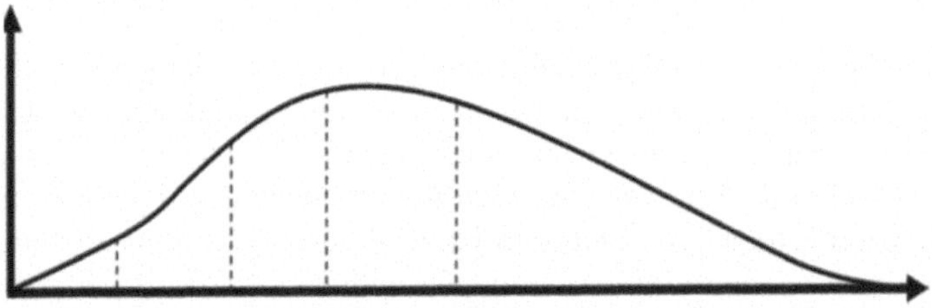

Fig. 18.1 Product lifecycle as described in this chapter shows evolution of sales over time

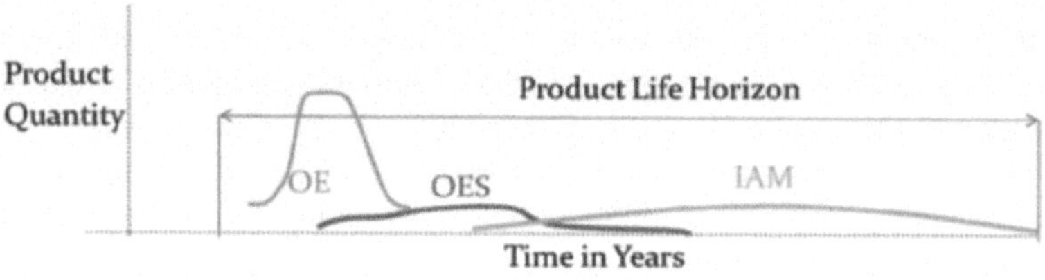

Fig. 18.2 Production and aftermarket business models of a manufacturing enterprise

initiated. Figure 18.3 shows the results of a case study conducted by the author during 2008–2009 for an automotive supplier [11]. During the business pursuit and product development phases, numerous remanufacturing opportunities – such as design for remanufacturing, intellectual property considerations, compliance with environmental regulations, and IAM potential – were largely ignored. This resulted in the creation of an engine control module, for example, that was either sealed or welded, making it challenging to *Remanufacture* once it entered the aftermarket service phase. The automotive industry in the U.S. hit a financial crisis during the same time period (2008/2009). The author conducted a survey to understand shifts in industry trends, the results of which were published in 2010 [12]. The survey results, shown in Table 18.1, indicated a growing interest in *Remanufacturing* and a shift toward more sustainable business models, driven by investor demand. A significant change in mindset around this time was also reported by Ioannou et al. [13], resulting in an accelerated need for Environment, Social and Governance (ESG) reporting starting in 2013.

Prior to the 2008 U.S. recession, the author distributed a survey to executives in the automotive supplier industry mainly from the USA. Another survey was conducted in 2010, after the recession ended, and was accompanied by an Analytic Hierarchy Process (AHP) analysis.

It was not surprising to see a high ranking for the financial impact. The green perception moved up in priority, going from seventh to fourth place. This aligns with findings by Ioannoau et al. [13] which noted a shift toward environmentally and socially responsible practices among U.S. companies following the financial crisis. Government and investor-funded green initiatives also played a role in encouraging

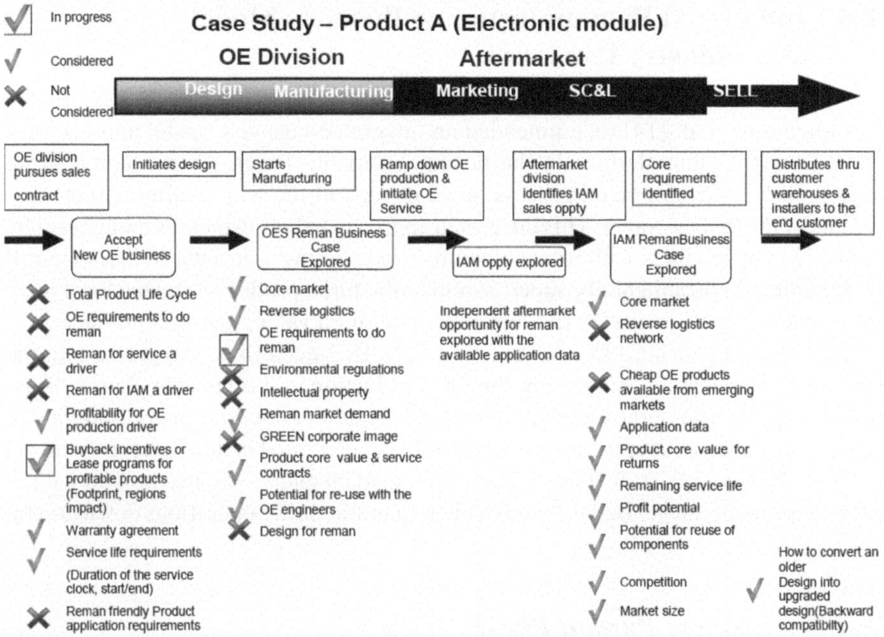

Fig. 18.3 Case study of an electronic automotive product A conducted by the author in 2008/2009

Table 18.1 Automotive industry survey conducted in 2008 and 2010

Key remanufacturing strategic factors	Industry survey ranking (2008)	AHP ranking (2010)
Financial impact of remanufacturing	1	1
Design for remanufacturing	2	8
Intellectual property of product	3	3
OE customer product specifications	4	5
Core management	5	2
Brand erosion	6	9
Green perception	7	4
Integrated organizational alignment	8	7
Government regulations	9	6

automotive companies to invest in sustainable business practices. The importance of managing used products, or "cores," rose in the rankings from fifth to second place, reflecting an increased focus on quality cores as companies expanded their *Remanufacturing* offerings. Meanwhile, the priority given to design for remanufacturing dropped from second to eighth place, as other factors took precedence and decision making processes for remanufactured products among OE suppliers were delayed.

18.4 Integrated Remanufacturing Business Model and Industry Use Cases

Subramoniam, et al. [14] recommended an integrated business model approach for remanufacturing that starts with the business pursuit, design for remanufacturing during product development, process development all the way to aftermarket sales and finally the end customer. This integrated approach helps businesses reduce cost by replacing new products with remanufactured ones during warranty, reduces landfill by providing environmentally superior remanufactured products, creates local jobs, and provides superior service for customers with high value, low cost products.

The remanufacturing PLC explained in Sect. 18.3 need to be integrated into the business model in Fig. 18.4 during the OE production and aftersales phases of the business cycle. The author learned during his experience with a major automotive supplier that a remanufacturing mindset should be integrated in the enterprise starting with the "Business Pursuit" phase. The subsequent chapters will discuss how a data-driven approach, can be used for improving remanufacturing operations downstream.

18.4.1 *Business Pursuit Phase*

In the business pursuit phase of the automotive supplier example, the vehicle manufacturer, who was the customer, failed to emphasize design for remanufacturing from the supplier. This resulted in a lack of design guidelines for remanufacturing in the supplier's product development process. In another use case, the OEM expressed the need to expand their stagnated remanufacturing revenue. There was a strong need to identify key stakeholders in the supply chain from a product/supplier database. The

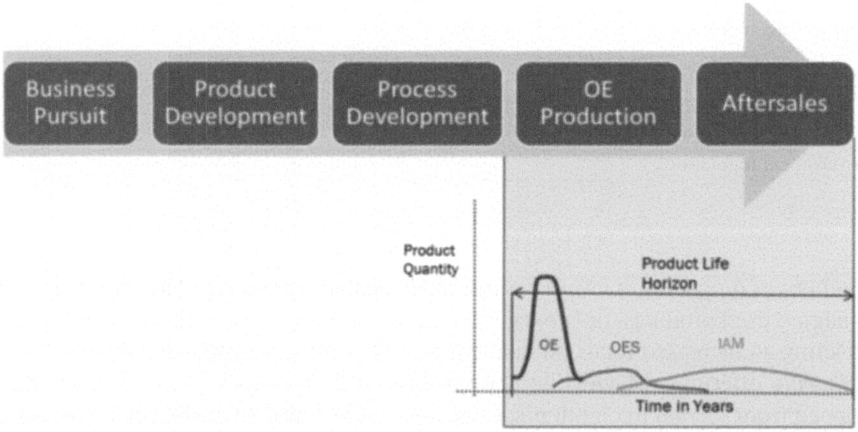

Fig. 18.4 Integrated business model with a PLC approach for a remanufacturing enterprise

author then looked for synergies with the supplier before they were recommended to the OEM for a potential joint venture and later for an acquisition. Many companies such as Renault have integrated leasing business models, where they engage with remanufacturing, recycling and reuse processes with their potential customers.

A major automotive supplier, for example, used the remanufacturing decision-making framework with product-specific data such as product design for remanufacturing, OE product specifications, core management process, etc. to develop a better selection process to decide which products to remanufacture and also improved the material flow process by working with the different stakeholders in the return process.

18.4.2 Product Development Phase

There is an urgent need to design products with a Product Service System (PSS) perspective, which integrates and optimizes the flow of products from a lifecycle perspective, significantly adding value for the customer and lowering costs for the manufacturer. In addition, the shorter lifecycle for many products requires manufacturers to get feedback about their products quickly so that they can fix them before they lose customers. Implementing a streamlined returns process that captures the reasons for product returns can provide valuable insights for manufacturers. The author observed that many companies now employ specialized teams focused on "Design for Remanufacturing" or "Design for Service". These teams work to optimize product designs for easier remanufacturing or servicing, thus positively contributing to the Circular Economy (CE).

A leading telecommunications service provider identified a significant issue with cell phone returns at their remanufacturing facility. The problem was traced back to the charging port's upward-facing design, which allowed water to seep into the phone. This feedback was relayed to the manufacturer, who then modified the design so that the port faced downward. This minor design adjustment led to a substantial reduction in product returns due to water damage and issues with the printed circuit board.

18.4.3 Process Development Phase

The uncertainty surrounding the condition of the returned products significantly affects the feasibility of *Remanufacturing* those items. This uncertainty also affects the scheduling of products for the *Remanufacturing* process. For instance, major automotive remanufacturers like John Deere don't offer remanufactured components for every part in their lineup. Various factors, such as volume and intellectual property, influence the decision to offer a remanufactured part, both from a financial and operational standpoint. The company also manages the core deposit value to regulate the availability of cores for remanufacturing. While a higher deposit value encourages more customers to return cores, it also requires balancing the influx of returned cores against the risk of those parts becoming obsolete.

A major telecommunications service provider leveraged the data of "customer comments" to gain insight into why a product was returned at their retail stores. This information was then relayed to the central returns plant responsible for remanufacturing. With this data, plant associates could plan their remanufacturing operations more efficiently, optimising resources such as labor and equipment. The author also observed that such streamlined routing led to a 5–10% reduction in processing costs for the service provider.

18.4.4 Original Equipment (OE) Production Phase

Many manufacturing operations feature pre-production or prototype facilities that produce items in limited quantities for design validation or process testing. These pre-production units often operate in isolation, sourcing functional test stands and other process equipment from prospective vendors. By adopting a coordinated approach that consolidates equipment needs for both pre-production and full-scale production, manufacturers can negotiate more favorable pricing with vendors, especially if those vendors are assured of supplying equipment for production, thereby reducing overall costs. Another opportunity for the manufacturer is to use scrap products that can be sorted for use from the production operations to prime the remanufacturing operation, particularly when product return volumes are initially low at the beginning stages of remanufacturing operations.

A major automotive supplier worked with their manufacturing plant and used the scrap from the plants to identify potential cores for remanufacturing. Many major OEMs, for example Detroit Diesel Remanufacturing, use digital technologies such as additive manufacturing to recover cores, which otherwise would have been discarded.

18.4.5 Aftersales Phase

During the aftersales phase, there is a level of uncertainty both in the volume of product returns and the demand for remanufactured products. The product visibility is better during the OES period. For example, in the automotive sector, dealerships keep track of products and can route them back to the OEM for remanufacturing. The rate of product returns is somewhat predictable, influenced by initial product defect rates and any shortcomings in communicating product features to customers, which can lead to increased returns. But after the product enters the IAM phase, the uncertainty of product returns and customer demand increases. These products are typically out of warranty, leaving the OEM with little to no visibility into their status. Many of these products do not get returned to the dealerships and so the availability of cores becomes an issue. Furthermore, the nature of defects becomes less clear as customers often opt for local repair shops once the product reaches a certain age. This results in a lack of product visibility not just for the dealerships but also for the OEM, complicating the remanufacturing process.

The author collaborated with a leading automotive OEM to facilitate partnerships with potential IAM retailers within the automotive supply chain. The OEM offered cost-effective product solutions to these retailers and also coordinated with them and their workshops to secure the return of cores for remanufacturing. Adopting this strategy necessitated that the OEM shift to an "All Makes, All Models" remanufacturing focus. This was because the relationships with independent automotive retailers encompassed a wider range of products, including those of competitors, compared to the more narrow product focus typical of a dealership partnership.

Before the pandemic, companies were already moving toward the development of more advanced forecasting tools to address uncertainties in product returns and demand. Post-pandemic, the emphasis on demand sensing models has grown, especially in the aftersales sector. Here, connectivity with OEMs products has become increasingly important, often facilitated through IoT technology or product sensors. These tools help track products and alert customers to service issues as needed. One telecommunications service provider took aftermarket remanufacturing to new heights by conducting pilot tests to explore cost-sharing models with manufacturers. This innovative approach aimed to consolidate remanufacturing operations into shared facilities, eliminating the need for separate specialized plants for both the service provider and the manufacturers to handle returned products.

18.5 Digital Transformation Trends for Remanufacturing

The coronavirus pandemic that began in 2020 served as a wake-up call for manufacturers to build the ability to quickly adapt to changes in customer demand. Remanufacturing is also affected by this trend and needs to be equally agile in response. In the future, digital capabilities will be key for remanufacturers to respond quickly to customer demands, reduce costs, and build supply chain resilience. There are many SMEs in the remanufacturing space, and they do not have the capital and infrastructure to support a digital transformation like the bigger OEMs. In addition, they do not own the design of these products, so OEMs have an advantage in developing remanufacturing capabilities faster in the immediate future.

Governments are also taking steps to boost digital transformation, such as in China, where the deployment of digital technologies such as big data, AI, and cloud computing is heavily encouraged especially to assess and ease the spread of the pandemic. In addition to that, cybersecurity and privacy concerns have increased in recent years, and many laws and rules have also surfaced to regulate them. Europe launched general data protection regulation in May 2018 with strict regulations for collecting and processing personal data. China implemented the China Cybersecurity Law, and noncompliance can lead to penalties and damage to brand reputation.

There is a convergence of data happening today from multiple data sources such as email, blogs and call center systems and remanufacturing organizations can build on their current Enterprise Resource Planning (ERP) systems to build a digital platform for enhanced customer experience and cost-effective operations [15]. Smart manufacturing is the process of converting data from manufacturing processes into actionable information. Cyber-physical system is the concept that connects networking, physical processes, and computation [16]. This type of integrated IT platform can result in a smart remanufacturing factory. Tracking the replacement of spare parts and consistently updating records for suppliers across all tiers is essential. Additionally, accounts must be reconciled to manage cores, or used products, for remanufacturing. This includes handling product returns, dealing with goods damaged during transport, and managing used or leased products. On top of these challenges, remanufacturers also face the complexity of managing core charges that come from multiple sources. Product tracking is a requirement for the remanufacturer to determine if and how often a returned product has already been remanufactured [17, 18].

Ragampeta [16] discussed the concept of a smart remanufacturing factory, where the use of IoT, Augmented Reality (AR) and Virtual Reality (VR) can provide the technological foundation for a remanufacturer. Also, the author warned about the need for increased caution from cyber attacks. Many authors have done research with these technologies for remanufacturing. Zheng et al. [19] applied IoT to automobile engines to form the Internet of Manufacturing Things (IoMT) environment. Using the IoMT environment, identification technology for disassembled engine parts was designed, real-time management of remanufacturing resources became possible, resulting in efficient management of resources, cost reduction, and a decrease in energy consumption by 34%, leading to sustainable development and greener production of automobile engines [14, 19]. Using AR technology, a process such as product disassembly, traditionally considered a non-value-added operation, can be optimized to be cost and time-efficient. An evaluation study by Harvard et al. [20] has concluded that AR technology can reduce maintenance time with improved quality compared to using hard copy manual or video instructions.

Praneetha et al. [14] discussed the remanufacturing challenges and provide technology solutions to the problems that are yet to be solved to meet the 2030 goals of the UN. Coupling technology with standards provides a unique perspective to solving these challenges. Several standards such as the RIC001.2-2021 [1] remanufacturing standard developed by the Remanufacturing Industries Council (RIC), can be integrated at different stages of the remanufacturing process and can help solve a remanufacturing challenge. The authors conducted an AHP analysis and found that 3D Printing (Table 18.2), followed by AI and cloud computing, were the most promising technology solutions based on a ranking by executives in the remanufacturing industry. Most of the digital technology surveys for forward supply chains have shown 3D printing to be a lower priority, and these results emphasize the importance of integrating reverse supply chain and remanufacturing in technology decision making. Failure to do so risks relegating remanufacturing to an afterthought.

The challenges of remanufacturing often begin with the formulation of a business case for OEMs or other manufacturers. Utilising IoT technology to track the volume

Table 18.2 Digital remanufacturing technologies ranked based on AHP analysis

Digital technologies	Rank
3D Printing	1
Inventory and network optimization	2
Prescriptive and predictive analytics	3
Artificial intelligence	4
Cloud computing	5
Augmented reality	6
Blockchain	7
Internet of Things	8
Autonomous vehicle and drones	9

of product returns is crucial. This enables accurate cost calculations that avoid duplicating material expenses for multiple returns, thereby making the business case for remanufacturing more viable.

Traditional remanufacturing has numerous challenges and barriers related to negative perception [12], acquisition, collection, evaluation, and reprocessing of cores [21], resulting in poor prediction of incoming cores (and spare parts). Additionally, ensuring the quality of the core and aligning its supply and demand presents significant challenges. Implementing quality management systems can improve customer confidence in remanufactured products. Adhering to established standards, such as the RIC standard [1] can further enhance trust and confidence in remanufactured items. Blockchain solutions can enhance the authenticity of remanufactured products, thereby increasing consumer confidence and willingness to purchase such items. Digital technologies such as IoT and Radio Frequency Identification (RFID) has the capability for core tracking, resulting in enhanced availability of cores, a key requirement for remanufacturing. Also, offering an app that allows OEMs and independent remanufacturers to locate core brokers in their specific regions can address RL challenges. In essence, core inventory must be transparent and easily accessible, enabling remanufacturers to source cores from available suppliers, regardless of location. In addition, collaborative Robots, Augmented Reality, and Virtual Reality help aid human skills while additive manufacturing assists in parts repair. AR can also be a valuable tool in manufacturing, offering design inputs that facilitate remanufacturing. For example, AR can guide choices related to materials and coatings, as well as improve the accessibility and layout of parts, making them more conducive to remanufacturing processes.

Blockchain technology has the capability to track products and raw materials back to their sources and can provide data to assess the energy consumption of products for ESG reporting. Also, blockchain can store cost, quality, date, location, and certification for efficient management of the circular network to provide authenticity and trust for customers who demand original remanufactured products and not counterfeit products. Also, blockchain can provide the capability with better smart contracts and the OEMs need to manage core charges.

The growing focus on data has led companies to leverage AI tools for analyzing large data sets to obtain meaningful insights. However, the quality of the data is just as crucial as the quantity. Blockchain technology can play a key role here by establishing a consensus among stakeholders regarding the reality of product location and ownership, thus ensuring data integrity [22]. According to Future Market Insights [23], the global market for blockchain AI will sky rocket to 3.5 bill. USD by 2032. In the realm of *Remanufacturing*, combating the use of fraudulent and counterfeit parts poses a significant challenge, particularly for automotive OEMs. A synergistic application of AI and blockchain technologies can significantly improve fraud detection measures, thereby increasing trust within the supply chain network.

While Industry 3.0 focused on the automation of individual machines, industry 4.0 extends digitization to encompass end-to-end supply chains and physical assets. Now, an emerging wave of Industry 5.0 [24] is bringing a human-centric, sustainable approach that integrates various technologies to add value across the remanufacturing lifecycle. Smart products, equipped with sensors and embedded software, offer added value for both manufacturers and consumers. Smart Product Lifecycle Management (SPLM) can seamlessly integrate with this embedded software, corporate backend systems, cloud services, and the Internet of Services (IoS) to provide comprehensive control over product data from inception to EOL. Ultimately, this smart and integrated SPLM approach significantly enhances servitization and product circularity, leading to more efficient and effective remanufacturing and RL processes.

Many manufacturers including Delphi, Xerox, Caterpillar, General Electric (GE), Volvo, Navistar and John Deere from diverse industrial sectors have established profitable remanufacturing business models [21]. Xerox established a cartridge remanufacturing program that saved 200 mill. USD in material costs in five years. General Electric moved from a manufacturer to a service provider by providing multi year service contracts and earning customer loyalty for many years. Traditional manufacturers such as GE and Delphi (now BorgWarner), as major auto suppliers, have joined the bigger digital ecosystem. A senior executive from a Tier 1 automotive supplier in the late 1990s stated that he works for a connected software company with the ability to collect data within the vehicle, transfer data to the cloud, and analyze the data in the cloud compared to a traditional mechanical automotive components supplier. This was a significant transformation from a traditional automotive supplier to a high-tech data-driven enterprise.

The transportation provider FedEx joined with Microsoft and launched a FedEx Surround program, which combines FedEx's IoT with Microsoft's suite of AI and cloud computing services to track packages and predict problems related to slow delivery, such as weather or clearance issues.

The most disruptive technologies in future will create more eco-friendly and human-centric methods of value creation. Data is the foundation for improved business processes today, and RFID and sensors can provide the data required for the digital twin in a real time or near real time manner. The increased ability to generate, store, and process data today can lead to network burden, and methodologies such as edge computing can reduce the burden and reduce data leakage. Also, real time transmission is possible today with 5G technology.

The author [25] recommends the following efforts with data available during the current digital transformation underway and to expand on it during future transformation waves using technologies and their applications in the remanufacturing PLC.

- Develop forward and reverse distribution channels to meet emerging customer expectations based upon survey and industry data.
- Focus on E-commerce with selected product portfolio data.
- Acquire/JV/partners to bridge capability gaps from industry data.
- Improve product design from returns/defects data (OEM).
- Monitor real time customer behaviour (forward and reverse supply chain) data, reduce response time to drive change.
- Locate products-in-use and motivate customer behaviour for increased core returns from point of use data.
- Continue to reduce "order to delivery" lead time by tracking real time logistics data.
- Reduce processing costs by using returns data.
- Track real time inventory based on customer demand data.
- Forecast/schedule effectively and efficiently to reduce inventory costs and improve customer experiences from point of sale data.

Table 18.3 links the digital technologies and its use cases to the produce lifecycle phases discussed in this chapter.

Table 18.3 Digital remanufacturing technologies linked to PLC phases

Product lifecycle phase	Enabling digital remanufacturing technologies
Business pursuit	Use of Quality management certifications, IoT for business case development & Blockchain Solutions to track fraud products & convince customers
Product development	AR to support design inputs for remanufacturing
Process development	IoT for product tracking that helps core availability
OE production	Additive manufacturing, 3D Printing, AI and Robotic systems to improve product availability and reduce inventory
Aftersales	Improved remanufacturing forecasting models, demand sensing models to deal with uncertainty in demand. ESG metrics using blockchain

18.6 Conclusion and Key Learnings

For many years, remanufacturing was largely an afterthought for OEMs, who often overlooked the opportunity to reclaim and remanufacture their own products. However, this began to change during the 1990s and into the early 2000s, as more OEMs started to embrace product take-back and remanufacturing initiatives. During this period, ERP systems gained prominence, allowing companies to realize the benefits of internal data visibility within their own organizational boundaries. The push for data visibility across organizations, led by companies like Amazon and the opportunity to monetize data, has introduced both challenges and opportunities for manufacturing firms. Around 2010–2013, the rise of ESG goals and an increased focus on sustainability led more companies to take control of their data. This was aimed at enhancing efficiency in return operations and remanufacturing processes. Although these efforts are still in their infancy, advances in product technologies like 3D printing are becoming more prevalent in the remanufacturing industry. These are currently overshadowing transformational technologies such as AI and blockchain, which have the potential to harness data for remanufacturing in more powerful ways.

The complexity of the RL processes, challenges in demand planning, and lack or inadequate management of core inventory have slowed the execution. As we embark on a new world of Industry 5.0 and a growing desire for a CE, the remanufacturing industry stands tall for a major transformation in the coming years.

Future success in remanufacturing will hinge on companies adopting a mindset open to change, backed by a robust technology infrastructure to facilitate the transition. Additionally, the capacity for innovation, demonstrated through creative pilot projects, will be essential for effective collaboration with other stakeholders in both the digital and remanufacturing supply chain ecosystems.

References

1. ANSI. (2021). *RIC001.2-2021 Specifications for the Process of Remanufacturing*.
2. Porter, M. E., & Heppelmann, J. E. (2014). How Smart, Connected Products Are Transforming Competition. *Harvard Business Review*. Section: Technology and analytics. ISSN: 0017-8012. https://hbr.org/2014/11/how-smart-connected-products-are-transforming-competition.
3. Porter, M. E., & Heppelmann, J. E. (2015). How Smart, Connected Products Are Transforming Companies. *Harvard Business Review*. Section: Business management. ISSN: 0017-8012. https://hbr.org/2015/10/how-smart-connected-products-are-transforming-companies.
4. Hayhoe, K., et al. (2018). *Chapter 2: Our Changing Climate. Impacts, Risks, and Adaptation in the United States: The Fourth National Climate Assessment, Volume II* (U.S. Global Change Research Program). https://nca2018.globalchange.gov/chapter/2/.
5. United Nation. (n. d.). *Goal 12—Department of Economic and Social Affairs*. Retrieved August 9, 2023 from https://sdgs.un.org/goals/goal12.
6. Lopes de Sousa Jabbour, A. B., Jabbour, C. J. C., Godinho Filho, M., & Roubaud, D. (2018). Industry 4.0 and the circular economy: a proposed research agenda and original roadmap for sustainable operations. *Annals of Operations Research, 270*, 273–286. ISSN: 0254-5330, 1572-9338.

7. Su, B., Heshmati, A., Geng, Y., & Yu, X. (2013). A review of the circular economy in China: moving from rhetoric to implementation. *Journal of Cleaner Production, 42*, 215–227. ISSN: 09596526.
8. *Design for Innovative Value Towards a Sustainable Society* (eds Matsumoto, M., Umeda, Y., Masui, K. & Fukushige, S.) (Springer Netherlands, Dordrecht, 2012). ISBN: 978-94-007-3010-6.
9. Simic, V. (2013). End-of-life vehicle recycling-A review of the state-of-the-art. *Tehnicki Vjesnik, 20*, 371–380.
10. Carrillo, P. (2016). *INDUSTRY 4.0: BUILDING THE DIGITAL ENTERPRISE - PWC*. https://www.readkong.com/page/industry-4-0-building-the-digital-enterprise-9807884.
11. Subramoniam, R., Huisingh, D., & Chinnam, R. B. (2009). Remanufacturing for the automotive aftermarket-strategic factors: literature review and future research needs. *Journal of Cleaner Production, 17*, 1163–1174. ISSN: 09596526.
12. Subramoniam, R., Huisingh, D., Chinnam, R. B., & Subramoniam, S. (2013). Remanufacturing Decision-Making Framework (RDMF): research validation using the analytical hierarchical process. *Journal of Cleaner Production. 40*, 212–220. ISSN: 09596526.
13. Ioannou, I., & Serafeim, G. (2010). The impact of corporate social responsibility on investment recommendations. *SSRN Electronic Journal*. ISSN: 1556-5068.
14. Pratapa, P., Subramoniam, R., & Gaur, J. (2022). Role of standards as an enabler in a digital remanufacturing industry. *Sustainability, 14*, 1643. ISSN: 2071-1050.
15. Budde, L., Hänggi, R., Laglia, L., & Friedli, T. (2024). Leading the transition to multi-sided platforms (MSPs) in a B2B context–The case of a recycling SME. *Industrial Marketing Management, 116*, 106–119. ISSN: 0019-8501. https://www.sciencedirect.com/science/article/pii/S0019850123002304.
16. Ragampeta, P. (2022). Industry 4.0 remanufacturing: a novel approach towards smart remanufacturing. *Masters Theses*. https://scholarsmine.mst.edu/masters_theses/8094.
17. Jensen, J., & Remmen, A. (2017). Enabling circular economy through product stewardship. *Procedia Manufacturing, 8*, 377–384. ISSN: 23519789.
18. Hayes, M. (2018). *Core Management Insight* Remanufacturing Industries Council. https://remancouncil.org/webinars/core-management-insight/.
19. Zheng, Y., Qureshi, A., & Ahmad, R. (2018). Algorithm for remanufacturing of damaged parts with hybrid 3D printing and machining process. *Manufacturing Letters, 15*, 38–41. ISSN: 22138463.
20. Havard, V., et al. (2016). In *Augmented Reality, Virtual Reality, and Computer Graphics* (eds De Paolis, L. T. & Mongelli, A.) Series Title: Lecture Notes in Computer Science, 302–320. Springer International Publishing, Cham. ISBN: 978-3-319-40621-3.
21. Teixeira, E. L. S., Tjahjono, B., Beltran, M., & Julião, J. (2022). Demystifying the digital transition of remanufacturing: A systematic review of literature. *Computers in Industry, 134*, 103567. ISSN: 01663615.
22. Nounou, H. (2023). *Crypto & AI: Competition, Convergence or Both?* Modern Consensus. https://modernconsensus.com/ai/crypto-ai-competition-convergence/.
23. FutureMarketInsights. (n.d.). *Market Research & Business Intelligence – Future Market Insights* Market Research & Business Intelligence – Future Market Insights. https://www.futuremarketinsights.com/.
24. Ghobakhloo, M., et al. (2023). Behind the definition of Industry 5.0: a systematic review of technologies, principles, components, and values. *Journal of Industrial and Production Engineering, 40*, 432–447. ISSN: 2168-1015.
25. Subramoniam, R., Sundin, E., Subramoniam, S., & Huisingh, D. (2021). Riding the Digital Product Life Cycle Waves towards a Circular Economy. *Sustainability, 13*, 8960. ISSN: 2071-1050.

Chapter 19
The Rapperswil Circular Economy Model – A Guide to Successful Implementation

Roman Hänggi, André Podleisek, Katharina Luban, and Nicolas Hofer

Abstract How can an industrial company effectively integrate the principles of Circular Economy into its company strategy and create an impact? In this chapter, we propose a 6-step model, the 'Rapperswil CE-Model'. This model is based on solid literature research, a survey with more than 500 participants, and 10 in-depth case studies. Our research reveals the practices and approaches that lead to the successful implementation of Circular Economy by providing an in-depth understanding of the essential drivers, enablers, and challenges. The 'Rapperswil CE-Model' starts with an analysis of the company's actual situation. Based on this, our findings support the development of a clear vision and specific goals for circular practices. This sets the basis for an open discussion to decide on the right approach to Circular Economy and its level of ambition. The subsequent steps consider the practical implementation: prioritising actions, crafting a detailed transition road map, and determining the pace of change. Critical to this model is the establishment of a continuous review and monitoring process that allows for ongoing oversight and necessary adjustments.

R. Hänggi (✉)
Brülisau, Switzerland
e-mail: roman.haenggi@ost.ch

A. Podleisek
Wettswil am Albis, Switzerland

K. Luban
Sevelen, Switzerland

N. Hofer
Zürich, Switzerland

A. Podleisek et al. (eds.), *Circular Economy: The Next Level of Company Success*,
https://doi.org/10.1007/978-3-031-66406-9_19

19.1 Strategic Direction Towards Circular Economy

In our research detailed in Sect. 4.4, we discovered that Circular Economy (CE) can significantly improve financial performance. This claim is based on a sample of 512 different respondents to a standard questionnaire, where 43% of the respondents were CEOs or business owners. In addition, case studies were conducted for 10 Retention Options (Rs) to further elaborate on this knowledge. Detailed occupation with companies significantly improves the understanding of how to implement CE in a company. Our research mirrors the actual situation and the reality of implementation in industrial companies. This is reason enough to discuss CE as an important strategic topic of increasing relevance for any company. In this chapter, we propose a practical approach for implementation.

19.1.1 Key Findings Concerning Drivers and Enablers of Successful Implementation of Circular Economy

Key Drivers

Our survey with 512 companies showed priorities with respect to the drivers (Sect. 4.4.3) and enablers (Sect. 4.4.4) of CE. A driver typically refers to a factor or variable that has a significant and often causal influence on a particular phenomenon or process. Drivers play a crucial role in understanding and explaining the behaviour or outcomes of a system. The relevance among the identified drivers varies significantly. As can be observed in Fig. 19.1, the drivers most frequently named by the respondents focus on reducing resource consumption (D.E1), carbon emissions (D.E2), and cost (D.F2). These findings underscore a dominant emphasis on efficient use of materials and resources, highlighting resource conservation and cost effectiveness as key motivators for current CE initiatives.

When allocating the number of driver mentions to the 10 Rs, we get a visual representation, a 'heat map' as illustrated in Fig. 19.2. This visual representation allows for an easy comparison, highlighting the varying degrees of relevance of each driver for different Rs. We observe variations in the relevance of different drivers for the Rs from Refuse (R0) to Remine (R9), but the key drivers D.E1, D.E2, and D.F2 mentioned above are important for all Rs.

To test the findings and hypothesis stemming from the heat map, for each R a detailed case study for the 10 Rs was conducted. Here again, the drivers can be represented in their relevance for different Rs (see Fig. 19.3).

Similarly to the survey participants, the companies analysed within the 10 case studies placed a high emphasis on the drivers related to resource management optimisation (D.E1 and D.E2). Notably, the driver 'Increasing competitiveness' (D.F3) received a higher ranking compared to its position in the survey results. We find

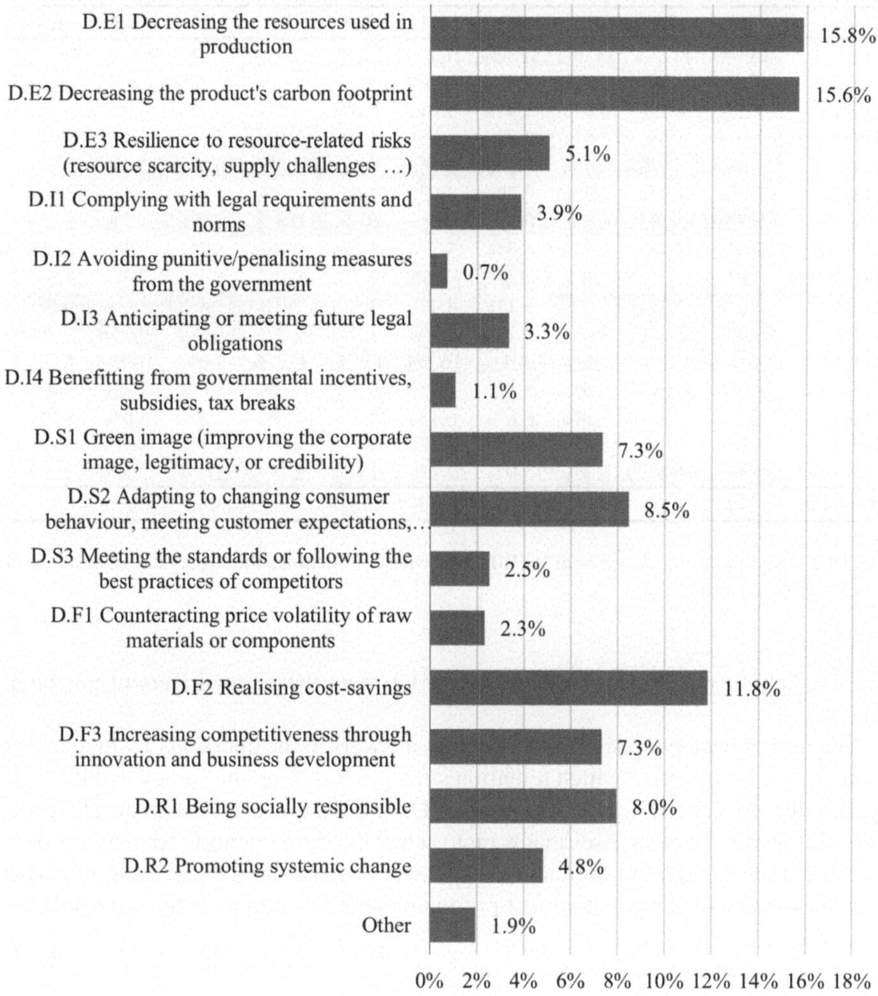

Fig. 19.1 Collective relevance for drivers ordered according to clusters (n = 5,259)

some more slight differences: For example, the driver 'Adapting to changing consumer behaviour' (D.S2) and 'increasing competitiveness through innovation and business development' (D.F3) was ranked higher in the practical case studies than in the survey, where 'Cost savings' (D.F2) held a more prominent position.

Additionally, the companies also stressed the importance of systemic change (D.R2) and social responsibility (D.R1), assigning these drivers more importance than was observed in the survey rankings. It should be noted that despite its applicability in these cases, the generic relevance of this fact must be questioned due to the particular nature of the enterprise analysed. But it is interesting to see that single

Driver	R0	R1	R2	R3	R4	R5	R6	R7	R8	R9
D.E1	13.0%	11.1%	18.2%	12.5%	16.6%	17.6%	18.0%	19.9%	15.1%	16.3%
D.E2	16.1%	17.4%	16.7%	9.4%	11.8%	13.9%	16.1%	17.4%	19.0%	18.5%
D.E3	5.2%	4.4%	4.6%	4.4%	5.0%	5.0%	4.1%	6.5%	6.9%	4.3%
D.I1	5.0%	4.6%	2.4%	3.8%	2.2%	2.9%	1.0%	5.6%	5.6%	5.4%
D.I2	1.8%	0.9%	0.3%	0.3%	0.7%	0.3%	0.5%	0.9%	0.4%	1.1%
D.I3	3.2%	4.9%	2.7%	2.2%	2.0%	2.6%	2.9%	4.3%	3.0%	5.4%
D.I4	0.8%	1.1%	0.2%	0.3%	1.5%	0.5%	0.5%	0.9%	1.7%	3.3%
D.S1	8.6%	10.0%	6.5%	6.0%	5.7%	5.2%	9.0%	8.4%	7.3%	6.5%
D.S2	6.6%	8.3%	7.6%	12.9%	11.6%	9.7%	9.0%	8.4%	7.3%	3.3%
D.S3	2.0%	2.3%	1.6%	5.1%	4.4%	3.1%	2.7%	1.4%	2.6%	0.0%
D.F1	2.0%	2.0%	3.3%	3.4%	2.6%	3.7%	2.2%	1.1%	0.9%	2.2%
D.F2	10.9%	8.1%	13.9%	14.1%	16.6%	15.2%	12.9%	7.0%	10.8%	8.7%
D.F3	8.3%	10.4%	6.6%	8.5%	7.7%	6.8%	7.1%	5.0%	7.3%	5.4%
D.R1	10.1%	8.3%	7.9%	8.8%	6.1%	7.1%	8.0%	7.4%	6.0%	9.8%
D.R2	5.6%	5.1%	5.5%	3.4%	3.3%	4.7%	4.9%	4.3%	3.9%	7.6%
Other	0.7%	1.0%	2.1%	5.0%	2.2%	1.6%	1.0%	1.6%	2.2%	2.2%
n =	713	1216	633	682	457	381	410	443	232	92

Fig. 19.2 Driver popularity heatmap for R0 to R9 where 'n' denotes the aggregate count of driver selections for a given R

cases see slightly different drivers for their CE than the focus drivers of our large survey.

The research we conducted now provides answers to the question of which driver is most closely related to which retention strategy. Based on the survey and the case studies, the drivers D.E1, D.E2, D.F2, and D.S2 get the overall the most relevance. This is essential for businesses that want to select the right retention strategy for their driving forces. Generally, companies map the four drivers to the particular retention strategies that will achieve the most promising results when they are achieved (see Fig. 19.4).

This summary is a high level generic result. It is understood, that this might differ for your specific company.

Driver	R0	R1	R2	R3	R4	R5	R6	R7	R8	R9
D.E1	+++	+++	+++	+	+++	++	0	0	+++	not applicable
D.E2	++	++	++	+	++	++	+	-	++	
D.E3	0	0	0	0	+	+	0	+	0	
D.I1	0	0	++	0	+	0	0	+++	++	
D.I2	0	0	++	0	0	0	0	+++	0	
D.I3	0	0	++	0	0	0	0	+	0	
D.I4	0	0	+	0	0	0	0	0	0	
D.S1	++	++	+	+	0	+	0	+	+++	
D.S2	+++	+++	+	+++	+	+++	++	+	+	
D.S3	++	++	0	0	++	0	0	0	0	
D.F1	0	0	0	0	+	0	0	+	+	
D.F2	+	+	+	0	0	++	+	++	+	
D.F3	+++	+++	0	++	++	++	+++	+	+++	
D.F4	0	0	0	0	++	++	+++	++	+	
D.R1	+	+	+++	+	++	+++	+++	0	0	
D.R2	++	++	+++	0	+++	++	+++	0	0	
Other										

Legend: Assessment of relevance:

+++	Critical
++	Important
+	Minor positive impact
0	Not relevant
-1	Negative impact

Drivers with higher relevance in the specific case studies

Fig. 19.3 Summary of drivers of eight case studies

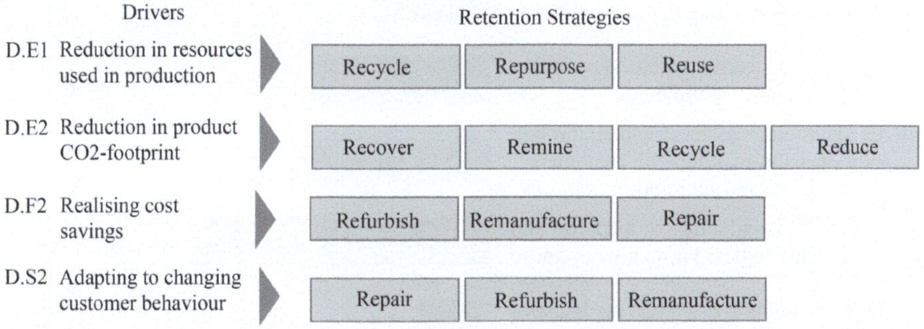

Fig. 19.4 Successful retention strategies for relevant drivers

Key Enablers

An enabler is a factor, element, or tool that facilitates or enhances a particular process, outcome, or capability. Enablers provide support or resources that contribute to the success or efficiency of a system or operation. Regarding enablers, in the survey, two key enablers emerged as the main priorities to transition to CE. As we can see in Fig. 19.5, the most critical enablers identified are the 'Establishment of close partnerships' (E.V1) and the 'Possession of technical and technological know-how' (ET3). The heat map in Fig. 19.6 further illustrates the varied level of importance each enabler has in relation to the different Rs.

We also reviewed how the answers of our case companies fit the overall survey in terms of enablers for the specific Rs (see Fig. 19.7).

In examining the enablers for transitioning to a CE, we found that the responses from companies in our case studies reflect the results of the broader survey. Both identified the establishment of close partnerships (E.V1) and the need for internal technical and technological know-how (E.T3) as the most critical enablers.

Although there are minor discrepancies between the case companies and the survey results, the overarching trends are consistent. The variations can be attributed to the specific selection of the companies for the specific retention strategy case study, the individual company's focus (market focus vs. cost focus), and the relatively smaller sample size in the case studies. It is important to note that the survey

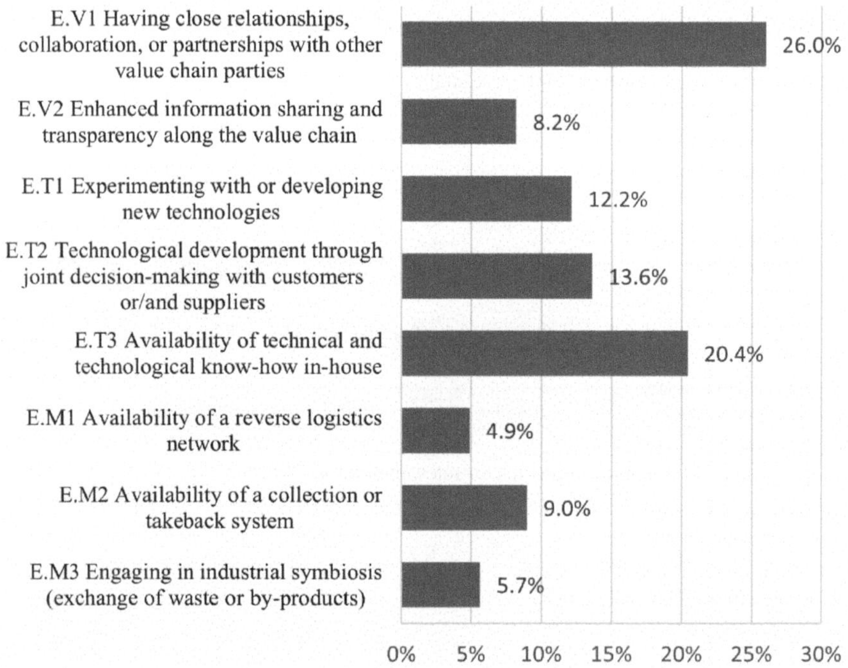

Fig. 19.5 Collective relevance for enablers across all Rs

Enabler	R0	R1	R2	R3	R4	R5	R6	R7	R8	R9
E.V1	23.4%	33.2%	26.3%	25.1%	24.2%	18.8%	25.4%	36.9%	20.0%	26.7%
E.V2	15.1%	11.4%	8.6%	6.2%	7.5%	7.5%	8.0%	5.0%	6.3%	6.7%
E.T1	20.6%	13.0%	7.2%	6.6%	5.6%	11.3%	13.0%	10.6%	13.8%	20.0%
E.T2	14.2%	15.5%	10.5%	12.8%	14.9%	17.3%	16.7%	11.3%	16.3%	6.7%
E.T3	15.6%	11.7%	22.0%	34.6%	34.8%	30.1%	19.6%	10.6%	8.8%	16.7%
E.M1	2.3%	2.8%	9.1%	5.8%	5.6%	5.3%	4.3%	4.3%	6.3%	3.3%
E.M2	4.1%	5.7%	11.0%	6.2%	6.2%	9.0%	7.2%	14.2%	16.3%	10.0%
E.M3	4.6%	6.6%	5.3%	2.9%	1.2%	0.8%	5.8%	7.1%	12.5%	10.0%
n =	221	318	212	247	165	135	139	143	84	32

Fig. 19.6 Enabler popularity heatmap for R0 to R9 where 'n' denotes the aggregate count of enabler selections for a given R

Enabler	R0	R1	R2	R3	R4	R5	R6	R7	R8	R9
E.F1	+++	+++	++	++	+++	+++	+++	+++	+++	not applicable
E.F2	++	++	+++	++	++	++	0	++	+++	
E.V1	+	+	+++	+	+	0	+++	0	+++	
E.V2	0	0	0	0	0	0	+++	0	-	
E.T1	+	+	0	0	+	0	+	0	++	
E.T2	0	0	+	+	0	++	+++	0	0	
E.T3	+++	+++	0	++	+	+++	++	0	0	
E.T4	0	0	0	0	0	0	0	0	+	
E.C1	++	++	+++	++	++	++	+	++	+	
E.M1	0	0	+	+	++	0	0	++	0	
E.M2	0	0	+++	0	+	0	+	-	0	
E.M3	0	0	++	0	++	++	++	-	0	

Legend: Assessment of relevance:

+++	Critical
++	Important
+	Minor positive impact
0	Not relevant
-1	Negative impact

Enablers with higher relevance in the specific case studies

Fig. 19.7 Summary of enablers of eight case study companies

encompasses all Rs in a large sample of 512 companies, offering a generic and averaged perspective, which may not capture the unique aspects of individual company scenarios. These company-specific enablers, as revealed in our case studies, can therefore vary depending on distinct business models and competitive environments.

In general, we have now developed a deep understanding of the elements of the transformation CE in practice. This will guide us later on towards our implementation model.

Impacts by Circularity

Having analysed drivers and enablers as *input factors* for the companies' CE journey, we also surveyed the companies in our case studies about the perceived *impacts* of applying CE principles to their business operations. So, in other words, what are the results of the implementation of CE? From the industry survey detailed in Sect. 4.4.1, several key impacts were deduced, including the following.

#1 Priority

- Improving company image
- Reduction of waste
- Realising cost savings in general

#2 Priority

- Improve sales
- Technological advancements
- Increase of profits
- Increase resilience against resource related risks

#3 Priority

- Reduction in manufacturing costs
- Reduction in material cost

Even for the priority item #3 'reduction in material cost', 33% of the respondents see a positive impact. Of course, the results will differ from company to company, but the overall positive impact has been widely confirmed. These findings are crucial for the success of circular business: Only when companies observe tangible benefits, especially financial ones, will CE be embraced as a long-term business concept.

The above results surveyed in our research are also supported in the literature [1–3]. Specifically, the following impacts, and therefore results, are often mentioned:

- **More attractive products to consumers** – Consumer behaviour is recently changing, therefore renewable, recyclable, and recycled input becomes an essential part of perceived value of the product.
- **Greater customer intimacy** – Focussing on the entire lifecycle of your product fosters a closer connection with your customer. The extent of this relationship may vary depending on the level of participation and the range of offerings provided during the service and usage phases.
- **Lower cost** – This is related not only to production costs, but also to costs throughout Product Lifecycle (PLC). Circularity can reduce costs in many areas, ultimately offering more attractive pricing to your customers. Using recycled materials, for example, can also reduce production costs.

- **Recurrent income sources** – Circularity bring you closer to your customer. Offering additional services can help increase the turnover of your company. This is nothing really new; many industrial companies, particularly in the investment goods sector, have been focusing on service elements from a sales and profit point of view for years. The focus on circularity further emphasises this view.
- **Develop new markets and customers** – Innovation through CE with regard to customer intimacy and product advantages can open new markets and attract new customers.
- **Greater resilience** – The better you understand your product and your supply chain, the better you can deal with risks. Furthermore, CE can reduce dependence on certain supply chains or, in the case of regional/local cycles, long transport routes.
- **Greater product and component utility drives product innovation** – Sharing economy models significantly advance the use of products as an extreme case. Every step towards increased serviceability and other technical enhancements increases the life of your products. This focus will also generate innovation in your products.
- **Reduced production time** – Focusing on refurbishment, reuse or repair can reduce the lead time to obtain the desired product.
- **Climate and environmentally friendly** – Circular practices contribute to better use of our globally available resources.
- **Increase attractiveness as an employer** – The change in consumer behaviour is only one facet of the changing values of your generation. The inclusion of circularity can increase your attractiveness in the job market.

This list shows many different concrete results by implementing a focus CE in a company. Although this list is extensive, our research also revealed unique and often very specific benefits – for example, balance of time, the ability to address specific challenges related to CE, and management within a limited investment budget pose significant challenges. However, these benefits did not appear as the main benefits for CE, but for the specific company, it is very important.

How to Drive the Change to Circularity

Our research revealed some useful insights into how to manage this transformation. In addition to the usual change elements such as broad team involvement or strong upper management support, we identified the following 'change facilitators' that should be taken into account for the successful implementation of CE practices:

- **Strengthening partnerships in the value chain** – As partnerships, we understand to increase the relationship in the value chain towards the supply chain as well as the customer side. In addition, consider involving other partners, such as technology or IT companies, to expand your product or service ecosystem.
- **Increasing and cultivating in-house technical and technological capabilities** – A deep understanding of your technologies and materials is vital. This knowledge forms the foundation for identifying and capitalising on future CE opportunities. Knowledge is the first step towards change.

- **Prioritising information sharing and transparency** – Specifically, when implementing material reduction strategy. Transparency and information sharing help you better understand the needs of other players in your ecosystem. It will also help them align their actions according to your own strategy.
- **Developing new business models** – Focus on areas such as Reverse Logistics (RL) or other services that match your specific R strategies. It is crucial to partner with other entities or to develop these services internally. Recall that operational excellence in delivery is a critical factor for success.

We also found that the integration of a mindset of CE into the core of the business is a priority objective. This means adapting the existing business model and products to align with the principles of CE. A disconnected peripheral addition of new circular services or products is inferior compared to embedding a mindset CE in the business strategy. The latter can significantly enhance and evolve the brand through a progressive, forward-looking concept, achieving greater business success. To do so, a deep understanding of the people involved in the innovation process is necessary.

As documented before, the research identified generic drivers for CE. When categorising them on a higher level, we see two distinct generic strategic dimensions that companies drive to implement CE – the companies drive on the one hand **operational excellence** (see Sect. 19.2) with their respective Rs, and on the other hand **innovate products, services up to new business models** (see Sect. 19.3) using the same or different Rs. These high-level dimensions are needed in practice when we discuss the concrete approach to implement circularity (see Sect. 19.5).

Our research showed that companies implement actions simultaneously for both generic strategic directions, but the focus lies first on the internal side, due to speed of results. Operational excellence approaches always come first. Innovation tends to take more time and effort, and reshaping existing business models involves greater risks due to many unknowns. These clear priority settings are also a critical change element. Too many activities and not achieved results can stop the change. Therefore, it is critical to focus where you get tangible results. Let us discuss the two different dimensions specifically for circularity.

19.2 Internal Circularity Focus: Operational Excellence Approaches

Our findings underscore that focusing on operational performance and cost savings is a top priority for industrial companies. Although the shift towards the introduction of new product offerings or even entirely new business models through CE is intriguing and relevant, it typically requires more investment and time. Often, such changes are made at a later stage.

Today's understanding of driving operational performance in an industrial company is grounded in the principles of Lean Management, supplemented by the application of digitalisation [4, 5]. The thinking of CE is adding an important piece to this current direction [6–8]. Our research with case companies showed that companies can further optimise their operational performance by applying CE principles in their operational excellence models. When focussing on strategies R0 to R9, new aspects of improvements can be brought to light. By applying these Rs within your own operations, you can use them as Narrowing R, to help decouple value creation from virgin material consumption and energy use. However, not all Rs are suitable for every organisation; it depends on the type of business, market access, and technological expertise.

To help companies select and implement the principles of CE, we have compiled a summary based on case studies from industrial companies. We have found that recapping the various Rs with simplified concrete examples is beneficial to achieving a shared understanding as a starting point. The following is a collection of improvements in operational excellence inspired by CE in traditional manufacturing-focused companies. This list of concepts is designed to inspire creativity, spur improvements, and generate new ideas; it is not exhaustive.

Short loop Rs: material reduction strategy

- Refuse (R0): *Refuse* is a perfect example of a retention strategy aimed at reducing manufacturing costs. Not doing so is the cheapest way to save money. The maximum amount of material, energy, time, and, therefore, cost can be reduced. Challenge your process, ideally also upstream and downstream, for the potential elimination of surplus activities or resources.
- Reduce (R1): In the production process, you can reduce the process steps or challenge operating instructions (e.g. heating up the machine for a manufacturing process)? Energy consumption in manufacturing is often a significant source of CO_2 emissions and also a cost driver. Challenge all the different parts of your products and their manufacturing steps. Can you reduce material thickness? Packaging is also a good area to scrutinise: Can you reduce and optimise it?

Medium loop Rs: product or part use cycle extending and renewing strategies

- Reuse (R2): Manufacturing processes are set up to consider reusing support materials such as transport boxes, tooling, etc. Optimise your workplace with reuse in mind, either internally as narrowing R or with a wider approach with your value chain neighbours.
- Repair (R3): Maintenance and cleaning of machines can extend the life of your production machines.
- Refurbish (R4): The same is true when it comes to refurbishing of machines. Have you talked to your machine supplier to see if a machine can be overhauled and its life extended?
- Remanufacture (R5): The rework process of products and parts falls into the category of remanufacturing. Instead of disposing of components, can you remanufacture them to meet quality standards? Material usage is optimised.
- Repurpose (R6): In production, you can find ways to repurpose materials, e.g., packaging materials from your vendors to support your production process. This can also be applied to excess or out-of-spec production material, finding other uses

(but, of course, only after solving the reason for this surplus material to exist in the first place).

Long loop Rs: material processing strategies

- Recycle (R7): The use of recycled materials instead of new (virgin) materials has gained significant momentum in recent times. For example, many plastic parts can be manufactured to the same quality standards using recycled materials.
- Recover (R8): Consider heat recovery in your own manufacturing environment, converting unused potential or kinetic energy into electricity, or recovering valuable resources in customers' or society's disposable waste; this increasingly requires thinking outside the box.
- Remine (R9): Using primary resources or recycled material is the classic form of input. Resources lost in landfills, oceans, or the atmosphere can be a substitute in some cases where the purity of the material is not an issue or where extraction or purification is reasonably possible.

Of course, the above list sometimes oversimplifies the concepts, e.g., for R0, where refusing to use a certain chemical or process could result in costs of change or increased material costs. Subsequently, it can result in increased sales or increased profits, as the survey in Chap. 4 showed.

The journey towards advanced digitalisation factories will also support the drive toward CE [5]. The availability of data will help you make fact-based decisions to focus on relevant retention strategies throughout your value chain. Additionally, circular economy drives the need to 'know' the entire lifecycle of your product and further challenge the status quo. A digital representation of your physical product helps significantly. This digital representation is based on a solid understanding of your processes and data. CE will further enhance the need for various track-and-trace technologies for product identification. As Budde et al. [5] stated, identification and asset tracking are a relevant use case for any smart factory. The focus was in the past more towards traceability for quality reasons or even for recall purposes. CE thinking will further enhance the case for track-and-trace technologies for an enlarged purpose and increase the relevance of implementation in any industrial company.

Therefore, the combination of **Lean Management** (stable processes), **Digitalisation** (data, technologies) and **Circular Economy** (sustainability) will bring operational performance to the next level. We believe that if an operational excellence performance programme is based on all three management concepts, operational excellence is addressed holistically and comprehensively.

We consider it important to highlight the priority setting. Our recommendation for your operational programme is first Lean Management, enhance it with Digitalisation as necessary, and continuously challenge and evolve your practices through Circular Economy (see Fig. 19.8). This 'Lean-Digital-Green' framework is often advocated not only for optimising profitability but also for increasing the robustness and resilience of a company.

Section 3.1.6 made the connection between *Green* through application of Rs, empowered by *Lean Management*'s focus on waste avoidance, resulting in specific

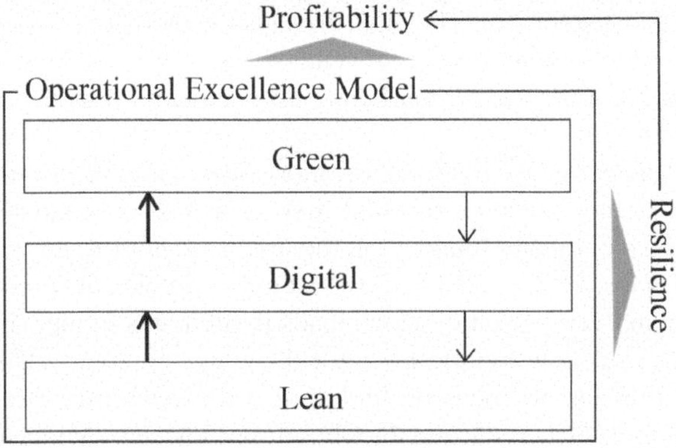

Fig. 19.8 Operational excellence model 'Lean-Digital-Green'

value propositions for better operational excellence for the different categories of waste and Lean Management approaches (see Table 3.2). Consequently, enhancing the existing waste and value perspective from *Lean* and *Digital* with a *green* point of view in operational excellence can further improve and stabilise your processes.

Stable, fact-based, and sustainable processes are the foundation of a resilient company [9]. CE will, in many cases, increase resilience through the following effects of the above mentioned CE strategies:

- Short loop Rs (material reduction strategys) can reduce the need for critical materials as well as energy consumption in manufacturing.
- Medium loop Rs (product or part use cycle renewing strategys) can make manufacturing less dependent on consumables, spare parts or availability of new machinery.
- Long loop Rs (material processing strategys) offer the potential to reduce the dependence on virgin material input through closed-loop internal recycling, as well as lower energy needs. Rethinking waste for recovery can further reduce the need for externally provided energy, and reduce criticality of waste disposal services.

In summary, you can conclude that a 'lean-digital-green' industrial company will increase your profitability and also be resilient to any possible future change. This strategic direction is key to driving financial performance and ensuring long-term stability.

19.3 External Circularity Focus: Value Proposition of Products, Services, and New Business Models

As a complementary strategy to the above analysis with an internal perspective, a mainly cost-driven direction, CE can be applied to products, services or even business models in a new ecosystem [10]. The main driver is the increase of volume of existing products, new products, or services, or revenue through new business models. These changes must be accompanied by marketing communication strategies and product repositioning strategies.

In fact, when discussing CE strategies, most often circular *product* concepts or even new business models come to mind first. A well-known case in Switzerland is FREITAG [11], a company founded on the unique idea of *Repurposing* old truck tarps (also see Chaps. 11.2.2 and 17.5.1). Naturally, circular business opportunities like this are rare and very specific, which leads participants in the CE workshops to deny the existence of such opportunities for their own company. This may or may not be true, but this narrow focus on finding that *a breakthrough circular product innovation* is not useful when starting the CE journey. Instead, a wider approach to CE should be adopted, challenging the company's business model from multiple angles. Especially, the external perspective can be challenging and time consuming due to its width and ambiguities in motivation and cooperation of value chain partners, including the customer. However, the examples discussed in the previous chapters demonstrated the significant potential for companies to excel in all the strategic priorities discussed in Sect. 19.1.

As a starting point for such discussions, we derived the following learnings from our research.

Short loop Rs: material reduction strategies

- Refuse (R0): *Refuse* can spark innovation for a completely new approach to your customers' needs. The maximum amount of material, energy, time, and therefore cost can be reduced. Challenge both your current offerings and their realisation. Can you, for example, switch from a physical to a virtual solution? This is definitely an extreme case, but should spark some thought.
- Reduce (R1): Similarly to internal efficiency strategies, reducing the amount of materials used in offered products can be a CE and a cost benefit. Consider also reducing the variety of materials and components, reducing energy, consumables, or the need for replacement parts during the product's use phase. Can you reduce or eliminate certain parts with a smarter design?

Medium loop Rs: product or part use cycle extending and renewing strategies

- Reuse (R2): User-specific features and extreme individualisation can be detrimental to *Reuse*. This can even be true for electronic or virtual products if (intentionally or unintentionally) not designed for resale or passing on to someone else. On the other hand, you might be able to generate additional income from a managed *Reuse* process that you can provide to users, e.g., quality control.
- Repair (R3): The offer of repair and maintenance is often highly valued by customers. Alternatively, especially in the absence of capacity or RL, repair manuals and spare parts can be made open source, creating a community of very loyal customers and sparking crowd knowledge for use extension, ultimately resulting in greater market share.

- Refurbish (R4): The offering of repair is often combined with a repair concept. If this is not an existing option, engage with other players in the value chain to build a wider network and share costs. For valuable or protected goods, implementing a buyback programme can be a strategic move. This approach not only ensures control over the second-life of the products but also potentially prevents competitors from offering similar refurbished goods.
- Remanufacture (R5): Are there components in your product that are of high value, but never break or are easily repaired? Check if it is possible to retrieve and reintroduce them into the manufacturing process. For products with a very long service life, check if recovery of spare parts from returned products can mitigate problems with electronic components' obsolescence.
- Repurpose (R6): Identify different uses for your product or some of its components. Or check if the functionalities of your product or your services could be realised by components or materials originating from other industries or uses.

Long loop Rs: material processing strategies

- Recycle (R7): Specify recycled content for the components you use and also for consumables. In addition, enable the recycling of components or consumables through design-for-recycling, for example, by making separation at End-of-Life (EOL) possible or by not mixing inseparable plastics.
- Recover (R8): It might be possible to tap into one of the current streams of waste and excess/lost energies. This goes beyond recycling. Question the form of energy your product needs: could it build on a different form of energy available in its environment? Or, thinking the other way round, could your product provide a form of energy to other processes or products in its environment, while in use or at EOL?
- Remine (R9): Lost resources can be an unexpected source of material input. Do any of the resources used for your processes or their precursors prevail in any of society's managed or unmanaged waste sinks or dump sites, landfills, waterways, oceans, or the atmosphere? Also, think of littering or wear and tear processes, to capture resources from. You could even solve an environmental problem by restoring an already lost resource to the material cycle.

Many of these ideas, due to the complexity of the change of CE in the value chain, will require the creation of partnerships, increasing transparency, or increasing customer engagement – these key enablers are discussed in Sect. 3.2.3.

Innovation projects for new products or redesigning existing products can benefit from early consideration of medium and long loop Rs, especially *Repair*, *Remanufacture* and *Recycle* as challengers and constraints.

19.4 Understanding Implementation Patterns

As presented in Sect. 4.4.2, our survey showed a decline in application of Rs from Reduce (R1) as the most common retention strategy to Remine (R9) as the least common strategy (see Fig. 4.21). On the other hand, our detailed analysis of R-cases of different companies in Part II found success examples also for the higher-numbered Rs. Apparently, companies find Short loop Rs more attractive than Long loop Rs, with the exception of Repair (R3) among unconscious CE practitioners. Also, Recycling (R7) among conscious CE practitioners is far outside the linear regression line. We found that these approaches are related to the specific business model and the motivation of the companies. In Sect. 4.4.5 we analysed the relationship between Rs and the different enablers. This analysis was complemented by creating a matrix of drivers relevant for each R above in Sect. 19.1.1.

In combination, this resulted in a rather surprising learning from the researched R-cases: the absence of a generic clear pattern for the implementation of retention strategies for every company, at least at first sight. The companies did not appear to undergo an evolution of options to remodel the company strategy through CE, e.g., starting with Recycle (R7) and then work their way up the Value Hill for each Retention Option until Reduce (R1) and Refuse (R0), nor the other way around. However, after further analysis of the survey and case studies, we found distinct implementation patterns shown in Fig. 19.9.

For further details, refer to Fig. 19.2 and Fig. 4.25 in Sect. 4.4.5 as well as the summary of case studies in Fig. 19.3. Environmental reasons were mentioned for every R in high percentages. We consider this driver to be common and not specific enough to understand the different patterns. Therefore, in the following analysis, we will only address particularly low or high results behind the motivation of the companies.

Within the Short loop Rs, Refuse (R0) and Reduce (R1) are common for nearly all industrial companies, as the company has a clear understanding of the product and its market: its main business objective is to optimise the design and increase the volume of the product in the respective markets through proactive communication and positioning of the product. A second common focus is *cost reduction* (e.g. reduction of materials used). In addition to cost reduction, in many cases, a key motivation

CE Retention Strategy		Company Strategy		Who
Refuse (R0), Reduce (R1)	→	Change of products (growth) and process (cost reduction)	→	Industrial companies
Repair (R3), Refurbish (R4), Remanufacture (R5)	→	Industrial service focus and new services (growth)	→	Industrial companies
Reuse (R2), Repurpose (R6)	→	New business models (growth)	→	Startup companies
Recycle (R7), Recover (R8), Remine (R9)	→	Extend recycle business, new products / services / business models (growth)	→	Recycle companies, construction companies, material focused industrial companies

Fig. 19.9 Implementation patterns through CE

and focus is *reacting to the change* in the product requirements, summarised in the motivational factor 'business development'.

Repair (R3), presumably due to its long history, exists in a plethora of variations, from exclusive repairs by company-owned service personnel through partnerships in multi-brand repair companies or free markets of professional repair services, to open source of repair manuals and spare parts for users themselves – especially for durable products. Many companies, especially in the Business-to-Business (B2B) segments, have an established repair and service business. In many cases, this is not only economically viable but also highly attractive, typically yielding higher margins than product sales. This is potentially the most common realisation of a *value retention* focus. The survey supports this with the highest results in business development (M.F2), growing customer demand (M.F3) and economic demand (M.F4); we see this as a strategy for *value retention* with new services.

Refurbish (R4) can be seen in the same group as *Repair* but with an additional hurdle: It requires companies to have access to their used products when their customers intend not to use them anymore. Although *Repair* is a pull process from the manufacturer's perspective (customers asking for repair), manufacturers generally do not receive information about EOL of their products, unless they replace them with newer generations. Expanding service offerings from simply focussing on *Repair* (which aims to maintain the original function of the product) to a more comprehensive strategy involving *Refurbishing* can provide significant benefits. *Refurbishing* not only ensures improved functionality or reduced energy or resource consumption, but also improves customer value and satisfaction. Similarly, Remanufacture (R5) further builds on this strategy for *value retention*. When it comes to enablers, this group R3 to R5 has the highest score for the availability of technical know-how in-house (see Fig. 4.27), showing potential for differentiation and securing market share through knowledge and patents.

Unlike the industrial service-orientated focus of Repair (R3) to Remanufacture (R5), the retention options Reuse (R2) and Repurpose (R6) require a change in focus towards the development of a *new business model*. Although *Reuse* as a concept is straightforward and easy to understand, its business logic often differs significantly from traditional models. For example, the production and sale of food containers represent a typical manufacturing and sales operation. However, the circulation of these containers involves an entirely different set of competencies, such as tracking, collecting, cleaning, testing, storing, and redistributing. We found that most companies that successfully establish a *Reuse* business are start-ups or new players in what used to be take-make-dispose markets. Therefore, the economic reasons shown in Sect. 4.4.5 in Fig. 4.30, where the combined score of *business development (M.F2)* and *economic reasons (M.F4)* are relatively high, are the economic potentials identified for the newly entered market. The attractiveness for established companies to introduce *Reuse* seems to be low; outside of specialised market places for reuse, only a few examples exist for companies starting to engage in *Reuse* in their own market (see Part II Sect. 7.1). This assumption is also supported by the very low score for future plans to introduce *Reuse* in Fig. 4.22. In the case of *Repurpose*, the examples we analysed were often niche products, e.g., furniture made from repurposed components or existing products because the source product is available at nearly zero cost, e.g., worn tyres used for embankments (see Chap. 11). In summary, this signifies the emergence of a new *business paradigm*. Existing market participants, or Original Equipment Manufacturer (OEM), are often observed trying to restrict new entrants by promoting product personalisation or exclusive usage rights, thus preventing opportunities for *Reuse*.

The remaining group of Recycle (R7), Recover (R8), and Remine (R9) is mainly focused on materials. Typically, it is not the company that creates the original products that implements these strategies itself. Instead, other market players specialise in material technologies, either for recycling to capitalise on residual value and create new source materials for other companies, or to tap into resources that are otherwise lost (R8) or already considered lost (R9). Interestingly, in the latter two R strategies, we found that industrial symbiosis is an important enabler, coupled with a notable emphasis on reducing production costs as another key motivator. In contrast, Recycle (R7) is carried out mainly by specialised recycling firms, with business growth as their central strategy. *Recover* and *Remine* businesses are more a niche business, where the key enablers are aspects of technology such as new (experimental) technologies, joint developments, or specific internal knowledge.

It showed that many companies are not comfortable adding another, currently unknown, market segment to their established market segment. Shifting to a new market is rare and moving into adjacent markets is also not very common. As an example, recycling companies typically do not start manufacturing products for end consumers, and fast-moving consumer goods companies typically do not leverage *Reuse* beyond their initial sales of new products. Bocken and Coffay coined the label 'gap exploiters' for new players who realise those circular business models where traditional companies are not active or are lagging behind [12, p. 1362]. This also proves that for a specific company, it might still make sense to reconsider existing cost structures and evaluate different business models, including niche markets or partnering in industrial symbiosis to rewrite the products' end-of-life.

19.5 Implementation Model for Industrial Companies

Now, we have summarised the relevant drivers, enablers, and results of the implementation of circularity. This understanding is crucial for discussing an implementation process for a specific company. This knowledge base is a critical part of an enhanced strategy CE. It is essential to begin by understanding the importance of CE for your business. This will significantly help to establish clear priorities and focus areas. Then you will also be able to achieve the best return and impact of a CE-oriented strategy.

This leads now to the following concrete questions: How does a specific company practically initiate this process? And how should the transformation process be managed?

Few process descriptions have been published on how CE can be implemented in a company [13, 14]. Knowing this, we have used our research to propose the following process.

1. Understanding of circular economy: Start by familiarising yourself and your team with the principles and concepts of CE. It is essential to understand how CE differs from the traditional linear economy and to understand the potential benefits it can bring to your business. In this initial phase, educating your employees is crucial. Only if you have a common understanding of the content of the CE model and its change, the transformation bears fruit.
2. Setting vision and goals: Shifting towards CE requires a clear vision and planning towards this vision. See and treat this shift as a journey. Set goals at different levels, financially or operationally, defining what you want to achieve. Furthermore, consider factors such as waste reduction, resource efficiency, extension of PLC and recycling goals. These goals will help guide your strategies and measure your progress.
3. Assess the current situation: Conduct a thorough evaluation of your company's current operations, supply chain, and business models. This step aims to gain an in-depth understanding of where your company is at this point. With this knowledge, you can identify areas where the principles of CE can be effectively integrated into your existing processes.
4. Develop a plan: Structure your future plan into three distinct sections: 'Products and services', 'Supply chain and operations', and 'Business models'. This separation will allow for a focused approach in each critical area of your business as you transition.
    a. Products and Services: Rethink your products on how to integrate a CE mindset into your products. Can you replace current materials with recyclable ones? Can you substitute and reduce materials according to the CE strategies? Also, consider additional services towards *Refurbish*, *Reuse* and other R-strategies.
    b. Supply Chain and Operation: Collaborate with your production team and suppliers to source materials more responsibly and focus on recyclability and reusability. Initiatives such as reducing material use can yield immediate results. Implement practices such as supplier take-back programmes RL, and the use of recycled input. Also, consider local sourcing to minimise transportation and its associated emissions.
    c. Business models: CE opens new avenues for innovative business models. The different R-strategies stipulate new ways of thinking about your business. You can add this new thinking and come up with new approaches that lead to fundamentally new ways of conducting business.
5. Engage stakeholders: Get the approval and participation of employees, suppliers, customers, and other relevant stakeholders. Foster a culture of innovation. Encourage employees to think creatively and find new ways of thinking.
6. Measure and communicate progress: Regularly track and measure your progress towards achieving your CE goals. Monitor Key Performance Indicators (KPI)

such as waste reduction, resource efficiency and recycling rates. Communicate your achievements transparently through different channels to all your stakeholders.

Through our research on the drivers and enablers of the various retention strategies from R1 to R9, we have established a solid foundation that can guide companies in their strategic approaches. In addition, we identified some additional good practices. Over the past three years, we also assisted various companies chart their course toward CE. These projects helped us refine and validate our approach through active participation in the development and implementation processes of the strategy with different companies. The learnings gained from both our research and these practical engagements have detailed and expanded the actual implementation models.

Our path of implementation is designed to include all businesses, with particular emphasis on Small and Medium-sized Enterprise (SMEs). Therefore, our approach is designed to be very concrete and can be managed with limited time and resource spending. Our Rapperswil Circular Economy Model is organised into six distinct steps, as depicted in Fig. 19.10.

19.5.1 Understand Your Current Situation

It all starts with understanding your current business situation with respect to circular economy (step 1). This step is crucial and sets the foundation for any advancement. As highlighted above, it is vital to involve a wide range of employees in this discussion. When starting such a process, circular economy is often not well understood, and in many cases it is reduced and oversimplified to only product repair or recycling. You should take the time to discuss *what a circular economy means*, as this understanding will be the basis for further progress.

Select Your Drivers

The first step is to assess which drivers are relevant to you. Go through the generic drivers for circularity and select those that are important to your business. Figure 19.11 gives an exemplary response from a company that selected eight drivers relevant to its transformation.

With the findings of our research, we were able to group generic retention strategies for different types of business (see Figs. 19.4 and 19.9). These generic categorisations will guide a first selection of your relevant Rs. The specific relevance of Rs will vary depending on your business's operational model. In our case example, the case company selected four Rs as the most relevant (see Fig. 19.12).

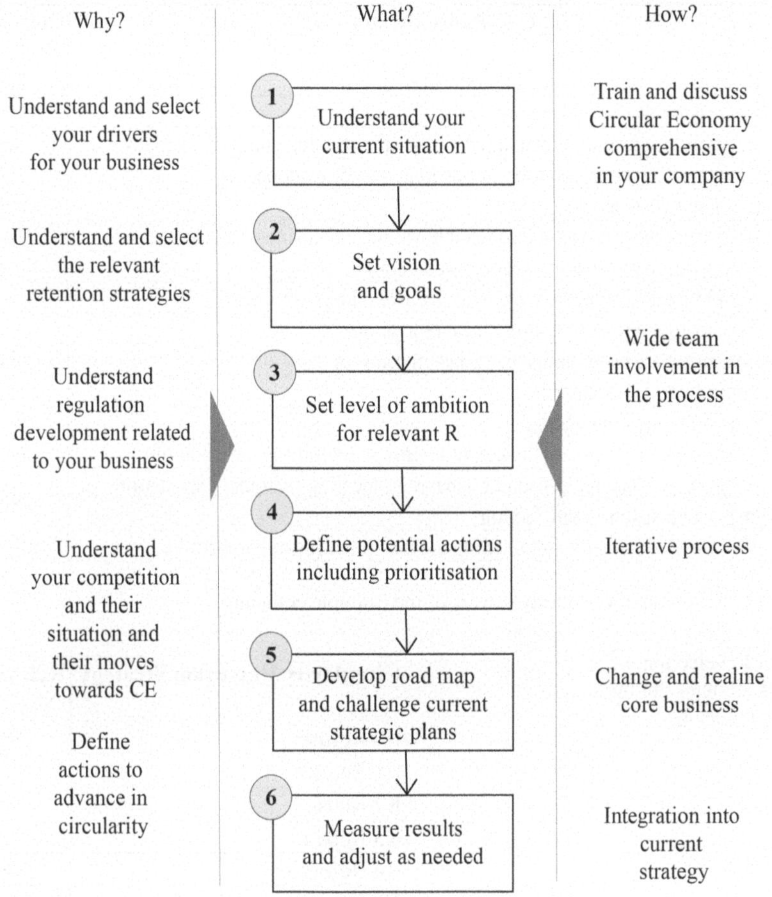

Fig. 19.10 The Rapperswil Circular Economy Model

Pair Your Rs and Drivers

Following this case-specific selection of drivers and Rs, you need to define which specific drivers influence which R in your company. We know from our research that not all drivers have the same impact on each R. This insight and your business understanding will guide the selection of which drivers influences which R (yes or no - answer).

As we have highlighted before, we propose to separate your strategic plans into two dimensions, namely *operational excellence* and *product/service/business model innovation*:

– **Operational Excellence:** Typical drivers are decreasing the resources needed (D.E1) or cost savings (D.F2). When you assess the situation, the typical team

Circularity Drivers		Relevant? Yes / No
D.E1	Decreasing resources used in production	Yes
D.E2	Decreasing product's footprint	No
D.E3	Resilience to resource-related risks (resource scarcity, supply challenges …)	Yes
D.F1	Counteracting price volatility of raw materials or components	Yes
D.F2	Realising cost-savings	Yes
D.F3	Increase competitiveness through innovation and business development	Yes
D.I1	Complying with legal requirements and norms	Yes
D.I2	Avoiding punitive/penalising measures from the government	No
D.I3	Anticipating or meeting future legal obligations	No
D.I4	Governmental tax benefits, subsidies, incentives	No
D.R1	Being socially responsible	No
D.R2	Promoting systemic change	No
D.S1	Green image (improving the corporate image, legitimacy, or credibility)	Yes
D.S2	Adapting to changing consumer behaviour, meeting customer expectations, or increasing customer satisfaction	No
D.S3	Meeting the standards or following the best practices of competitors	Yes

Fig. 19.11 Selection of circularity drivers of the example company

Fig. 19.12 Selection of retention strategies of the example company

Circularity Retention Strategy	Relevant? Yes / No
R0 Refuse	No
R1 Reduce	Yes
R2 Reuse	No
R3 Repair	Yes
R4 Refurbish	Yes
R5 Remanufacture	Yes
R6 Repurpose	No
R7 Recycle	No
R8 Recover	No
R9 Remine	No

compromising people from production, logistics, and facility management. They need IT support.

– **Product/Service/Business Model Innovation:** Here you change products, services, or business models to adapt to changing customer behaviour (D.S2). Team members are often coming from R&D, product management, sales and marketing.

Figure 19.13 represents the pairing of Rs with relevant drivers for each of the two dimensions. This separation enables companies to start with their internal processes and leverage CE strategies for the first low-hanging fruit improvements. It

Actual Situation for Operational Excellence

	Circularity Drivers	Relevant? Yes / No	Operational Excellence R1 Reduce	R3 Repair	R4 Refurbish	R5 Remanu-facture
D.E1	Decreasing resources used in production	Yes	✓	✗	✗	✗
D.E3	Resilience to resource-related risks (resource scarcity, supply challenges …)	Yes	✓	✗	✗	✓
D.F1	Counteracting price volatility of raw materials or components	Yes	✓	✗	✗	✗
D.F2	Realising cost-savings	Yes	✓	✓	✗	✗
D.F3	Increase competitiveness through innovation and business development	Yes	✗	✗	✗	✗
D.I1	Complying with legal requirements and norms	Yes	✓	✗	✗	✗
D.S1	Green image (improving the corporate image, legitimacy, or credibility)	Yes	✓	✗	✗	✗
D.S3	Meeting the standards or following the best practices of competitors	Yes	✓	✗	✗	✗

Actual Situation for Product/Service/Business Model Innovation

	Circularity Drivers	Relevant? Yes / No	Product/Service/Business Model Innovation R1 Reduce	R3 Repair	R4 Refurbish	R5 Remanu-facture
D.E1	Decreasing resources used in production	Yes	✗	✗	✗	✗
D.E3	Resilience to resource-related risks (resource scarcity, supply challenges …)	Yes	✗	✗	✗	✗
D.F1	Counteracting price volatility of raw materials or components	Yes	✓	✓	✗	✗
D.F2	Realising cost-savings	Yes	✗	✗	✗	✗
D.F3	Increase competitiveness through innovation and business development	Yes	✓	✓	✓	✓
D.I1	Complying with legal requirements and norms	Yes	✗	✗	✗	✗
D.S1	Green image (improving the corporate image, legitimacy, or credibility)	Yes	✗	✓	✓	✓
D.S3	Meeting the standards or following the best practices of competitors	Yes	✗	✓	✓	✓

Fig. 19.13 Defined impact of circularity drivers for resilience strategy

also gives focus to the appropriate teams and challenges the path towards 'internal' and 'external' perspectives simultaneously.

Calculating the Actual Level of Circularity

For the fields of impact, that is, the combinations of drivers and Rs that you selected, you can now calculate your actual circularity (CE) score. You need to assess the dimensions 'Time' (Time Assessment) and 'Impact' (Impact Assessment).

We propose a method that is as fact-based as possible.

Time Assessment

For the Time Assessment factor, you rate yourself with (1) when you have just started. (5) you select when you have been driving your programme for 3 years and a (10) you rank yourself when you have been implementing your actions already for 5 years or more. It should be mentioned that you can choose numbers from 1 to 10, the distinct definition of (1), (5), or (10) gives you some guidance.

Impact Assessment

For the factor Impact Assessment, we propose the following levels: (1) no impact yet, (5) 50% of impact reached, (10) full impact realised.

Circularity

To calculate the combined number of your actual situation for circularity (CE), you multiply Time Assessment by Impact Assessment. If you do this for the dimension 'Operational Excellence', you get the values for Operational Excellence Circularit, and if you do this for the dimension 'Product/Service/Business Model Innovation' you get the values for Product/Service/Business Model Innovation Circularity.

By multiplying the two factors 'Time Assessment' and 'Impact Assessment', you will quantify your actual level of circularity for operational excellence (OE-CE) and product/service/business model innovation (PM-CE) (see Fig. 19.14).

Normalization

For comparison reasons, you need to normalise the figures. Divide the calculated circularity score and divide it by the number of relevant drivers, multiply by 100 and you will get the normalised score in percent. Calculate your normalised percentages for Operational Excellence Circularity (sum of all OE-CE / number of circularity drivers x 100%) and Product/Service/Business Model Innovation Circularity (sum of all PM-CE / number of circularity drives x 100%). Note that the situation for *operational excellence* and for *innovation of the product/service/business model* can be different for each R. However, we recommend that they be displayed on a combined chart. This will help you visualise the current state, identify gaps, and uncover potential opportunities across both dimensions (see Fig. 19.15).

Actual Situation for Operational Excellence

	Circularity Drivers	Relevant? Yes / No	Operational Excellence Level for Circularity (OE-CE) R1 Reduce	R3 Repair	R4 Refurbish	R5 Remanu-facture
D.E1	Decreasing resources used in production	Yes	25			
D.E3	Resilience to resource-related risks (resource scarcity, supply challenges …)	Yes	9			80
D.F1	Counteracting price volatility of raw materials or components	Yes	1			
D.F2	Realising cost-savings	Yes	30	70		
D.F3	Increase competitiveness through innovation and business development	Yes				
D.I1	Complying with legal requirements and norms	Yes	100			
D.S1	Green image (improving the corporate image, legitimacy, or credibility)	Yes	20			
D.S3	Meeting the standards or following the best practices of competitors	Yes	70			
		OE-CE absolut (total)	255	70		80
		OE-CE (normalised)	36%	70%		80%

Time Assessment (TA): (1) just started, (5) implemented 3 years ago, (10) implemented > 5 years ago
Impact Assessment (PA): (1) no impact yet, (5) 50% of impact reached, (10) full impact realised
OE-CE = TA x PA, Normalisation calculated considered number of circularity drivers, normalized on 100%

Actual Situation for Product/Service/Business Model Innovation

	Circularity Drivers	Relevant? Yes / No	Product/Service/Business Model Innovation Level for Circularity (PM-CE) R1 Reduce	R3 Repair	R4 Refurbish	R5 Remanu-facture
D.E1	Decreasing resources used in production	Yes				
D.E3	Resilience to resource-related risks (resource scarcity, supply challenges …)	Yes				
D.F1	Counteracting price volatility of raw materials or components	Yes	50	50		
D.F2	Realising cost-savings	Yes				
D.F3	Increase competitiveness through innovation and business development	Yes	30	1	1	1
D.I1	Complying with legal requirements and norms	Yes				
D.S1	Green image (improving the corporate image, legitimacy, or credibility)	Yes		10	10	10
D.S3	Meeting the standards or following the best practices of competitors	Yes		80	80	80
		PM-CE absolut (total)	80	141	91	91
		PM-CE (normalized)	40%	35%	30%	30%

Time Assessment (TA): (1) just started, (5) implemented 3 years ago, (10) implemented > 5 years ago
Impact Assessment (PA): (1) no impact yet, (5) 50% of impact reached, (10) full impact realised
PM-CE = TA x PA, Normalisation calculated considered number of circularity drivers, normalised on 100%

Fig. 19.14 Actual situation of the example company's circularity

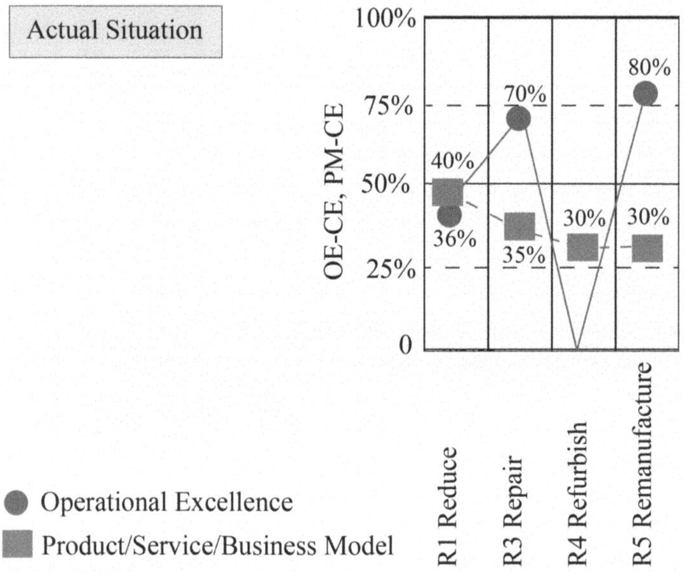

Fig. 19.15 Actual situation of the example company's circularity in a diagram

19.5.2 Set Vision and Goals

The evaluation of the situation of drivers and related Rs is the basis for defining the vision and goals toward CE (step 2). Given the importance and increased relevance of this topic, a long-term vision of your company is relevant and plots the way forward. Be ambitious and as concrete as possible! You need to decide whether you want to be a leader and become widely recognised for CE or if you prefer to follow industry standards for your company. This vision must be collectively embraced by the entire management team, as it will significantly influence the strategic path of your company. Typically, companies focus on a limited number of Rs where they see the greatest potential and feasibility ([15], p. 296). We recommend starting with one, with a maximum of three Rs.

19.5.3 Define Level of Ambition for Relevant Rs

The agreed-upon vision lays the foundation for step 3 in our model. With a clear understanding of your current situation and a well-defined vision in place, you are now ready to move forward.

Set Direction

Where do you want to go from here? For example, you might aim to significantly strengthen the *Remanufacturing* service within your existing service business to meet your customers' growing need to extend the useful life of your existing product installations. Our research study has found an indication of the relevance levels of certain drivers and specific retention strategies. In addition, different enablers for the change to CE were also identified and prioritised. Consider the study and research in Chap. 4 as a source of support and guidance to define your ambitious level; not all findings may apply one-to-one to your business.

Identify Dependencies

Furthermore, when defining your level of ambition, you should be aware of the dependencies that could influence your level of ambition for a certain Rs. Some dependencies are universal, e.g., for any remanufacturing to take place, a RL process is needed. Otherwise, no products or components can circulate back into the material flow of the factory. In Chap. 16 the RL had to be installed from scratch, which caused a lot of effort, leading to the ambitions towards the highest possible level of value retention.

However, there might also be a business case for a third party company to establish such a RL flow instead of OEMs, e.g., to reach the economy of scale. If such services exist, you need to understand who is offering them and potentially partner with them. To understand and navigate these dependencies early in the process, it is vital to set your goals.

Other dependencies are more tied to the specifics of your circular setup, e.g., an industrial symbiosis setup for *Recover* (R8). These networks must exist; creating them independently will be difficult. The robustness of these networks plays a key role in determining your level of ambition.

Your Value Hill Position

Throughout our research, we built on the concept of the Value Hill (see Sect. 3.1.2), a strategic framework that proposes a categorisation and valuation based on the lifecycle phases of a product: preuse, use, and post-use. The Value Hill model then proposes strategies and actions to maximise retention of value and to increase circularity, e.g., re-use of components is preferred over recycling of materials. This understanding can help define your level of ambition. Your position for specific products on this 'hill' defines your option towards CE. To better define your position, consider the following questions:

1. Where are you on Value Hill? Assess your position from different angles: from the perspective of your output (e.g., your product as a component in someone else's machine); from the final use perspective (the machine that incorporates your component); and from the end-user value perspective (what the machine with your component is producing).
2. Check your neighbouring Value Hill positions: Do you have access or partnership to your suppliers (e.g., component manufacturer) or your customers?

a. Can you establish returns within your own space?
b. Can you create a short loop with your customer?
c. Can you create a short loop with your supplier?

Discussing your position on the Value Hill within your management team is a powerful tool. It not only fosters a deeper understanding of the principles of CE among your team members, but it also encourages them to think beyond the existing boundaries of your business.

Challenge Your Ambition

Drawing on our experience, an additional simple method can help to significantly open up the discussion. Table 19.1 illustrates the provoker sentences for each R aimed at creating agreement or opposition and finding an appropriate resolution to the conflict. Extreme positions are a good base for defining realistic levels of ambition because they open up the horizon. Especially the increased focus on CE creates tension in many companies, because it is an additional partially new requirement for many department managers.

Calculate Your Ambition Level Outlook

Through all these discussions, companies can now complement their actual situation with their level of ambition (see Fig. 19.16) based on a detailed and solid discussion of each driver for the relevant R. We use the same methods as we have used to define your actual situation for your circularity (see step 1). The only small change is that the evaluation criteria or levels for Time Assessment (TA) and Impact Assessment (PA) have changed to the nature of this analysis. The outlook or ambition level needs here a change. Specifically, we propose for TA the following levels: (1) implemented in 5 years, (5) implemented in 3 years, and (10) implemented this year. For PA, we see the following levels: (1) first impact (e.g., one process, one product, one plant), (5) 50% scope, (10) full scope (e.g., all processes, all products, globally all plants). A detailed and fact-based assessment of your level of ambition will also help you align actions to get approval. Therefore, the effort is well spent.

Comparing this level of ambition with the actual situation visualises to you where the biggest change is aspired (see Fig. 19.17 for 'Operational Excellence' and Fig. 19.18 for 'Product/Service/Business Model Innovation').

Table 19.1 R provokers to generate conflicts about ambitions

R	Positive challenge	Negative challenge
Refuse (R0)	We can eliminate the product or service through a virtual substitute	Not a single aspect of this product or service has any potential to be removed
Reduce (R1)	We can reduce the use of materials and energy by at least 90%	We have already exploited every smallest reduction potential in the manufacturing or use of our product
Reuse (R2)	We can buy back and resell our machines to other customer segments	We do not know the status of the used machine and cannot guarantee its function
Repair (R3)	We design the product so that parts that are critical to failure can be repaired	It is too time consuming, replacing the complete product is much easier
Refurbish (R4)	The design of our product allows us to replace the control unit, then the machine runs faster	We would cannibalise the sales of our new product line
Remanufacture (R5)	All main modules of our product are designed for remanufacturing	The need for specialised skills and training for remanufacturing will increase our labour costs
Repurpose (R6)	Our machines or its modules can serve in a different application, e.g. in someone else's machines	Our modules are too specific for anyone to use them for anything other than what we do
Recycle (R7)	All the plastics used in our product can be recycled	The specifications of plastics components and using inseparable composites are non-negotiable
Recover (R8)	We can recover energy from unused waste heat or non-recyclable scrap materials	We do not want to engage in any new processes or invest in any such technology as energy costs might fluctuate
Remine (R9)	We can reclaim and make use of waste materials lost in the ocean or in landfills	We don't want to invest in relationships and technologies but stick to virgin materials use

19.5.4 Define and Prioritise Actions

With these strategic changes in mind, concrete actions that align with the ambition must be defined (step 4). Let us illuminate this with an example.

During our discussion with a case company, a specific implementation topic arises. This example highlights the additional thinking and topics that emerge when we focus on CE. And particularly brings totally new aspects to the business that you have never thought of before. The discussion circulated around modularisation of a product and service that enable the removal of personalisation for remanufacture

Ambition Level for Operational Excellence

	Circularity Drivers	Relevant? Yes / No	Operational Excellence Level for Circularity (OE-CE) R1 Reduce	R3 Repair	R4 Refurbish	R5 Remanu-facture
D.E1	Decreasing resources used in production	Yes	100			
D.E3	Resilience to resource-related risks (resource scarcity, supply challenges …)	Yes	50			80
D.F1	Counteracting price volatility of raw materials or components	Yes	30			
D.F2	Realising cost-savings	Yes	100	80		
D.F3	Increase competitiveness through innovation and business development	Yes				
D.I1	Complying with legal requirements and norms	Yes	100			
D.S1	Green image (improving the corporate image, legitimacy, or credibility)	Yes	40			
D.S3	Meeting the standards or following the best practices of competitors	Yes	100			
		OE-CE absolut (total)	520	80		80
		OE-CE (normalized)	74%	80%		80%

Time Assessment (TA): (1) implemented in 5 years, (5) implemented in 3 years, (10) implemented this year
Impact Assessment (PA): (1) first impact, (5) 50% of scope (10) full scope (eg. all process, globally)
OE-CE = TA x PA, Normalisation calculated considered number of circularity drivers, normalized on 100%

Ambition Level for Product/Service/Business Model Innovation

	Circularity Drivers	Relevant? Yes / No	Product/Service/Business Model Innovation Level for Circularity (PM-CE) R1 Reduce	R3 Repair	R4 Refurbish	R5 Remanu-facture
D.E1	Decreasing resources used in production	Yes				
D.E3	Resilience to resource-related risks (resource scarcity, supply challenges …)	Yes				
D.F1	Counteracting price volatility of raw materials or components	Yes	80	80		
D.F2	Realising cost-savings	Yes				
D.F3	Increase competitiveness through innovation and business development	Yes	30	1	1	1
D.I1	Complying with legal requirements and norms	Yes				
D.S1	Green image (improving the corporate image, legitimacy, or credibility)	Yes		10	10	10
D.S3	Meeting the standards or following the best practices of competitors	Yes		80	80	80
		OE-PM absolut (total)	110	171	91	91
		OE-PM (normalized)	55%	43%	30%	30%

Time Assessment (TA): (1) implemented in 5 years, (5) implemented in 3 years, (10) implemented this year
Impact Assessment (PA): (1) first impact, (5) 50% of scope (10) full scope (eg. all products, serives)
OE-CE = TA x PA, Normalisation calculated considered number of circularity drivers, normalised on 100%

Fig. 19.16 Level of ambition separated for operational excellence and product/service/business model

or reuse. In case of digital products or services, this could be offering a secure data wipe or storing user-specific information and usage data in different containers. This might be requisite for re-use of the product, because you want to erase the usage data, before you re-use the product again in a different format. This question with respect to CE for traced products is often not discussed and, therefore, overlooked. We know that these additional requirements can further increase the complexity of product and service design. But if this topic is not addressed early, some R strategies will not

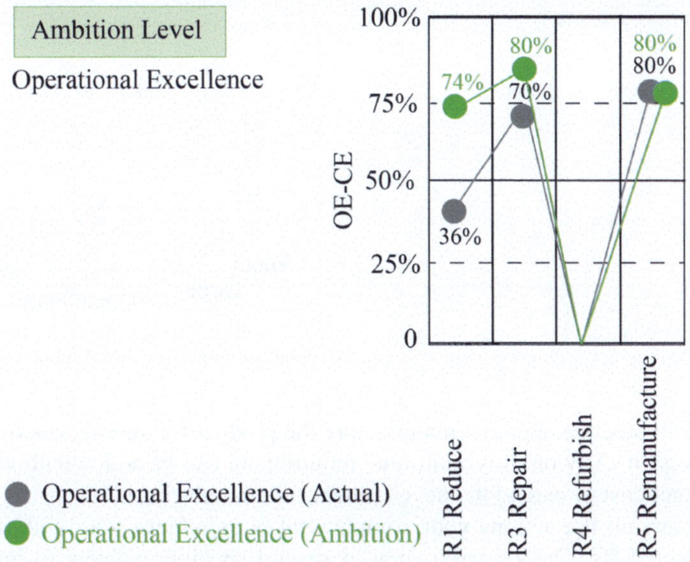

Fig. 19.17 Level of ambition for operational excellence

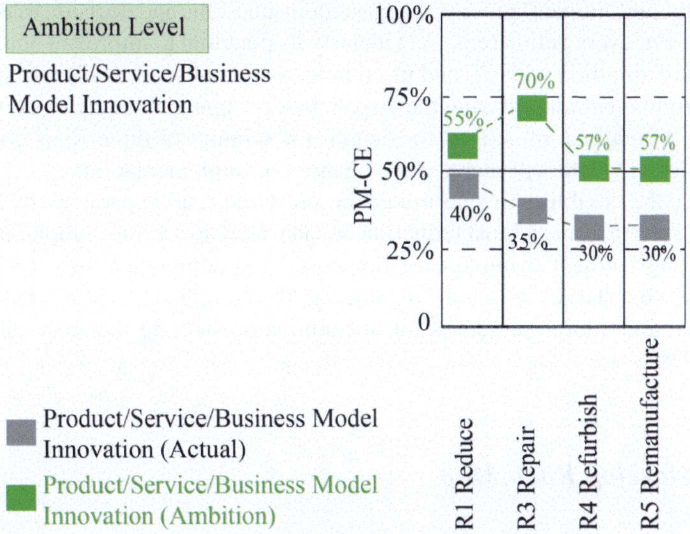

Fig. 19.18 Level of ambition for product/service/business model innovation

Fig. 19.19 Priority matrix

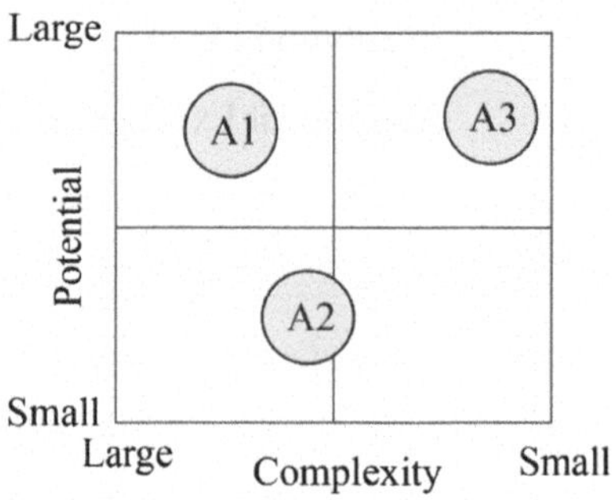

be possible in practice, e.g. to remanufacture the product for another customer. And if you address it early on, this additional requirement can be added without adding much product cost or extend the development cost massively.

Define now all the actions your company takes to achieve your ambition level for the selected Rs. Our research showed several specific enablers to implement circularity (see Sects. 3.2.3 and 4.4.4). They can help you as a reference to define your actions from a research point of view. The comprehensive lists of actions you have developed must then be prioritised. Prioritisation according to the dimensions of 'complexity' and 'impact' gives you a base for discussions and decision making (see Fig. 19.19). For every action (e.g., A1) identify its potential to improve your situation to the level of ambition with CE and its complexity to realise. Prioritise those actions in the top-right quadrant, offering the largest gain compared to complexity. Limiting yourself to a small set of actions for one or a maximum of three Rs is considered good practice [15] and will increase the chances of implementation.

You have the flexibility to determine your preferred scaling parameters. Potential can be assessed in terms of cost reduction or sales increase, while complexity can be gauged through project or implementation costs. You define then for your company perspective what 'large' or 'small' means, e.g. the complexity will be defined with project cost, and a large project is for this company anything that costs more than 500,000 EUR.

19.5.5 Develop Road Map

Step 5 is to align these actions into a road map for CE and integrate them into your strategy. Ideally, CE actions are integrated into your existing road map for your general management and company strategy. The better you align CE actions with your current strategic actions, the better and faster the results will be. If you need to redesign products, it is vital that all CE redesign requirements are linked to the

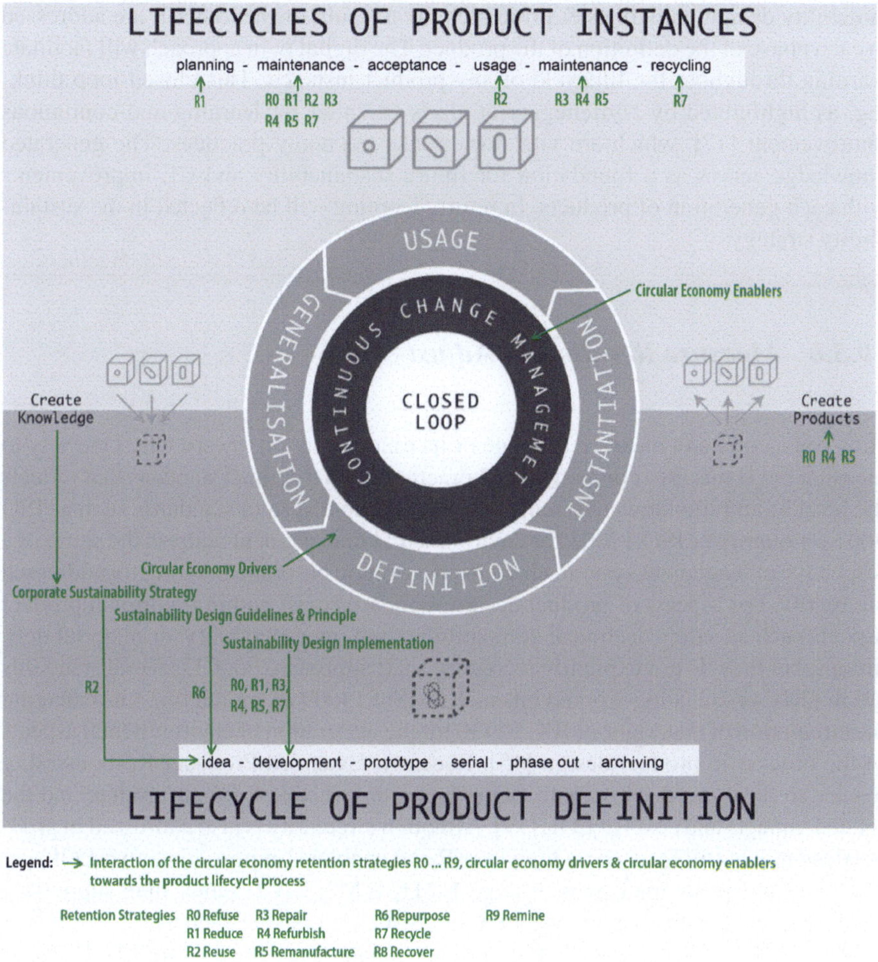

Fig. 19.20 PLM extended by contribution to Circular Economy and related Retention Options based on [17, 18]

already defined R&D-road map for your product portfolio. Realisation efforts will be significantly reduced if you integrate any running plans.

In this context, integration into the Product Lifecycle Management (PLM) system can help realise advantages, both in identifying improvements and in managing them [16], as presented in Sect. 3.1.5 to identify the value of waste. Specifically, the roadmap for CE benefits from a PLM view towards the innovation process (product definition lifecycle) and the lifecycle of the product (product instance lifecycle) to address Rs and their related actions at the right time (see Fig. 19.20). Strategic decisions must be aligned with elements of the corporate sustainability strategy, sustainability design guidelines & principles. As a result, the actions CE are addressed in each phase of the definition of the product. The digital twin approach will facilitate learning throughout the lifecycle of any product instance. This closed-loop thinking, as highlighted by Nyffenegger et al., is essential for learning and continuous improvement [17], which are vital for circular economy practices. The generated knowledge serves as a foundation for future sustainability and CE improvements with each generation of products. In return, learning will be reflected in the sustainability strategy.

19.5.6 Measure Results and Adjust

How can a company measure the state of its transformation toward CE? One way to measure the status is to compare actual practices with a defined standard that reflects the level of ambition toward circularity. Management system standards such as ISO 9001 for quality or ISO 14001 for environmental management address the setup of a respective management system, albeit from a generic basis and not directly addressed circularity. For aspects of product design, standards exist within traditional product aspects such as safety, technical compatibility and interoperability, or material definitions, but they do not explicitly recognise a CE approach. For CE several standards are in place or are being worked on, such as ISO 14009 as a guideline for managing the circulation of materials, or IEC 62430 for the integration of environmental aspects of the process in product design [15]. Recently, Extended Producer Responsibility has led to the concept of manufacturers being responsible for their product and the related impacts and costs of EOL [19]. Although circularity is also addressed in product design, it is still focused on long loop R and enablers for post-consumer EOL and materials and waste management (e.g., E.M1, E.M2, E.V2) rather than supporting Medium loop R with enablers we identified with high impact such as E.V1, E.T2 or E.M3 (see Sect. 4.4.4). Specific CE product standards are fewer, with EN 4555x as a promising addition for material efficiency. More development is needed to make companies with CE strategies benefit more compared to traditional Linear Economy (LE) focused companies.

In addition to measuring against norms or standards, several models and tools for measuring circularity have been developed in the past, quantitative and qualitative, for nano, micro or macro areas (see [20] for an overview). In this list by Valls-Val et al. several tools were found to evaluate products, but it was noted that research on 'indicators, metrics and methodologies to assess the implementation of CE is still under development' [20, p. 12].

We found the existing sets of metrics and approaches to be very complex, time consuming and too difficult to assess for SMEs that are starting their CE transition. The practical implementation and relevance of these approaches tend to be questionable. Depending on the focus areas and actions for CE, in later stages one of these complex tools might be valuable to maintain good standard practices or to calculate circularity indices for reporting purposes.

Our approach to measuring results is to measure progress toward level of ambition in 'Operational Excellence' and 'Product/Service/Business Model Innovations'. We want to have an impact. Therefore, we are proposing to track the status towards the defined actions. This concept is intended to provide a more of a 'guidance for change towards CE'than it is about exact arithmetically measuring the grade of CE realisation. The measurement of results and progress is also to allow adjustments during realisation and to stay on track.

Due to its novelty and dynamic environment for CE, it is important to continuously assess the situation, track the results and adjust accordingly (step 6). We recommend reassessing your actual level of circularity in both dimensions 'Operational Excellence (OE-CE)' and 'Product/Service/Business Model Innovation (PM-CE)' at least once a year. To conduct this assessment effectively, review your defined actions, and, collaboratively with your team, assess your current position relative to your level of ambition. Here, we highlight the fact that you measure results in relation to your set project goals, which represents your level of ambition for CE. Nothing is different from what you do for every other project. The open discussion will help you define your actual situation. Then mark it on the already proposed chart. This will show you the gap to your established level of ambition (see Fig. 19.21). This reflection will create transparency and allow you to reshape or refocus your actions.

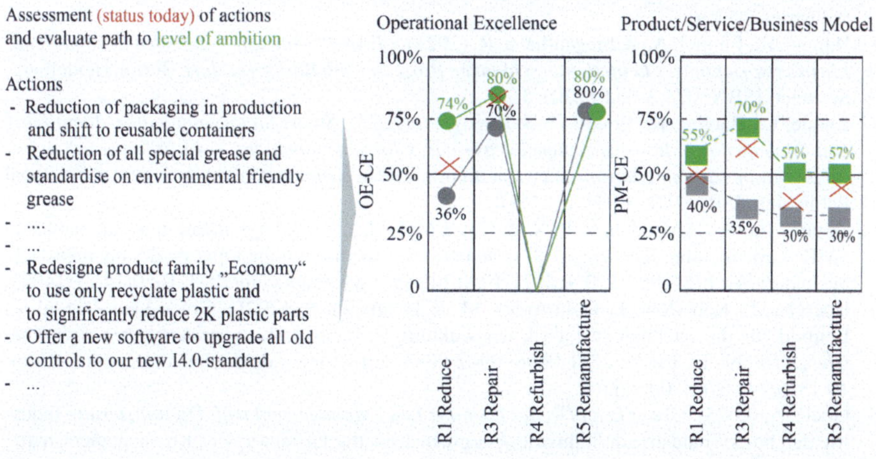

Fig. 19.21 Assessment and evaluation of path to level of ambition

19.5.7 Conclusion on Implementation

Our Rapperswil Circular Economy Model gives us a clear guide to a fact-based approach for a company-specific implementation approach based on research on drivers and enablers. This comprehensive foundation comes from practice.

We believe that only the fact-based alignment and transparency of the results will help drive CE in practice. Tangible results are the essential part of the change in the integration of new CE thinking. This fact base and constant tracking of the initiatives will also close the gap to your level of ambition.

Integrating CE into your strategy can have a significant effect on your business and CE. In this way, circularity is an essential part of business, not a separate issue that needs to be addressed separately.

References

1. Jensen, H. H. (2023). *8 ways the circular economy outperforms linear businesses* World Economic Forum. https://www.weforum.org/agenda/2023/03/8-ways-the-circular-economy-outperforms-traditional-business-models/.
2. Weetman, C. (2023). *7 reasons why the circular economy is better for business* Rethink Global. https://www.rethinkglobal.info/7-circular-economy-benefits-for-business/.
3. Baker, R. (2023). *The circular economy & how it benefits businesses - Terran industries* Terran Industries Pty Ltd.
4. Hänggi, R., Fimpel, A., & Siegenthaler, R. (2021). *LEAN Production – einfach und umfassend: Ein praxisorientierter Leitfaden zu schlanken Prozessen mit Bildern erklärt*. Berlin, Heidelberg: Springer. ISBN: 978-3-662-62701-3.
5. Budde, L., Hänggi, R., Friedli, T. & Rüedy, A. (2023). *Smart factory navigator: Identifying and implementing the most beneficial use cases for your company – 44 use cases that will drive your operational performance and digital service business*. Cham: Springer International Publishing. ISBN: 978-3-031-17254-0.
6. Queiroz, G. A., Alves Junior, P. N. & Costa Melo, I. (2022). Digitalization as an enabler to SMEs implementing lean-green? A systematic review through the topic modelling approach. *Sustainability, 14*, 14089. ISBN: 2071-1050. https://www.mdpi.com/2071-1050/14/21/14089.
7. Roesch, M., Köberlein, J., Goldmanns, M. & Hohmann, A. (2021). Green-Lean-Digital als Leitbild für die nachhaltige Fabrik der Zukunft. *Zeitschrift für wirtschaftlichen Fabrikbetrieb, 116*, 29–33. ISBN: 2511-0896, 0947-0085. https://www.degruyter.com/document/doi/10.1515/zwf-2021-0006/html.
8. Goelz, A. (2023). *Green Lean Digital Factory: Lean Management trifft Digitalisierung* Industrie.de. https://industrie.de/industrie-4-0/green-lean-digital-factory-lean-management-trifft-digitalisierung/.
9. Luban, K. & Hänggi, R. (Eds.). (2023). *Erfolgreiche Unternehmensführung durch Resilienzmanagement*. Berlin, Heidelberg: Springer. ISBN: 978-3-662-64023-4.
10. Christopher, L. (2023). *BMW's lean, green and digital difference* Automotive Logistics. https://www.automotivelogistics.media/supply-chain-management/bmws-lean-green-and-digital-difference/43943.article.
11. FREITAG. (2023). *One-off pieces made from recycled truck tarps*. FREITAG. https://www.freitag.ch/en.

12. Bocken, N. & Coffay, M. (2023). The circular experimentation workbench – a lean and effectual process. *Circular Economy and Sustainability, 3*, 1361–1383. ISBN: 2730-597X, 2730-5988. https://link.springer.com/10.1007/s43615-022-00239-w.
13. impacX team (2023). *3 Ways to Implement the Circular Economy In Your Business* impacX. https://impacx.io/blog/circular-economy-business/.
14. Elffers, D. (2014). A guide to implementing the circular economy in your business. *The Guardian*. ISBN: 0261-3077. https://www.theguardian.com/sustainable-business/how-to-implement-circular-economy-business.
15. Dworak, C. (2022). In Schwager, B. (Ed.) *CSR und Nachhaltigkeitsstandards* (pp. 289–299). Berlin, Heidelberg: Springer. ISBN: 978-3-662-64912-1.
16. Cholewa, M. & Minh, L. H. B. (2021). PLM solutions in the process of supporting the implementation and maintenance of the circular economy concept in manufacturing companies. *Sustainability, 13*, 10589. ISBN: 2071-1050.
17. Nyffenegger, F., Hänggi, R. & Reisch, A. (2018). In Chiabert, P., Bouras, A., Noël, F. & Ríos, J. (Eds.) *Product lifecycle management to support industry 4.0* (pp. 358–366). Cham: Springer International Publishing. ISBN: 978-3-030-01614-2.
18. Reike, D., Vermeulen, W. J. & Witjes, S. (2018). The circular economy: New or refurbished as CE 3.0? – exploring controversies in the conceptualization of the circular economy through a focus on history and resource value retention options. *Resources, Conservation and Recycling, 135*, 246–264. ISBN: 0921-3449.
19. Brown, A., Laubinger, F. & Börkey, P. (2023). *New Aspects of EPR: Extending producer responsibility to additional product groups and challenges throughout the product lifecycle* OECD Environment Working Papers 225. https://www.oecd-ilibrary.org/environment/new-aspects-of-epr-extending-producer-responsibility-to-additional-product-groups-and-challenges-throughout-the-product-lifecycle_cfdc1bdc-en.
20. Valls-Val, K., Ibáñez-Forés, V. & Bovea, M. D. (2022). How can organisations measure their level of circularity? A review of available tools. *Journal of Cleaner Production, 354*, 131679. ISBN: 09596526. https://linkinghub.elsevier.com/retrieve/pii/S0959652622012938.

Glossary

Amazon is an US-American, globally active online retailer with a broadly diversified product range.

Analytic Hierarchy Process is a method for organising and analysing complex decisions, using maths and psychology.

Apeel Sciences is a company that has developed an edible coating for fruits and vegetables. This coating, made from plant-derived materials like lipids and glycerolipids, aims to extend the freshness of produce by reducing oxygen penetration and moisture loss.

Artificial Intelligence is the theory and development of computer systems able to perform tasks normally requiring human intelligence, such as visual perception, speech recognition, decision-making, and translation between languages.

Asphalt is a material used to pave roads. It is a mixture of aggregates of different sizes mixed with a binder, typically long-chain hydrocarbons, derived from natural deposits or as residue from the refining of crude oil.

Asphalt Suisse Swiss Asphalt Industry Association.

Augmented Reality images produced by a computer and used together with a view of the real world ; achieved through the use of digital visual elements, sound, or other sensory stimuli and delivered via technology.

avec is a convenience store at highly frequented locations such as train stations or gas stations with an extensive range of fresh food and regional products [1].

Barrier are factors that hinder or obstruct the implementation and practice of retention option. It is important to note that enablers can be challenged by their own barriers, and their absence is often perceived as a barrier in itself. Barriers are discussed analogously to enablers in Sect. 3.2.3.

Bio-chemical recovery is a process that involves the extraction of energy through the breakdown of organic waste, utilising biological agents such as microbes or enzymes to facilitate bio-decomposition.

Biocharcoal is charcoal produced by pyrolysis of biomass.

A. Podleisek et al. (eds.), *Circular Economy: The Next Level of Company Success*,
https://doi.org/10.1007/978-3-031-66406-9

Biodiesel is produced from oil crops, lipidic wastes, and algae. The oils or fats of these feed-stocks are brought to reaction with short-chained alcohols with or without a catalyst.

Biogas is a gas mixture that is produced by the breakdown of organic matter in the absence of oxygen.

Bitumen refers to a substance produced through the distillation of crude oil. Bitumen is known for its waterproofing and adhesive properties and is commonly used in the construction industry, particularly for roads and highways.

Brezelkönig is a bakery chain in Switzerland that specialises in pretzel pastries [2].

Building Information Modeling is a process concerning generating and managing data of certain structures throughout their life cycles.

Burberry is a British luxury fashion brand, known for its iconic check pattern.

Business Performance Factor evaluation criteria that is generally used to measure the performance of a company or business unit. It is often made up of various performance indicators.

Business-to-Business is a form of transaction between businesses, such as one involving a manufacturer and wholesaler, or a wholesaler and a retailer. It refers to business that is conducted between companies.

Business-to-Consumer refers to the process of selling products and services directly between a business and consumers who are the end-users of its products or services.

Caffé Spettacolo is a coffee chain. They offer a variety of products, including organic and fairtrade coffee beans [3].

carbon capture utilisation is the process of capturing carbon dioxide and using it for various purposes, e. g. for producing synthetic fuels, chemicals or plastics.

Carbon footprint is a measure of the total amount of carbon dioxide emissions caused directly and indirectly by an activity or generated over the life stages of a product.

Caterpillar is a US company that provides equipment and machinery for construction and other industries, they introduced their first remanufacturing program in 1973.

Certification can either refer to a company fulfilling defined standards (e.g. management system certification such as ISO 14001) or to products fulfilling specifications defined in a standard (e.g. certification for installation in drinking water, like NSF/ANSI 61). Official certificates are usually issued after a third-party review, test, or audit by an accredited body.

Chemical recovery is a term often used in the context of specific industries where certain chemicals are recovered from the waste stream for reuse.

Chemicals recovery In contrast to chemical recover this process is to recover the chemicals directly from a previously lost state such as a sludge. It is more general and can refer to the recovery of any type of chemicals from a waste stream. It is often used in the context of waste management and recycling, where the goal is to recover as much of the valuable materials as possible and reduce the amount of waste that needs to be disposed of.

CHF Swiss Francs, the official currency of Switzerland. At the time of writing this book, 1 CHF was approximately 1.00 EUR or 1.15 USD.

Circular Economy (CE) is a system based on three principles: reduce, reuse and recycle. The aim is to keep the original material in the economy for as long as possible and implies reducing waste to a minimum.

Circular Economy Business Models are specific types of business models that aim to close the loop on resource use by reusing, refurbishing, and recycling materials and products to create value by using resources in multiple cycles.

Circularity refers to the adoption and integration of Circular Economy principles into various aspects of a business.

Closed-Loop Supply Chains refers to the design, operation, and control of the system to ensure maximum value from returning products of different breeds and quantities.

Collaborative Robot are a form of robotic automation built to work safely alongside human workers in a shared, collaborative workspace. In most applications, a collaborative robot is responsible for repetitive and menial tasks while a human worker completes more complex and thought-intensive tasks.

Combustion is the process of burning.

Composting is a biological degradation of organic matter usually containing various polymeric organic compounds. It is a dynamic process that includes the enzymatic breakdown of these polymers into soluble organic carbon and sugars.

Construction creating buildings (residential, commercial, public) and infrastructure (roads, waterways, train tracks, airports). Also referred to as public works, building construction, or civil engineering.

Consumer-to-Consumer is a business model whereby customers can trade with each other, typically in an online environment. Two implementations of C2C markets are auctions and classified advertisements. C2C marketing has gained in popularity with the arrival of the internet and companies such as eBay and Etsy.

Coop is a leading Swiss supermarket that offers a wide range of products [4].

COVID-19 infectious disease caused by the SARS-CoV-2 coronavirus.

Cyber-physical system is a system that combines computer-based algorithms with physical processes for real-time monitoring, control, and optimisation.

Cybersecurity refers to any technology, measure or practice for preventing cyber-attacks or mitigating their impact. Cybersecurity aims to protect the systems, applications, computing devices, sensitive data, and financial assets of individuals and organisations against simple and annoying computer viruses, sophisticated and costly ransomware attacks, and everything in between.

DACH region refers to the economic region of Germany (**D**eutschland), **A**ustria and Switzerland (country code **CH**).

Delphi today BorgWarner; BorgWarner is a leading global automotive supplier that designs and manufactures systems for mobility solutions.

DESSO is a company that manufactures high quality and innovative carpet tiles for office and commercial use. It was acquired by Tarkett in 2015.

Digitalisation is the use of digital technologies to change a business model and provide new revenue and value-producing opportunities; it is the process of moving to a digital business.

Driver typically refers to a factor or variable that has a significant and often causal influence on a particular phenomenon or process. Drivers play a crucial role in understanding and explaining the behaviour or outcomes of a system. In the context of CE and our study, drivers refer to the key factors or motivations that encourage Circular Economy options. The Drivers are described in Sect. 3.2.2.

E-bike also known as an electric bicycle, is a single-track vehicle with an electric engine.

e-waste is discarded Electrical and Electronic Equipment (EEE); it can potentially harm health and the environment if not properly disposed of, treated or recycled.

Eastern Switzerland University of Applied Sciences is the University of Applied Science. The university offers bachelor and master degree courses, as well as a continuing education programme in six departments [5].

eBay is an online auction and shopping website where people and businesses can buy and sell a wide variety of goods and services worldwide.

Ecodesign also known as Design-for-Environment, Sustainable Design or Green Design. It refers to the practice of designing products with special consideration for the environmental impacts of the product during its whole lifecycle.

Eileen Fisher is a US-based fashion brand that offers recycled and organic clothing.

Ekokem is a waste management company based in Finland; it has been acquired by Fortum and is now known as Fortum Waste Solutions.

Electrical and Electronic Equipment is defined as equipment which is dependent on electric currents or electromagnetic fields in order to work properly and equipment for the generation, transfer and measurement of such currents and fields.

Empa is short for Swiss Federal Laboratories for Materials Science and Technology (in German: Eidgenössische Materialprüfungs- und Forschungsanstalt), is a Swiss research institution for application-oriented materials science and technology.

Enabler is a factor, element, or tool that facilitates or enhances a particular process, outcome, or capability. Enablers provide support or resources that contribute to the success or efficiency of a system or operation. In the context of CE and our study, enablers are defined as making application of a retention option successful; for further detail, see Sect. 3.2.3.

End-of-life is the end of a product's useful life, i. e. when a product is no longer supported by its manufacturer, or when it is no longer able to meet the needs of its users. When a product reaches its EOL, it is usually disposed of.

Energy price refers to the cost of energy in various forms such as electricity, natural gas, kerosene, fuel oil, coal and more.

Energy recovery is a process which converts waste materials into usable heat, electricity, or fuel through a variety of methods including combustion, gasification and anaerobic digestion.

Enterprise Resource Planning supports all of a company's business processes. It contains modules for the areas of procurement, production, sales, asset management, human resources, finance, and accounting, etc., which are linked through a common database.

Environment, Social and Governance refers to a comprehensive set of rules for assessing the sustainable and ethical practices of companies.

Environmental Impact Points are a measure used in Life Cycle Assessment (LCA) studies to quantify the potential environmental impact of a product or service. The concept is based on the principle that all human activity has some impact on the environment.

Environmental Protection Agency is an agency of the United States federal government whose mission is to protect human and environmental health [6].

EUR Euro, the official currency within certain countries of the European Union. At the time of writing this book, 1 EUR was approximately 1.00 CHF or 1.15 USD.

European Union is an organisation of European countries which have joint policies on matters such as trade, agriculture, and finance.

Eutrophication is a process by which pollution from such sources as sewage effluent or leachate from fertilised fields causes a lake, pond, or fen to become overrich in organic and mineral nutrients, so that algae and cyanobacteria grow rapidly and deplete the oxygen supply.

Extended Producer Responsibility is a policy approach to make producers responsible for their products at the post-consumer stage of the life cycle, i.e. waste from consumer products and costs for its waste management.

FedEx is a multinational delivery services company.

Filson is a US-based manufacturer of outdoor clothing, accessories, and luggage.

FREITAG is a Swiss company famous for its bags and other items made of used truck tarps, car seat belts and bicycles inner tubes.

Gasification is the process of turning a solid biomass into usable gases.

GBP British Pound Sterling, the official currency of the United Kingdom. At the time of writing this book, 1 GBP was approximately 1.15 EUR or 1.25 USD.

GE General Electric Company (GE) is an American multinational conglomerate. It operates in several divisions, including aerospace, power, renewable energy, digital industry, additive manufacturing, venture capital, and finance.

General data protection regulation is a legal framework that sets guidelines for the collection and processing of personal information from individuals who live in and outside of the European Union (EU).

Georg Fischer AG is an industrial company, that offers products and solutions for the safe transport of liquids and gases, lightweight cast components and high-precision manufacturing technologies.

Global Warming Potential is a term used to describe the relative potency, molecule for molecule, of a greenhouse gas, taking account of how long it remains active in the atmosphere.

Green washing refers to the attempt by organisations to achieve a 'green imag' through communication, marketing and individual measures without having systematically embedded corresponding measures in the operational business.

GreenToGo is a service that provides reusable containers for take-out food.

Gross Domestic Product is the total monetary value of all goods and services produced within a country's borders in a specific time period, typically measured annually or quarterly.

H&M is a multinational fashion retailer, founded in Sweden.

Heat recovery is the process of collecting and re-using heat that would normally be lost. It refers to the use of heat energy that is released from some industrial processes and that would otherwise dissipate into the immediate environment unused.

Henkel AG & Co. KGaA is a German multinational company, known for its consumer goods and adhesive technologies.

Hewlett-Packard is a US company which manufactures PCs and printers.

Home Depot Inc. is an American multinational home improvement retail corporation.

Ikea is a Swedish furniture store.

Image is how a company, product or individual is perceived by others, e.g. as a green company protecting the environment or as a polluter.

Impact Assessment is a systematic process for identifying the potential positive or negative impacts (consequences) of activities, projects or policies. In this context, it is used to measure the effect realised.

Independent Aftermarket means the market composed of purchasers of systems, equipment and components but excluding the original equipment market and the original equipment service market.

Industry 4.0 refers to the fourth industrial revolution, characterised by the integration of digital technology into the industrial sector's fabric. It embodies the transition to smart factories and interconnected systems, fundamentally altering operational and innovation processes within industries. Central to this concept are technologies such as the Internet-of-Things (IoT) Artificial Intelligence (AI) big data, and automation.

Instant Pot is a revolutionary cooking appliance that has become the number one selling brand on Amazon.com, acclaimed for its exceptional customer ratings.

Institut für Wissen, Energie und Rohstoffe Zug WERZ is a competence center for energy and resource efficiency at Eastern Switzerland University of Applied Sciences (OST) [7].

Institute for Materials Science and Plastics Processing is a research facility of the OST – Eastern Switzerland University of Applied Sciences for component design and production in plastics technology.

Institute of Product Development and Design in German 'Institut für Produktdesign, Entwicklung und Konstruktion', a research and services institute at Eastern Switzerland University of Applied Sciences (OST) deals with all aspects of the product lifecycle, especially innovation and development, manufacturing, supply chain, after sales; it is located at the campus in Rapperswil-Jona [8].

Interface is a carpet manufacturer and commercial flooring company from the US, they launched their first circular economy 'Mission Zero' approach in 1995 and have since then been covered by many articles and reports.

Internet of Manufacturing Things is an open network system that combines advanced manufacturing, Internet-of-Things (IoT) information, and modern management.

Internet-of-Things refers to a system of networked machines, systems and devices via and with the Internet.

ISO 14001 is an international standard applicable to any organisations, regardless of type, size or location, as a framework for policies, procedures, and practices to manage their environmental impact. Through independent third-party auditors, organisations can be awarded ISO 14001 certification if their environmental management system meets the requirements of the ISO 14001 standard.

k kiosk is a convenience store with the core product ranges tobacco, lottery, snacks and press [9].

Key Performance Indicators Key figures that relate to the success, performance or capacity utilisation of the company, its individual organisational units or a machine.

Kinetic recovery refers to recovering kinetic energy by converting it into another form for immediate use or storage instead of disposing of it in the form of heat, for example in a braking system.

kooky is a startup based in Zürich, Switzerland, that provides a circular infrastructure for reusable food packaging. They offer smart cups, smart return boxes, and washing services [10].

Kyburz Switzerland was founded in 1991 and is based in Freienstein-Taufen, Switzerland; it employs around 170 employees; it develops and produces electric vehicles for delivery and industrial companies as well as for private individuals.

Lean Management is a set of management practices that aim to deliver value to customers quickly by focusing on reducing delays and eliminating waste, which results in increased quality and lower cost.

Lidl is an internationally renowned supermarket chain from Germany. In Switzerland, Lidl offers a wide range of products, including food, household goods and more [11].

Life Cycle Assessment is a systematic method for investigating the environmental impact of products and processes throughout their entire life cycle.

Linear Economy is a system in which people produce a product, buy it, use it and throw it away. There is no thought along the line regarding recycling or reuse.

LinkedIn is a global social network for business professionals, with about 930 mill. members in 2022.

Liquefaction is a subcategory of pyrolysis for extraction of high-grade bio-oil.

Lista Office is a Swiss company that provides high-quality office furniture, founded in 1945.

Lithium-Ion Battery rechargeable battery that uses lithium ions as the main component of their electrolyte, also known as Li-ion battery. Among other applications, it is widely used for mobile devices such as portable power tools, mobile phones, laptop computers, consumer electronics, and for electric vehicles such as e-cars, e-scooters, e-bikes.

Long loop R R7 to R9 close the cycle of materials or energy at the end of use back to the production stage.

Mahonitas is a Spanish online retailer of vegan and recyclable footwear.

Manufacturers of Convenience Products companies that produce goods designed for consumption, such as foods, beverages, tobacco, etc.

Manufacturers of Durable Products companies that produce long-lasting goods, such as machinery, furniture, vehicles, electrical and electronic equipment, appliances, etc.

Manufacturers of Non-Durable Products companies that manufacture goods with a shorter lifespan, such as chemicals, cleaning products, cosmetics, textiles, apparel, etc.

Manufacturers of Parts and Components companies that produce components or parts for other products, such as automotive parts, replacement parts, etc.

Material processing strategy represents retention option R7 to R9. These are waste management techniques to close the material cycle. This involves the processing of discarded products or parts into materials that can be used and reused in the production of the same or other products to create new forms of value [12].

Material reduction strategy epresents retention option R0 and R1. They aim to reduce the use of virgin materials and the generation of waste generated during production. As such, they affect product design and production processes. Companies that are successful in these strategies can reduce their demand for virgin materials, increase operational efficiency, and reduce their environmental impact [12].

Medium loop R R2 to R6 are slowing the flow of resources by extending the existing use cycle (R2, R3) or allowing additional use cycles (R4 to R6).

Microsoft is a multinational technology company that develops, manufactures, licenses, supports, and sells computer software, consumer electronics, personal computers, and related services.

Migros is the leading Swiss supermarket that offers a wide range of products [13].

Minergie is a Swiss construction standard for new and modernised buildings, targeting comfort, efficiency and climate protection. It is known for certifying low-energy, passive and plus-energy buildings [14].

Mobility is a Swiss company that provides mobility solutions. It provides rental vehicles to its customers at different locations in Switzerland in various vehicle categories 24 hours a day, usually used for short-term rentals of several hours.

Märtplatz Foundation is an educational institution for young people with psychological and social difficulties, offering training in ten different fields and residential places with individual living assistance.

Narrowing R aim to reduce resource use, improve the efficiency of production processes, and ultimately decouple value creation from virgin material consumption. It encompasses strategies such as lean management, increasing material and energy efficiency, and reuse of materials or waste heat in the process.

Near Field Communication is a protocol enabling short-range communication between compatible devices, commonly used for contactless payments and data exchange.

Operational Excellence Circularity involves the efficient and effective implementation of business strategies to ensure that core processes are continuously optimised. This includes focusing on quality improvement, customer focus, process management, speed, and learning and adaptability.

Original Equipment components or parts for the final product.

Original Equipment Manufacturer is a company whose goods are used as components in the products of another company.

Original Equipment Service is a term used in the manufacturing industry. It refers to the market where spare parts and replacements are sold. These parts are not meant for the initial assembly of a product, but rather for maintenance or repair purposes.

Oxford University is located in England and is a prestigious institution known for its academic excellence and most renowned university globally.

Partnership is an agreement between organisations, people, etc. to work together.

Patagonia is an American company for outdoor clothing and gear [15].

Physical recovery refers to the process of reclaiming materials from waste and reintroducing them into the production cycle. An example of physical waste recovery is the recovery of oil and water along with a dry solid from an oily sludge of a refinery.

Polybutylene Terephthalate is a thermoplastic material that is used in technical applications for industry, in household appliances, in the electronics industry and in automotive engineering.

Polycyclic Aromatic Hydrocarbons are a class of organic compounds produced by incomplete combustion or high-pressure processes. They form when complex organic substances are exposed to high temperatures or pressures. PAHs comprise the largest class of cancer-causing chemicals. PAHs are classified as persistent organic pollutants. They are found almost everywhere (air, water, food, soil, etc.).

Polyethylene is a plastic, being made into products ranging from clear food wrap and shopping bags to detergent bottles and automobile fuel tanks.

Polyethylene Terephthalate PET plastics is a polyester and belongs to the group of polycondensates. PET is used as a material for the manufacture of a wide range of products, including textile fibres, films and beverage bottles.

Polylactic Acid is a biodegradable plastic derived from renewable resources like corn starch or sugarcane. PLA is commonly used in packaging, disposable tableware, and 3D printing.

Polyoxymethylene is a thermoplastic formed by polymerising formaldehyde, with high strength and rigidity.

Polypropylene is a plastic made from the polymerisation of propene. It is often used for packaging, textiles, car parts and other products.

Polyvinylchloride is a thermoplastic material that is used in numerous industrial sectors.

Pressure Regulating Valve is a valve for fluids; it enables a precise and optimised pressure management, ensures a balance within the water supply network and provides a large contribution to eliminate leaks and water losses in distribution networks.

Product as a Service is a business model where customers pay to use a product for a certain period instead of owning it. The focus is on the manufacturer's responsibility for product maintenance and performance, with examples including leasing products or subscribing to services.

Product Lifecycle is the length of time from a product first being introduced to consumers until it is removed from the market.

Product Lifecycle Management refers to the handling of a good as it moves through the typical stages of its product life: development and introduction, growth, maturity/ stability and decline.

Product Service System is a business model where companies maintain ownership of their products through renting, sharing, or leasing them out. This grants the company more control over the product's life cycle and facilitates the implementation of CE practices.

Product use cycle extending strategy represents retention option R2 and R3. These strategies slow the flow of resources in the economy and allow companies to extract value from products by extending their use cycle [12]. This means that they aim to extend the period during which a product performs its original function.

Product use cycle extending strategy Represents retention option R4 to R6. These strategies also slow the flow of resources and allow companies to extract value from products or parts thereof [12] that have reached the end of their lifecycle. When these strategies are implemented, the lifecycle is renewed to a new use cycle.

Product/Service/Business Model Innovation Circularity is a process that involves creating, modifying, or defining the fundamental structure and components of a business model to create new value propositions, capture new market opportunities, and gain a competitive advantage. It involves developing innovative ways to generate revenue, deliver products or services, and create and capture customer value.

ProQuest is a company that provides research, learning, and insight tools. It offers a wide range of products and services for academic, corporate, government, public, and school libraries.

Pyrolysis turns solid biomass into usable liquids (bio-oil or pyrolysis oil – synthetic fuel which could substitute crude oil) and solids (biocharcoal – charcoal produced by pyrolysis of biomass).

QR Code is a two-dimensional barcode that encodes information in a square-shaped grid of black squares on a white background. It is used to store data such as website URLs, text, or other information, and is readable by QR scanners and smartphones with camera capabilities.

R Framework is a combination of a set of Rs which, from the view of the framework's authors, together form the comprehensive representation of all meaningful retention options to create a CE with closed loops.

R-Technology refers to a technology that supports circular strategies to practise an Retention Option (R) and encompasses the availability of technical and technological know-how to apply it.

Radio Frequency Identification is a technology for contactless identification of objects using radio waves.

Rapperswil Circular Economy Model is an empirically supported approach for implementing CE to improve market situation and profitability, reduce resource consumption and waste and increase resilience.

reCIRCLE distributes reusable packaging in the gastronomy sector, which can be reused several hundred times before being recycled. It was founded to replace single-use packaging with reusable and ecological packaging [16].

Reclaimed Asphalt also known as recycled asphalt, is reprocessed pavement containing asphalt and aggregates. These materials are created when asphalt pavements are removed for various activities, such as reconstruction, resurfacing, or to obtain access to buried utilities.

Reclaimed Asphalt Pavements is reprocessed pavement material containing asphalt and aggregates from installed asphalt pavements removed for reconstruction or repair.

Recover R8, extracting elements or materials from end-of-life composites that are not recycled, and energy recovery from incineration.

Recycle R7, extending the material lifespan by processing mixed streams of products to obtain the same high-grade or lower-grade quality. It involves technological equipment to shred, melt and capture materials, and the structure of the product is not maintained. Recycled material is often referred to as secondary material that can be reintroduced in production for the same or alternative initial purpose.

recycled thermoplastic polyurethane is a form of TPU made from recycled materials. It is a multipurpose plastic that is used in many applications, from electronic devices to sporting goods.

Recycling Returning production and consumption waste to the economic cycle.

Reduce R1, eliminating waste in production by using less material per unit produced; reduce natural resources and energy input.

Refurbish R4, extending to new use cycles of products or parts by repairing or replacing defective components of a multi-component product to restore functionality that may be inferior to the original specification. It can also refer to upgrading the product with newer/advanced components and may involve partial disassembly. Synonyms: recondition, retrofit, refresh, remodel.

Refuse R0, rejecting virgin or hazardous materials in product design or eliminating physical products by offering the same function or combined functions; typically enabled by radically different products, technologies, or both. Synonyms: reinvent, prevent.

REI is a US-based outdoor company for gear, apparel, rental equipment and outdoor experiences.

Remanufacture R5, extending to new use cycles of products or parts and using them for the original function with at least the same OEM performance specification and quality; synonyms: rebuild, overhaul, remake.

Remanufacturing Industries Council is a strategic alliance located in the United States of businesses and academic institutions, supporting the remanufacturing industry through collaboration, education, advocacy and research; it published the

ANSI standards RIC001.1-2016 and its successor RIC001.2-2021 which defines and provides a benchmark for the process of remanufacturing.

Remine R9, extracting valuable materials and minerals (e. g. gold, copper) stored in landfills for revalorisation purposes. Synonyms: landfill mining, urban mining.

Repair R3, extending the existing use cycle of products or parts through maintenance or replacing broken parts to retain or restore the original functionality. Synonyms: Corrective-, condition-based-, predictive- and prescriptive maintenance.

Repair and Installation of Machinery and Equipment companies that specialise in repairing, maintaining, and installing machinery and equipment.

Repurpose R6, extending to new use cycles of products and parts by using them for other than their original function. Synonym: alternate use.

Research Question is a specific inquiry which the research seeks to provide a response to. It resides at the core of systematic investigation and it helps to clearly define a path for the research process.

Resilience describes the strength of a company of being able to return quickly to a previous good condition after problems.

Resource recovery is the activity of separating materials from waste that can be recycled into new products or used as an energy alternative to fossil fuels and is actioned with the goal of diverting as much waste from landfill as possible. It is done mostly for sludges that remain after waste handling processes and industrial processes.

Retail and Wholesale companies involved in the sale of goods to consumers or other businesses, including supermarkets, department stores, and distributors.

Retention option refers to the various methods and strategies that can be used to store, reuse and recycle materials and resources.

Return on Investment is a financial metric used to evaluate the efficiency and profitability of an investment. It measures the return of an investment relative to its cost.

Reuse R2, extending the existing use cycle of products or parts by using them again for the same purpose for which it was conceived without reconditioning. It can involve a minimum amount of condition monitoring (cleaning or repackaging), but involves no disassembly. Synonyms: redistribution, resell.

Revendo is a Swiss company that buys and sells used smartphones, tablets, and laptops and offers repair services.

Reverse Logistics collecting and transporting used products and materials from their final destination back to the manufacturer or other designated point in the supply chain.

reWood is a Swiss company that specialises in the production of upcycled furniture. Their products are not made from new wood, but exclusively from wood that has already been used elsewhere.

SBB The Swiss Federal Railways (Schweizerische Bundesbahnen SBB) is Switzerland's national railway company [17].

Service Industry companies that primarily offer services to customers or other businesses, including consulting, financial services, public service, IT, etc.

Service life is the total period during which it remains in use, or ready to be used, in a productive process.

Short loop R R0 and R1 narrow the flow of resources, mitigating the need for virgin material extraction and reducing waste from the outset.

Small and Medium-sized Enterprises refers to companies that have revenues, assets, or a number of employees below a certain threshold. The EU recommendation 2003/361 defines it as companies with 10 to 250 headcount and 2 to 50 million EUR turnover.

Smart Product Lifecycle Management refers to the integration of advanced technologies and processes to manage the entire life cycle of a product efficiently. It typically involves the use of digital tools, such as Internet-of-Things (IoT) devices, Artificial Intelligence (AI), and data analytics.

Social housing refers to dwellings for people with no or low income, oftentimes subsidised by government, foundations or social welfare.

Starbucks Starbucks Corporation is an internationally operating US-based retail company and franchiser specialising in coffee products.

Super Guud is a small, trend-setting snacking concept in Switzerland [18].

SurveyCircle is a website that helps to find participants for online surveys and experiments, representing a large community for mutual support in online research, with over 2 mill. study participations from 100+ countries.

Sustainable Development Goal also known as the Global Goals, were adopted by the United Nations in 2015 as a universal call to action to end poverty, protect the planet, and ensure that by 2030 all people enjoy peace and prosperity.

Swiss Post provides letter and parcel delivery services across Switzerland, it was founded in 1849 has since become one of Switzerland's largest employers.

Tarkett is a company manufacturing carpets that meet design and functional requirements.

Thermal recovery refers to the use of heat energy that is released from some industrial processes and would otherwise dissipate into the immediate environment unused. Heat recovery is the most common form of thermal recovery.

Thermo-chemical recovery is a process wehre energy is extracted from waste through high temperatures such as in combustion, gasification, pyrolysis, or liquefaction.

Time Assessment is a process that involves evaluating how time is spent. In this context, time assessment is a self rating system for evaluating the duration of programme and action implementation.

Total Cost of Ownership is an estimation of the expenses associated with purchasing, deploying, using and retiring a product or piece of equipment. TCO quantifies the cost of the purchase across the product's entire life cycle.

Transparency in this context refers to the availability of information. It's often used interchangeably with terms such as openness, publicity, or visibility.

Tritan is a thermoplastic polyester. It is often used for the manufacture of drinking bottles and other products.

United Nations Environmental Programme is the leading environmental authority in the United Nations system. UNEP uses its expertise to strengthen environmental standards and practices while helping implement environmental obligations at the country, regional and global levels [19].

USD US Dollar, the official currency of the United States of America. At the time of writing this book, 1 USD was approximately 0.90 EUR.

Use cycle extending and renewing strategy represents retention options R2 to R6, a combination of product use cycle extending strategy and product use cycle extending strategy.

UVC Partners is a venture capital firm based in Munich and Berlin. They specialise in partnering with B2B startups across Europe in their initial stages up to Series A [20].

Valora Holding AG is a Swiss retail and food service holding company [21].

Value Hill is a circular business strategy tool that provides companies with an understanding of how to position their business in a circular context and to develop future strategies for a circular economy. It provides an overview of the circular partners and collaborations essential to the success of a circular value network [22].

Verve Ventures is a network and technology-driven venture capital firm based in Switzerland [23].

Virtual Reality is a computer-generated simulation of a three-dimensional image or environment that can be interacted with in a seemingly real or physical way by a person using special electronic equipment, such as a helmet with a screen inside or gloves fitted with sensors.

Vytal s a company in Germany that offers reusable alternatives for food and beverages. It was founded to replace single-use packaging with reusable and ecological packaging [24].

Waste from Electrical and Electronic Equipment (WEEE) Directive (2012/19/EU) refers to the European Community Directive 2012/19/EU. It mandates the treatment, recovery, and recycling of e-waste.

Wastewater Water that is not clean because it has already been used in homes, businesses, factories, etc.

Web of Science is a web portal developed by Clarivate Analytics. It provides access to various databases, which can be searched individually or collectively.

Wind Turbine Blade are key components of wind turbines that convert wind energy into mechanical energy. They are designed to capture the kinetic energy from the wind and cause the rotor to turn.

World Wide Fund for Nature is an organisation focused on wildlife preservation, environmental conservation, and reducing human impact on nature [25].

YORMA's is a German food service company [26].

References

1. *avec - handmade with love* avec. https://www.avec.ch (2023).
2. *Leidenschaft für Laugengebäck* Brezelkönig. https://brezelkoenig.ch/de/ (2023).
3. *Kaffeekultur seit 1999* Caffè Spettacolo. https://spettacolo.ch:443/de (2023).
4. *Der Online-Supermarkt* Coop. https://www.coop.ch/de.html (2023).
5. *OST - Ostschweizer Fachhochschule*. https://www.ost.ch (2023).
6. *U.S. Environmental Protection Agency-US EPA*. https://www.epa.gov/ (2023).
7. *WERZ Institut für Wissen, Energie und Rohstoffe Zug* OST. https://www.ost.ch/werz (2023).
8. *IPEK* OST. https://www.ipek.ch (2023).
9. *Gönn dir was!* k kiosk. https://www.kkiosk.ch (2023).
10. *kooky - Das Mehrwegsystem für dein Büro* kooky. https://www.kooky2go.com/ (2023).
11. *Lidl Schweiz*. https://www.lidl.ch/ (2023).
12. Vermunt, D., Negro, S., Verweij, P., Kuppens, D., & Hekkert, M. (2019). Exploring barriers to implementing different circular business models. *Journal of Cleaner Production, 222*, 891–902. ISSN: 0959-6526.
13. *Migros, der führende Schweizer Supermarkt* Migros. https://www.migros.ch/de (2023).
14. *Schweizer Baustandard* Minergie. https://www.minergie.ch/de/ (2023).
15. *Patagonia Outdoor Clothing & Gear* Patagonia. https://eu.patagonia.com/de/de/home/ (2023).
16. *reCIRCLE Wir gehen den Mehrweg!* reCIRCLE. https://www.recircle.ch/en/ (2023).
17. *DasOnline-Portal der SBB für Zug undÖV* SBB. https://www.sbb.ch/de (2023).
18. *SuperGuud* Valora Holding AG. https://www.valora.com:443/de/brands/superguud/ (2023).
19. UNEP. *Circularity* UNEP - United Nations Environment Programme. http://www.unep.org/circularity (2019).
20. UVC Partners. *UVC Partners* https://www.uvcpartners.com/ (2023).
21. Valora Holding AG. *Valora ist die führende Foodvenience-Anbieterin* Valora Holding AG. https://www.valora.com:443/de/ (2023).
22. Achterberg, E., Hinfelaar, J., & Bocken, N. (2016). *Master Circular Business with the Value Hill*.
23. Verve Ventures. *Verve Ventures is a leading pan-european venture capital investor* verve.vc. https://www.verve.vc/ (2023).
24. *Vytal - Das nachhaltige Mehrwegsystem*. https://www.vytal.org/ (2023).
25. *Organisation für Natur- und Umweltschutz* WWF Schweiz. https://www.wwf.ch/de (2023).
26. *Yorma's Kaffeespezialitäten, Baguettes & Snacks am Bahnhof* Yorma's AG. https://www.yormas.de/ (2023).

The manufacturer's authorised representative in the EU is Springer Nature Customer Service Centre GmbH, Europaplatz 3, 69115 Heidelberg, Germany. If you have any concerns regarding our products, please contact ProductSafety@springernature.com

Printed and bound by CPI Group (UK) Ltd, Croydon, CR0 4YY
01/10/2026
02225932-0015